17th Workshop on Biomedical Natural Language Processing (BioNLP 2018)

Held at ACL 2018

Melbourne, Australia
19 July 2018

ISBN: 978-1-5108-6710-9

BioNLP 2018

**SIGBioMed Workshop on
Biomedical Natural Language Processing**

Proceedings of the 17th BioNLP Workshop

July 19, 2018
Melbourne, Australia

Biomedical natural language processing in 2018:
Spotlight on Deep Learning

Dina Demner-Fushman, Kevin Bretonnel Cohen, Sophia Ananiadou, and Jun-ichi Tsujii

The number of community challenges, corpora and publicly available tools in the domain continues to grow rapidly. The past year has seen several hackathons, a variety of shared tasks and growing numbers of workshops dedicated to specific biomedical and clinical sublanguages and tasks. The BioNLP meeting has now been ongoing for 17 years. BioNLP continues to stay the flagship and the generalist in biomedical language processing, accepting noteworthy work independently of the tasks and sublanguages studied. The quality of submissions continues to impress the program committee and the organizers. BioNLP 2018 received 28 submissions, of which 13 were accepted for oral presentation and 12 as poster presentations. This year, Deep Learning approaches are explored in the overwhelming majority of the papers, with focus on interesting new models and in-depth exploration of the state-of-the-art publicly available tools. As for the past several years, the themes in this year's papers and posters continue to focus equally on clinical text and biological language processing, as well as reveal growing interest in consumer language processing. The morning session presents clinical text processing for extraction of causes of death, risk factors identification and named entity recognition, among others. The next session presents work on fundamental NLP problems, such as ontology alignment and key-phrase extraction, whereas the afternoon session presents exceptionally strong work on complex text mining tasks, such as event extraction and question answering.

The invited talk and the invited presentation reflect thus growing interest in automated support for systematic reviews of the literature. In the invited talk, professor Paul Glasziou discusses progress and challenges in automating systematic reviews. Paul Glasziou, FRACGP, PhD is Professor of Evidence-Based Medicine at Bond University and a part-time General Practitioner. He was the Director of the Centre for Evidence-Based Medicine in Oxford from 2003-2010. His key interests include identifying and removing the barriers to using high quality research in everyday clinical practice. He is the author of six books related to evidence based practice: Systematic Reviews in Health Care, Decision Making in Health Care and Medicine: integrating evidence and values, An Evidence-Based Medicine Workbook, Clinical Thinking: Evidence, Communication and Decision-making, Evidence-Based Medicine: How to Practice and Teach EBM, and Evidence-Based Medical Monitoring: Principles and Practice. He has authored over 160 peer-reviewed journal articles and his h-index is currently 43. He is the recipient of an NHRMC Australia Fellowship which he commenced at Bond University in July, 2010.

The invited presentation follows suit by bringing to our attention a new corpus of about 5,000 abstracts of randomized control trials annotated with granular information regarding the study populations, interventions, comparators and outcomes.

Organizers:

Kevin Bretonnel Cohen, University of Colorado School of Medicine, USA
Dina Demner-Fushman, US National Library of Medicine
Sophia Ananiadou, National Centre for Text Mining and University of Manchester, UK
Jun-ichi Tsujii, National Institute of Advanced Industrial Science and Technology, Japan

Program Committee:

Sophia Ananiadou, National Centre for Text Mining and University of Manchester, UK
Emilia Apostolova, Language.ai, USA
Eiji Aramaki, University of Tokyo, Japan
Asma Ben Abacha, US National Library of Medicine
Olivier Bodenreider, US National Library of Medicine
Leonardo Campillos Llànos, LIMSI - CNRS, France
Brian Connolly, Kroger Digital, USA
Dina Demner-Fushman, US National Library of Medicine
Filip Ginter, University of Turku, Finland
Graciela Gonzalez-Hernandez, University of Pennsylvania, USA
Travis Goodwin, The University of Texas at Dallas, USA
Cyril Grouin, LIMSI - CNRS, France
Tudor Groza, The Garvan Institute of Medical Research, Australia
Antonio Jimeno Yepes, IBM, Melbourne Area, Australia
Halil Kilicoglu, US National Library of Medicine
Robert Leaman, US National Library of Medicine
Ulf Leser, Humboldt-Universität zu Berlin, Germany
Zhiyong Lu, US National Library of Medicine
Timothy Miller, Children's Hospital Boston, USA
Makoto Miwa, Toyota Technological Institute, Japan
Danielle L Mowery, VA Salt Lake City Health Care System, USA
Yassine M'Rabet, US National Library of Medicine
Aurelie Neveol, LIMSI - CNRS, France
Claire Nedellec, INRA, France
Mariana Neves, Hasso Plattner Institute and University of Potsdam, Germany
Nhung Nguyen, The University of Manchester, UK
Naoaki Okazaki, Tohoku University, Japan
Sampo Pyysalo, University of Cambridge, UK
Francisco J. Ribadas-Pena, University of Vigo, Spain
Fabio Rinaldi, University of Zurich, Switzerland
Kirk Roberts, The University of Texas Health Science Center at Houston, USA
Angus Roberts, The University of Sheffield, UK
Hagit Shatkay, University of Delaware, USA
Pontus Stenetorp, University College London, UK
Karin Verspoor, The University of Melbourne, Australia
Byron C. Wallace, University of Texas at Austin, USA
W John Wilbur, US National Library of Medicine
Pierre Zweigenbaum, LIMSI - CNRS, France

Additional Reviewers:

Pramod Chandrashekar, University of Pennsylvania, USA
Nicolas Fiorini, US National Library of Medicine
Arjun Magge, University of Pennsylvania, USA
Yijia Zhang, US National Library of Medicine

Invited Speaker:

Paul Glasziou, Bond University, Australia

Table of Contents

Conference Program

Thursday July 19, 2018

9:00–9:15   **Opening remarks**

9:15–10:30   **Session 1: Clinical NLP**

9:15–9:30   *Embedding Transfer for Low-Resource Medical Named Entity Recognition: A Case Study on Patient Mobility*
Denis Newman-Griffis and Ayah Zirikly

9:30–9:45   *Multi-task learning for interpretable cause of death classification using key phrase prediction*
Serena Jeblee, Mireille Gomes and Graeme Hirst

9:45–10:00   *Identifying Risk Factors For Heart Disease in Electronic Medical Records: A Deep Learning Approach*
Thanat Chokwijitkul, Anthony Nguyen, Hamed Hassanzadeh and Siegfried Perez

10:00–10:15   *Keyphrases Extraction from User-Generated Contents in Healthcare Domain Using Long Short-Term Memory Networks*
Ilham Fathy Saputra, Rahmad Mahendra and Alfan Farizki Wicaksono

10:15–10:30   *Identifying Key Sentences for Precision Oncology Using Semi-Supervised Learning*
Jurica Ševa, Martin Wackerbauer and Ulf Leser

10:30–11:00   *Coffee Break*

11:00–12:30   Session 2: Foundations

11:00–11:15   *Ontology alignment in the biomedical domain using entity definitions and context*
Lucy Wang, Chandra Bhagavatula, Mark Neumann, Kyle Lo, Chris Wilhelm and Waleed Ammar

11:15–11:30   *Sub-word information in pre-trained biomedical word representations: evaluation and hyper-parameter optimization*
Dieter Galea, Ivan Laponogov and Kirill Veselkov

11:30–11:45   *PICO Element Detection in Medical Text via Long Short-Term Memory Neural Networks*
Di Jin and Peter Szolovits

11:45–12:00   *Coding Structures and Actions with the COSTA Scheme in Medical Conversations*
Nan Wang, Yan Song and Fei Xia

12:00–13:30   *Lunch break*

13:30–14:30   Invited Talk: "Automating systematic reviews: progress and challenges" – Paul Glasziou

14:30–15:30   Session 3 Literature mining and retrieval; Question Answering

14:30–14:45   *A Neural Autoencoder Approach for Document Ranking and Query Refinement in Pharmacogenomic Information Retrieval*
Jonas Pfeiffer, Samuel Broscheit, Rainer Gemulla and Mathias Göschl

14:45–15:00   *Biomedical Event Extraction Using Convolutional Neural Networks and Dependency Parsing*
Jari Björne and Tapio Salakoski

15:00–15:15   *BioAMA: Towards an End to End BioMedical Question Answering System*
Vasu Sharma, Nitish Kulkarni, Srividya Pranavi, Gabriel Bayomi, Eric Nyberg and Teruko Mitamura

15:15–15:30   *Phrase2VecGLM: Neural generalized language model–based semantic tagging for complex query reformulation in medical IR*
Manirupa Das, Eric Fosler-Lussier, Simon Lin, Soheil Moosavinasab, David Chen, Steve Rust, Yungui Huang and Rajiv Ramnath

15:30–16:00   *Coffee Break*

16:00–16:15   **Invited Presentation: "A Corpus with Multi-Level Annotations of Patients, Interventions and Outcomes to Support Language Processing for Medical Literature" – Ben Nye**

16:15–18:00   **Poster Session**

Convolutional neural networks for chemical-disease relation extraction are improved with character-based word embeddings
Dat Quoc Nguyen and Karin Verspoor

Domain Adaptation for Disease Phrase Matching with Adversarial Networks
Miaofeng Liu, Jialong Han, Haisong Zhang and Yan Song

Predicting Discharge Disposition Using Patient Complaint Notes in Electronic Medical Records
Mohamad Salimi and Alla Rozovskaya

Bacteria and Biotope Entity Recognition Using A Dictionary-Enhanced Neural Network Model
Qiuyue Wang and Xiaofeng Meng

SingleCite: Towards an improved Single Citation Search in PubMed
Lana Yeganova, Donald C Comeau, Won Kim, W John Wilbur and Zhiyong Lu

A Framework for Developing and Evaluating Word Embeddings of Drug-named Entity
Mengnan Zhao, Aaron J. Masino and Christopher C. Yang

MeSH-based dataset for measuring the relevance of text retrieval
Won Gyu KIM, Lana Yeganova, Donald comeau, W John Wilbur and Zhiyong Lu

CRF-LSTM Text Mining Method Unveiling the Pharmacological Mechanism of Off-target Side Effect of Anti-Multiple Myeloma Drugs
Kaiyin Zhou, Sheng Zhang, Xiangyu Meng, Qi Luo, Yuxing Wang, Ke Ding, Yukun Feng, Mo Chen, Kevin Cohen and Jingbo Xia

Prediction Models for Risk of Type-2 Diabetes Using Health Claims
Masatoshi Nagata, Kohichi Takai, Keiji Yasuda, Panikos Heracleous and Akio Yoneyama

On Learning Better Embeddings from Chinese Clinical Records: Study on Combining In-Domain and Out-Domain Data
Yaqiang Wang, Yunhui Chen, Hongping Shu and Yongguang Jiang

Investigating Domain-Specific Information for Neural Coreference Resolution on Biomedical Texts
Long Trieu, Nhung Nguyen, Makoto Miwa and Sophia Ananiadou

Toward Cross-Domain Engagement Analysis in Medical Notes
Sara Rosenthal and Adam Faulkner

Embedding Transfer for Low-Resource Medical Named Entity Recognition: A Case Study on Patient Mobility

Denis Newman-Griffis[1,2] and **Ayah Zirikly**[1]

[1]Rehabilitation Medicine Department, Clinical Center, National Institutes of Health, Bethesda, MD
[2]Department of Computer Science and Engineering, The Ohio State University, Columbus, OH
{denis.griffis, ayah.zirikly} @nih.gov

Abstract

Functioning is gaining recognition as an important indicator of global health, but remains under-studied in medical natural language processing research. We present the first analysis of automatically extracting descriptions of patient mobility, using a recently-developed dataset of free text electronic health records. We frame the task as a named entity recognition (NER) problem, and investigate the applicability of NER techniques to mobility extraction. As text corpora focused on patient functioning are scarce, we explore domain adaptation of word embeddings for use in a recurrent neural network NER system. We find that embeddings trained on a small in-domain corpus perform nearly as well as those learned from large out-of-domain corpora, and that domain adaptation techniques yield additional improvements in both precision and recall. Our analysis identifies several significant challenges in extracting descriptions of patient mobility, including the length and complexity of annotated entities and high linguistic variability in mobility descriptions.

1 Introduction

Functioning has recently been recognized as a leading world health indicator, joining morbidity and mortality (Stucki and Bickenbach, 2017). Functioning is defined in the International Classification of Functioning, Disability, and Health (ICF; WHO 2001) as the interaction between health conditions, body functions and structures, activities and participation, and contextual factors. Understanding functioning is an important element in assessing quality of life, and automatic extraction of patient functioning would serve as a useful tool for a variety of care decisions, including rehabilitation and disability assessment (Stucki et al., 2017). In healthcare data, natural language processing (NLP) techniques have been successfully used for retrieving information about health conditions, symptoms and procedures from unstructured electronic health record (EHR) text (Soysal et al., 2018; Savova et al., 2010). As recognition of the importance of functioning grows, there is a need to investigate the application of NLP methods to other elements of functioning.

Recently, Thieu et al. (2017) introduced a dataset of EHR documents annotated for descriptions of patient mobility status, one area of activity in the ICF. Automatically recognizing these descriptions faces significant challenges, including their length and syntactic complexity and a lack of terminological resources to draw on. In this study, we view this task through the lens of named entity recognition (NER), as recent work has illustrated the potential of using recurrent neural network (RNN) NER models to address similar issues in biomedical NLP (Xia et al., 2017; Dernoncourt et al., 2017b; Habibi et al., 2017).

An additional strength of RNN models is their ability to leverage pretrained word embeddings, which capture co-occurrence information about words from large text corpora. Prior work has shown that the best improvements come from embeddings trained on a corpus related to the target domain (Pakhomov et al., 2016). However, free text describing patient functioning is hard to come by: for example, even the large MIMIC-III corpus (Johnson et al., 2016) includes only a few hundred documents from therapy disciplines among its two million notes. While recent work suggests that using a training corpus from the target domain can mitigate a lack of data (Diaz et al., 2016), even a careful corpus selection may not produce suffi-

Proceedings of the BioNLP 2018 workshop, pages 1–11
Melbourne, Australia, July 19, 2018. ©2018 Association for Computational Linguistics

cient data to train robust word representations.

In this paper, we explore the use of an RNN model to recognize descriptions of patient mobility. We analyze the impact of initializing the model with word embeddings trained on a variety of corpora, ranging from large-scale out-of-domain data to small, highly-targeted in-domain documents. We further explore several domain adaptation techniques for combining word-level information from both of these data sources, including a novel nonlinear embedding transformation method using a deep neural network.

We find that embeddings trained on a very small set of therapy encounter notes nearly match the mobility NER performance of representations trained on millions of out-of-domain documents. Domain adaptation of input word embeddings often improves performance on this challenging dataset, in both precision and recall. Finally, we find that simpler adaptation methods such as concatenation and preinitialization achieve highest overall performance, but that nonlinear mapping of embeddings yields the most consistent performance across experiments. We achieve a best performance of 69% exact match and over 83% token-level match F-1 score on the mobility data, and identify several trends in system errors that suggest fruitful directions for further research on recognizing descriptions of patient functioning.

2   Related work

The extraction of named entities in free text has been one of the most important tasks in NLP and information extraction (IE). As a result, this track of research has matured over the last two decades, especially in the newswire domain for high resource languages such as English. Many of the successful existing NER systems use a combination of engineered features trained using conditional random fields (CRF) model (McCallum and Li, 2003; Finkel et al., 2005). NER systems have also been widely studied in medical NLP, using dictionary lookup methods (Savova et al., 2010), support vector machine (SVM) classifiers (Kazama et al., 2002), and sequential models (Tsai et al., 2006; Settles, 2004). In recent years, deep learning models have been used in NER with successful results in many domains (Collobert et al., 2011). Proposed neural network architectures included hybrid convolutional neural network (CNN) and bi-directional long-short term

```
Evaluation:
[Scoring: 1=totally dependent,
2=requires assistance,
3=requires appliances, 4=totally
independent] ScoreDefinition .

[Ambulation: 4] Mobility
Observations:
Pt is weight bearing: [she
ambulates independently w/o
use of assistive device] Mobility .
Limited to very brief
examination.
```

Figure 1: Synthetic document with examples of ScoreDefinition (in blue) and Mobility (in orange).

memory (Bi-LSTM) as introduced by Chiu and Nichols (2015). State-of-the-art NER models use the architecture proposed by Lample et al. (2016), a stacked bi-directional long-short term memory (Bi-LSTM) for both character and word, with a CRF layer on the top of the network. In the biomedical domain, Habibi et al. (2017) used this architecture for chemical and gene name recognition. Liu et al. (2017) and Dernoncourt et al. (2017a) adapted it for state-of-the-art note deidentification. In terms of functioning, Kukafka et al. (2006) and Skube et al. (2018) investigate the presence of functioning terminology in clinical data, but do not evaluate it from an NER perspective.

3   Data

Thieu et al. (2017) presented a dataset of 250 deidentified EHR documents collected from Physical Therapy (PT) encounters at the Clinical Center of the National Institutes of Health (NIH). These documents, obtained from the NIH Biomedical Translational Research Informatics System (BTRIS; Cimino and Ayres 2010), were annotated for several aspects of patient mobility, a subdomain of functioning-related activities defined by the ICF; we therefore refer to this dataset as BTRIS-Mobility. We focus on two types of contiguous text spans: descriptions of mobility status, which we call Mobility entities, and measurement scales related to mobility activity, which we refer to as ScoreDefinition entities.

Two major differences stand out in BTRIS-Mobility as compared with standard NER data. The entities, defined for this task as contiguous text spans completely describing an aspect of mobility, tend to be quite long: while prior NER datasets such as the i2b2/VA 2010 shared task data (Uzuner et al., 2012) include fairly short entities (2.1 tokens on average for i2b2), Mobility entities

Entity	Train	Valid	Test
Mobility	1,533	467	947
ScoreDefinition	82	24	48

Table 1: Named entity statistics for training, validation, and test splits of BTRIS-Mobility. Due to the rarity of ScoreDefinition entities, we use a 2:1 split of training to test data, and hold out 10% of training data as validation.

are an average of 10 tokens long, and ScoreDefinition average 33.7 tokens. Also, both Mobility and ScoreDefinition entities tend to be entire clauses or sentences, in contrast with the constituent noun phrases that are the meat of most NER. Figure 1 shows example Mobility and ScoreDefinition entities in a short synthetic document. Despite these challenges, Thieu et al. (2017) show high (> 0.9) inter-annotator agreement on the text spans, supporting use of the data for training and evaluation.

These characteristics align well with past successful applications of recurrent neural models to challenging NLP problems. For our evaluation on this dataset, we randomly split BTRIS-Mobility at document level into training, validation, and test sets, as described in Table 1.

3.1 Text corpora

In order to learn input word embeddings for NER, we use a variety of both in-domain and out-of-domain corpora, defined in terms of whether the corpus documents include descriptions of function. For in-domain data, with explicit references to patient functioning, we use a corpus of 154,967 EHR documents shared with us (under an NIH Clinical Center Office of Human Subjects determination) from the NIH BTRIS system.[1] A large proportion of these documents comes from the Rehabilitation Medicine Department of the NIH Clinical Center, including Physical Therapy (PT), Occupational Therapy (OT), and other therapeutic records; the remaining documents are sampled from other departments of the Clinical Center.

Since BTRIS-Mobility is focused on PT documents, we also use a subset of this corpus consisting of 17,952 PT and OT documents. Despite this small size, the topical similarity of these documents makes them a very targeted in-domain corpus. For clarity, we refer to the full corpus as

BTRIS, and the smaller subset as PT-OT.

3.1.1 Out-of-domain corpora

As the BTRIS corpus is considered a small training corpus for learning word embeddings, we also use three larger out-of-domain corpora, which represent different degrees of difference from the in-domain data. Our largest data source is pretrained FastText embeddings from Wikipedia 2017, web crawl data, and news documents.[2]

We also make use of two biomedical corpora for comparison with existing work. PubMed abstracts have been an extremely useful source of embedding training in biomedical NLP (Chiu et al., 2016); we use the text of approximately 14.7 million abstracts taken from the 2016 PubMed baseline as a high-resource biomedical corpus. In addition, we use two million free-text documents released as part of the MIMIC-III critical care database (Johnson et al., 2016). Though smaller than PubMed, the MIMIC corpus is a large sample of clinical text, which is often difficult to obtain and shows significant linguistic differences with biomedical literature (Friedman et al., 2002). As MIMIC is clinical text, it is the closest comparison corpus to the BTRIS data; however, as MIMIC focuses on ICU care, the information in it differs significantly from in-domain BTRIS documents.

4 Methods

We adopt the architecture of Dernoncourt et al. (2017a), due to its successful NER results on CoNLL and i2b2 datasets. The architecture, as depicted in Figure 2, is a stacked LSTM composed of: i) character Bi-LSTM layer that generates character embeddings. We include this in our experimentations due to its performance enhancement; ii) token Bi-LSTM layer using both character and pre-trained word embeddings as input; iii) CRF layer to enhance the performance by taking into account the surrounding tags (Lample et al., 2016). We use the following values for the network hyperparameters, as they yielded the best performance on the validation set: i) hidden state dimension of 25 for both character and token layers. In contrast to more common token layer sizes such as 100 or 200, we found the best validation set performance for our task with 25 dimensions; ii) learning rate = 0.005; iii) patience = 10; iv) optimization with stochastic gradient de-

---

3

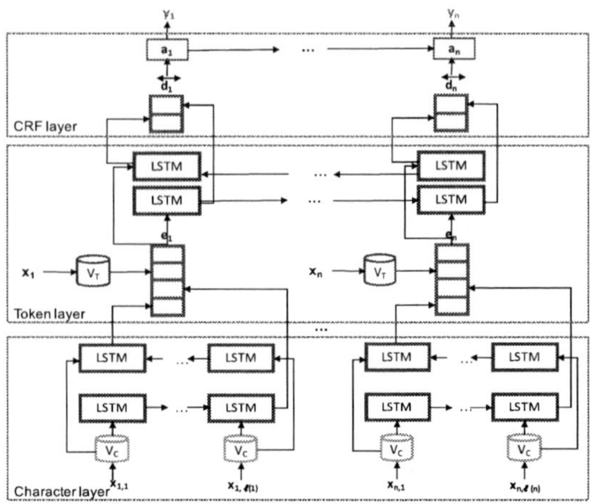

Figure 2: Bi-LSTM-CRF network architecture

scent (SGD) which showed superior performance to adaptive moment estimation (Adam) optimization technique (Kingma and Ba, 2014).

4.1 Embedding training

We use two popular toolkits for learning word embeddings: word2vec[3] (Mikolov et al., 2013) and FastText[4] (Bojanowski et al., 2017). We run both toolkits using skip-gram with negative sampling to train 300-dimensional embeddings, and use default settings for all other hyperparameters.[5]

4.2 Domain adaptation methods

We evaluate several different methods for adapting out-of-domain embeddings to the BTRIS corpus.

Concatenation In addition to the original embeddings, we concatenate out-of-domain and BTRIS/PT-OT embeddings as a baseline, allowing the model to learn a task-specific combination of the two representations.

Preinitialization Recent work has found benefits from retraining learned embeddings on a target corpus (Yang et al., 2017). We pre-initialize both word2vec and FastText toolkits with embeddings learned on each of our three reference corpora, and retrain on the BTRIS corpus using an initial learning rate of 0.1. Additionally, we use the regularization-based domain adaptation approach introduced by Yang et al. (2017) as another baseline, due to its successful results in improving

NER performance. Their method aims to help the model to differentiate between general and domain specific terms, using a significance function ϕ of a word w. ϕ is dependent on the definition of w's frequency, where in our implementation it is the word frequency in the target corpora.

Linear transform However, these approaches suffer from the same limitations as training BTRIS embeddings directly: a restricted vocabulary and minimal training data, both due to the size of the corpus. We therefore also investigate two methods for learning a transformation from one set of embeddings into the same space as another, based on a reference dictionary. Given an out-of-domain source embedding set and a target BTRIS embedding set, we use all words in common between source and target as our training vocabulary.[6] We adapt this to the linear transformation method successfully applied to bilingual embeddings by Artetxe et al. (2016), using this shared vocabulary as the training dictionary.

Non-linear transform As all of our embeddings are in English, but from domains that do not intuitively seem to have a linear relationship, we also extend the method of Artetxe et al. to a non-linear transformation. We randomly divide the shared vocabulary into ten folds, and train a feed-forward neural network using nine-tenths of the data, minimizing mean squared error (MSE) between the learned projection and the true embeddings. After each epoch, we calculate MSE on the held-out set, and halt when this error stops decreasing. Finally, we average the learned projections from each fold to yield the final transformation function. Following Artetxe et al. (2016), we apply this function to all source embeddings, allowing us to maintain the original vocabulary size.

Our model is a fully-connected feed-forward neural network, with the same hidden dimension as our embeddings. We evaluate with both 1 and 5 hidden layers, and use either tanh or rectified linear unit (ReLU) activation throughout. Model structure is denoted in the result; for example, "5-layer ReLU" refers to nonlinear mapping using a 5-layer network with ReLU activation. We train with Adam optimization (Kingma and Ba, 2014) and a minibatch size of 5.[7]

---

[3] We use word2vec modified to support pre-initialization, from `github.com/drgriffis/word2vec-r`.

[4] `github.com/facebookresearch/fastText`

[5] For PT-OT embeddings, due to the extremely small corpus size, we use an initial learning rate of 0.05, keep all words with minimum frequency 2, and train for 25 iterations.

[6] We evaluated using subsets of 1k, 2k, or 10k shared words most frequent in BTRIS, but the best downstream performance was achieved using all pivot points.

[7] Source implementation available at `github.com/drgriffis/NeuralVecmap`

Corpus	Size	Toolkit	Mobility						ScoreDefinition					
			Exact match			Token match			Exact match			Token match		
			Pr	Rec	F1	Pr	Rec	F1	Pr	Rec	F1	Pr	Rec	F1
Random initialization			67.7	61.8	64.6	84.0	75.9	79.7	86.5	93.4	90.0	97.7	98.9	98.3
WikiNews	16B	FT	67.0	64.0	65.4	83.0	80.0	81.5	83.3	93.4	88.2	96.8	99.3	98.0
PubMed	2.6B	FT	68.7	**65.9**	67.2	82.0	84.5	83.2	**93.6**	91.7	92.6	**98.1**	97.8	97.9
		w2v	64.9	64.7	64.8	77.4	**87.7**	82.2	90.0	93.8	91.8	97.8	99.6	**98.7**
MIMIC	497M	FT	37.7	10.6	16.5	78.9	21.7	34.0	86.0	90.0	87.8	97.9	97.7	97.8
		w2v	**71.9**	64.9	**68.2**	84.3	83.0	**83.6**	91.7	91.7	91.7	96.5	99.6	98.0
BTRIS	74.6M	FT	66.8	63.8	65.3	80.6	83.4	82.0	90.2	**95.8**	92.9	95.9	99.0	97.4
		w2v	69.7	63.7	66.7	**86.0**	79.2	82.4	88.2	93.8	90.9	96.7	**99.9**	98.3
PT-OT	4.2M	FT	68.8	62.5	65.5	84.5	80.2	82.3	92.0	**95.8**	**93.9**	97.1	97.7	97.4
		w2v	70.0	63.4	67.0	85.8	79.4	82.3	88.3	91.7	88.9	96.3	98.9	97.6

Table 2: Exact and token-level match results on BTRIS-Mobility, using randomly-initialized embeddings as a baseline and unmodified word2vec (w2v) and FastText (FT) embeddings from different corpora. *Size* is the number of tokens in the training corpus.

5 Results

We report exact match results, calculated using CoNLL 2003 named entity recognition shared task evaluation scoring (Tjong Kim Sang and De Meulder, 2003), which requires that all tokens of an entity are correctly recognized. Additionally, given the long span of Mobility and ScoreDefinition entities (see Section 3), we evaluated partial match performance using token-level results. For simplicity, we report only performance on the test set; however, validation set numbers consistently follow the same trends observed in test data. We denote embeddings trained using FastText with the subscript $_{FT}$, and word2vec with $_{w2v}$.

5.1 Embedding corpora

Exact and token-level match results for both Mobility and ScoreDefinition entities are given for embeddings from each corpus in Table 2. By and large, the in-domain BTRIS and PT-OT embeddings yield higher precision than out-of-domain embeddings, though this comes at the expense of recall. word2vec embeddings consistently achieve better NER performance than FastText embeddings from the clinical corpora, although this was reversed with PubMed, suggesting that further research is needed on the strengths of different embedding methods in biomedical data. The unusually poor performance of MIMIC_{FT} embeddings persisted across multiple experiments with two embedding samples, manifesting primarily in making very few predictions (less than 30% as many Mobility entities other embeddings yielded).

Most notably, despite a thousand-fold reduction in training corpus size, we see that PT-OT embeddings match the performance of PubMed embeddings on Mobility mentions and achieve the best overall performance on ScoreDefinition entities. Together with the overall superior performance of PT-OT embeddings even to the larger BTRIS corpus, our findings support the value of using input embeddings that are highly representative of the target domain. Nonetheless, MIMIC embeddings have both the best precision and overall performance on Mobility data, despite the domain mismatch of critical care versus therapeutic encounters. This indicates that there is a limit to the benefits of in-domain data that can be outweighed by sufficient data from a different but related domain.

Token-level results follow the same trends as exact match, with clinical embeddings achieving highest precision, while PubMed embeddings yield better recall. As many entity-level errors are only off by a few tokens, token-level scores are generally 15-20 absolute points higher than their corresponding entity-level scores. At the token level, it is clear that ScoreDefinition entities are effectively solved in this dataset, with all F1 scores are above 97.4%. This is primarily due to the regularity of ScoreDefinition strings: they typically consist of a sequence of single numbers followed by explanatory strings, as shown in Figure 1.

5.2 Mapping methods

Table 3 takes a single representative source/target pair and compares the different results obtained on recognizing Mobility entities when the NER model is initialized with embeddings learned using different domain adaptation methods. In this case, as with several other source/target pairs we evaluated, the concatenated embeddings give the best overall performance, stemming largely from

Target	Source	Concat			Preinit			Linear			5-layer tanh		
		Pr	Rec	F1	Pr	Rec	F1	Pr	Rec	F1	Pr	Rec	F1
BTRIS_{FT}	WikiNews_{FT}	**72.2**	**65.3**	**68.6**	55.0	59.2	57.0	65.1	61.9	63.5	69.3	64.2	66.7
	PubMed_{FT}	**69.5**	65.8	**67.6**	64.2	**66.5**	65.4	65.6	60	62.7	66.1	64.5	65.3
	PubMed_{w2v}	65.3	65.3	**65.3**	64.8	**65.4**	65.1	70.3	65.8	68	**66.3**	62.6	64.4
	MIMIC_{FT}	35.0	10.4	16.0	37.8	15.5	22.0	63.7	**62.9**	63.3	**70.3**	61.3	**65.5**
	MIMIC_{w2v}	67.4	**67.6**	**67.5**	68.5	64.6	66.5	66.8	60.3	63.4	**69.2**	64.3	66.7
PT-OT_{FT}	WikiNews_{FT}	67.5	**63.9**	65.6	54.5	57.9	56.1	68.9	63.8	66.2	**68.5**	63.4	**65.8**
	PubMed_{FT}	62.8	**65.1**	**63.9**	61.3	50.2	55.2	62.6	62.6	62.6	**68.3**	60.1	**63.9**
	MIMIC_{w2v}	64.1	**66.1**	**65.1**	59.9	61.8	60.8	57.9	54.1	55.9	**67.3**	63.2	**65.1**

Table 4: Exact match precision and recall for Mobility entities with word embeddings mapped from each source to BTRIS_{FT} embeddings, using four selected domain adaptation methods. The best-performing embeddings from each source corpus were also mapped to PT-OT_{FT} embeddings. The best precision, recall, and F1 achieved with each source/target pair is marked in bold.

Method	Exact match			Token match		
	Pr	Rec	F1	Pr	Rec	F1
WikiNews_{FT}	67.0	64.0	65.4	83.0	80.0	81.5
BTRIS_{w2v}	70.0	63.7	66.6	**86.0**	79.2	81.5
Concatenated	68.6	**66.7**	**67.6**	84.3	81.8	**83.0**
Preinitialized	66.8	64.5	65.6	78.4	**86.4**	82.2
Linear	**72.5**	58.9	65	79.1	83	81
1-layer ReLU	69.2	63.2	66.0	83.4	76.9	80.0
1-layer tanh	70.6	61.0	65.5	84.9	75.7	80.1
5-layer ReLU	67.3	61.9	64.5	83.5	76.6	79.9
5-layer tanh	67.9	62.1	64.9	82.1	77.0	79.4

Table 3: Comparison of mapping methods, using WikiNews_{FT} as source and BTRIS_{w2v} as target. Results are given for exact entity-level match and token-level match for test set Mobility entities.

Source	Target	Method	Pr	Rec	F1
WikiNews_{FT}	PT-OT_{w2v}	Preinit	72.1	66.1	**69.0**
WikiNews_{FT}	BTRIS_{w2v}	Linear	**72.5**	58.9	65
MIMIC_{w2v}	BTRIS_{FT}	Concat	67.4	**67.6**	67.5

Table 5: Best precision, recall, and F1 (exact) for test set Mobility mentions, with the source/target pair and domain adaptation method used.

an increase in recall over the baselines. However, we see that the nonlinear mapping methods tend to yield high precision: all settings improve over WikiNews embeddings alone, and the 1-layer tanh mapping beats the BTRIS embeddings as well. Reflecting the earlier observed trends of in-domain data, this is offset by a drop in recall, often of several absolute percentage points.

These differences are fleshed out further in Table 4, comparing four domain adaptation methods across several source/target pairs. Concatenation typically achieves the best overall performance among the adaptation methods, but nonlinear mappings yield highest precision in 6 of the 8 settings shown. Concatenation is also more sensitive to noise in the source embeddings, as shown with MIMIC_{FT} results, and preinitialization varies widely in its performance. By contrast, linear and nonlinear mapping methods are less affected by the choice of source embeddings, yielding more consistent results than preinitialization or concatenation for a given target corpus. Nonlinear mappings exhibit this stability most clearly, producing very similar results across all settings. The

regularization-based domain adaptation method of Yang et al. (2017) consistently yielded similar results to preinitialization: for example, an F1 score of 65% when PubMed_{w2v} embeddings are adapted to BTRIS, as compared to 65.4% using pre-initialization with word2vec. We therefore omit these results for brevity.

Comparing both Tables 3 and 4 to the performance of unmodified embeddings shown in Table 2, we see a surprising lack of overall performance improvement or degradation. While the different adaptation methods exhibit consistent differences between one another, only 12 of the 32 F1 scores in Table 4 represent improvements over the relevant unmapped baselines. Many adaptation results achieve notable improvement in precision or recall individually, suggesting that different methods may be more useful for downstream applications where one metric is emphasized over the other. However, several of our results indicate failure to adapt, illustrating the difficulty of effectively adapting embeddings for this task.

5.3 Source/target pairs

Table 5 highlights the source/target pairs that achieved the best exact match precision, recall, and F1 out of all the embeddings we evaluated, both unmapped and mapped. Though each source/target pair produced varying downstream results among the domain adaptation methods, a couple of broad trends emerged from our analysis. The largest performance gains over unmapped baselines were found when adapting high-resource WikiNews and PubMed embeddings to in-domain representations; however, these pairings also had the highest variability in results. The most consistent gains in precision came from using MIMIC embeddings as source, and these were mostly achieved through the nonlinear mapping approach.

There was no clear trend in the domain-adapted results as to whether word2vec or FastText embeddings led to the best downstream performance: it varied between pairs and adaptation methods. word2vec embeddings were generally more consistent, but as seen in Tables 4 and 5, FastText embeddings often achieved the highest performance.

5.4 Error analysis

Several interesting trends emerge in the NER errors produced in our experiments. Most generally, punctuation is often falsely considered to bound an entity. For example, the following string is part of a continuous Mobility entity:[8]

```
supine in bed with elevated leg,
and was left sitting in bed
```

However, most trained models separated this at the comma into two Mobility entities. Unsurprisingly, given the length of Mobility entities, we find many cases where most of the correct entity is tagged by the model, but the first or last few words are left off, as in

```
[he exhibits compensatory gait
patterns]Pred as a result]Gold
```

This behavior is illustrated in the large performance difference between entity-level and token-level evaluation discussed in Section 5.1.

We also see that descriptions of physical activity without specific evaluative terminology are often missed by the model. For example, working out in the yard is a Mobility entity ignored by the vast majority of our experiments, as is negotiate six steps to enter the apartment.

5.4.1 Corpus effects

Within correctly predicted entities, we see some indications of source corpus effect in the results. Considering just the original, non-adapted embeddings as presented in Table 2, we note two main differences between models trained on out-of-domain vs in-domain embeddings. In-domain

embeddings lead to much more conservative models: for example, PT-OT$_{w2v}$ only predicts 850 Mobility entities in test data, and BTRIS$_{w2v}$ predicts 863; this is in contrast to 922 predictions from MIMIC$_{w2v}$ and 940 from PubMed$_{w2v}$. This carries through to mapped embeddings as well: adding PT-OT embeddings into the mix decreases the number of predictions across the board.

Several predictions exhibit some degree of domain sensitivity, as well. For example, "fatigue" is present at the end of several Mobility mentions, and both PubMed and MIMIC embeddings typically end these mentions early. PubMed embeddings also append more typical symptomatic language onto otherwise correct Mobility entities, such as no areas of pressure noted on skin and numbness and tingling of arms. MIMIC and the heterogeneous in-domain BTRIS corpus append similar language, including and chronic pain. WikiNews embeddings, by contrast, appear oversensitive to key words in many Mobility mentions, tagging false positives such as my wife (spouses are often referred to as a source of physical support) and stairs are within range.

5.4.2 Changes from domain adaptation

Domain-adapted embeddings fix some corpus-based issues, but re-introduce others. Out-of-domain corpora tend to chain together Mobility entities separated by only one or two words, as in

```
[He ambulates w/o ad]Mobility, no
walker observed, [antalgic gait
pattern]Mobility
```

While source PubMed and WikiNews embeddings often collapse these to a single mention, adapting them to the target domain fixes many such cases. However, some of the original corpus noise remains: PT-OT$_{w2v}$ correctly ignored and chronic pain after a Mobility mention, but MIMIC$_{w2v}$ mapped to PT-OT$_{w2v}$ re-introduces this error.

The most consistent improvement obtained from domain adaptation was on Mobility entities that are short noun phrases, e.g. gait instability, and unsteady gait. Non-adapted embeddings typically miss such phrases, but mapped embeddings correctly find many of them, including some that in-domain embeddings miss.

5.4.3 Adaptation method effects

The most striking difference we observe when comparing different domain adaptation methods is that preinitialization universally leads to longer

---

[8] Several examples in this section have been edited for deidentification purposes and brevity.

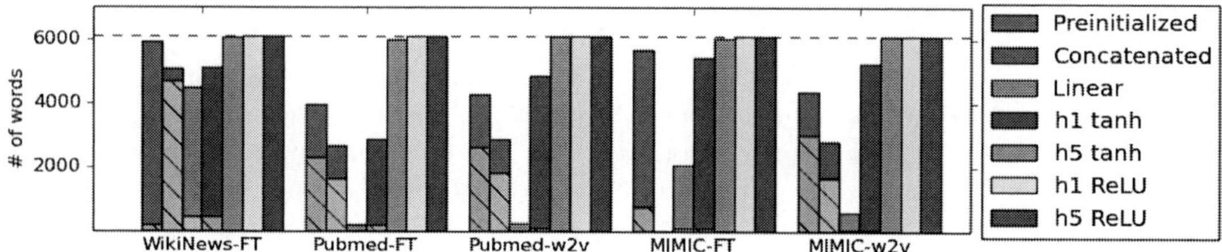

Figure 3: Number of words in shared vocabulary with different nearest neighbors in source and domain-adapted embeddings, using BTRIS$_{FT}$ as target. Light hatched bars indicate the number of words whose new nearest neighbor matches BTRIS$_{FT}$. The dashed line indicates shared vocabulary size.

Source set	Source	Target	Preinit	Concat	Linear	h1 tanh	h5 tanh
PubMed$_{FT}$	ambulating	ambulating	ambulating	ambulating	ambulating	ambulating	worsening
	ambulate	ambulate	ambulate	ambulate	ambulate	ambulate	wearing
	crutches	ambulatory	walker	crutches	crutches	crutch	complaints
WikiNews$_{FT}$	ambulating	ambulating	pos	ambulating	cardiopulmonary	robotic	respiratory
	ambulate	ambulate	76	ambulate	neurosurgical	overhead	sclerotic
	extubation	ambulatory	acuity	ambulatory	resuscitation	ambulating	acupuncture

Table 6: Top 3 nearest neighbors of *ambulation* in embeddings mapped to BTRIS$_{FT}$ using different adaptation methods. Source and Target are neighbors in the original source and BTRIS$_{FT}$ embeddings.

Mobility entity predictions, by both mean and variance of entity length. Though preinitialized embeddings still perform well overall, many predictions include several extra tokens before or after the true entity, as in the following example:

```
(now that her leg is healed [she
is independent with wheelchair
transfer]Gold and using her
shower bench)Pred
```

Preinitialized embeddings also have a strong tendency to collapse sequential Mobility entities. Both of these trends are reflected in the lower token-level precision numbers in Table 3.

Comparing nonlinear mapping methods, we find that a 1-layer mapping with tanh activation consistently leads to fewer predicted Mobility entities than with ReLU (for example, 814 vs 859 with WikiNews$_{FT}$ mapped to BTRIS$_{w2v}$, 917 vs 968 with MIMIC$_{w2v}$ mapped to PT-OT$_{w2v}$). However, this difference disappears when a 5-layer mapping is used. Despite their consistent performance, nonlinear transformations seem to re-introduce a number of errors related to more general mobility terminology. For example, he is very active and runs 15 miles per week is correctly recognized by concatenated WikiNews$_{FT}$ and BTRIS$_{w2v}$, but missed by several of their nonlinear mappings.

6 Embedding analysis

To further evaluate the effects of different domain adaptation methods, we analyzed the nearest neighbors by cosine similarity of each word before and after domain adaptation. We only considered the words present both in the dataset and in each of our original sets of embeddings, yielding a vocabulary of 6,201 words. We then took this vocabulary and calculated nearest neighbors within it, using each set of out-of-domain original embeddings and each of its domain-adapted transformations.

Figure 3 shows the number of words whose nearest neighbors changed after adaptation, using BTRIS$_{FT}$ as the target; all other targets display similar results. We see that in general, the neighborhood structure of target embeddings is well-preserved with concatenation, sometimes preserved with preinitialization, and completely disposed of with the nonlinear transformation. Interestingly, this reorganization of words to something different from both source and target does not lead to the performance degradation we might expect, as shown in Section 5.

We also qualitatively examined nearest neighbors before and after adaptation. Table 6 shows nearest neighbors of *ambulation*, a common Mobility word, for two representative source/target pairs. Preinitialization generally reflects the neighborhood structure of the target embeddings, but can be noisy: in WikiNews$_{FT}$/BTRIS$_{FT}$, other words such as *therapy* and *fatigue* share *ambulation*'s less-than-intuitive neighbors.

Reflecting the changes seen in Figure 3, the linear transformation preserves source neighbors in the biomedical PubMed corpus, but yields a neighborhood structure different from source or target with highly out-of-domain WikiNews embeddings. Nonlinear transformations sometimes yield sensible nearest neighbors, as in the single-layer tanh mapping of PubMed$_{FT}$ to BTRIS$_{FT}$. More often, however, the learned projection significantly shuffles neighborhood structure, and observed neighbors may bear only a distant similarity to the query term. In several cases, large swathes of the vocabulary are mapped to a single tight region of the space, yielding the same nearest neighbors for many disparate words. This occurs more often when using a ReLU activation, but we also observe it occasionally with tanh activation.

7 Conclusions

We have conducted an experimental analysis of recognizing descriptions of patient mobility with a recurrent neural network, and of the effects of various domain adaptation methods on recognition performance. We find that a state-of-the-art recurrent neural model is capable of capturing long, complex descriptions of mobility, and of recognizing mobility measurement scales nearly perfectly. Our experiments show that domain adaptation methods often improve recognition performance over both in- and out-of-domain baselines, though such improvements are difficult to achieve consistently. Simpler methods such as preinitialization and concatenation achieve better performance gains, but are also susceptible to noise in source embeddings; more complex methods yield more consistent performance, but with practical downsides such as decreased recall and a non-intuitive projection of the embedding space. Most strikingly, we see that embeddings trained on a very small corpus of highly relevant documents nearly match the performance of embeddings trained on extremely large out-of-domain corpora, adding to the recent findings of Diaz et al. (2016).

To our knowledge, this is the first investigation into automatically recognizing descriptions of patient functioning. Viewing this problem through an NER lens provides a robust framework for model design and evaluation, but is accompanied by challenges such as effectively evaluating recognition of long text spans and dealing with complex syntactic structure and punctuation within relevant mentions. It is our hope that these initial findings, along with further research refining the appropriate framework for representing and approaching the recognition problem, will spur further research into this complex and important domain.

Acknowledgments

The authors would like to thank Elizabeth Rasch, Thanh Thieu, and Eric Fosler-Lussier for helpful discussions, the NIH Biomedical Translational Research Information System (BTRIS) for their support, and our anonymous reviewers for their invaluable feedback. This research was supported in part by the Intramural Research Program of the National Institutes of Health, Clinical Research Center and through an Inter-Agency Agreement with the US Social Security Administration.

References

Mikel Artetxe, Gorka Labaka, and Eneko Agirre. 2016. Learning principled bilingual mappings of word embeddings while preserving monolingual invariance. In *Proceedings of the 2016 Conference on Empirical Methods in Natural Language Processing*, pages 2289–2294, Austin, Texas. Association for Computational Linguistics.

Piotr Bojanowski, Edouard Grave, Armand Joulin, and Tomas Mikolov. 2017. Enriching Word Vectors with Subword Information. *Transactions of the ACL*, 5:135–146.

Billy Chiu, Gamal Crichton, Anna Korhonen, and Sampo Pyysalo. 2016. How to Train Good Word Embeddings for Biomedical NLP. *Proceedings of the 15th Workshop on Biomedical Natural Language Processing*, pages 166–174.

Jason PC Chiu and Eric Nichols. 2015. Named entity recognition with bidirectional lstm-cnns. *arXiv preprint arXiv:1511.08308*.

James J. Cimino and Elaine J. Ayres. 2010. The clinical research data repository of the US National Institutes of Health. *Studies in Health Technology and Informatics*, 160(PART 1):1299–1303.

Ronan Collobert, Jason Weston, Léon Bottou, Michael Karlen, Koray Kavukcuoglu, and Pavel Kuksa. 2011. Natural language processing (almost) from scratch. *Journal of Machine Learning Research*, 12(Aug):2493–2537.

Franck Dernoncourt, Ji Young Lee, and Peter Szolovits. 2017a. NeuroNER: an easy-to-use program for named-entity recognition based on neural networks. In *Proceedings of the 2017 Conference on Empirical Methods in Natural Language Processing: System Demonstrations*, pages 97–102, Copenhagen, Denmark. Association for Computational Linguistics.

Franck Dernoncourt, Ji Young Lee, Ozlem Uzuner, and Peter Szolovits. 2017b. De-identification of patient notes with recurrent neural networks. *Journal of the American Medical Informatics Association*, 24(3):596–606.

Fernando Diaz, Bhaskar Mitra, and Nick Craswell. 2016. Query Expansion with Locally-Trained Word Embeddings. In *Proceedings of the 54th Annual Meeting of the Association for Computational Linguistics (Volume 1: Long Papers)*, pages 367–377, Berlin, Germany. Association for Computational Linguistics.

Jenny Rose Finkel, Trond Grenager, and Christopher Manning. 2005. Incorporating non-local information into information extraction systems by gibbs sampling. In *Proceedings of the 43rd Annual Meeting of the Association for Computational Linguistics*, pages 363–370. Association for Computational Linguistics.

Carol Friedman, Pauline Kra, and Andrey Rzhetsky. 2002. Two biomedical sublanguages: A description based on the theories of Zellig Harris. *Journal of Biomedical Informatics*, 35(4):222–235.

Maryam Habibi, Leon Weber, Mariana Neves, David Luis Wiegandt, and Ulf Leser. 2017. Deep learning with word embeddings improves biomedical named entity recognition. *Bioinformatics*, 33(14):i37–i48.

Alistair E W Johnson, Tom J Pollard, Lu Shen, Li-Wei H Lehman, Mengling Feng, Mohammad Ghassemi, Benjamin Moody, Peter Szolovits, Leo Anthony Celi, and Roger G Mark. 2016. MIMIC-III, a freely accessible critical care database. *Scientific data*, 3:160035.

Jun'ichi Kazama, Takaki Makino, Yoshihiro Ohta, and Jun'ichi Tsujii. 2002. Tuning support vector machines for biomedical named entity recognition. In *Proceedings of the ACL-02 Workshop on Natural Language Processing in the Biomedical Domain-Volume 3*, pages 1–8. Association for Computational Linguistics.

Diederik P Kingma and Jimmy Ba. 2014. Adam: A method for stochastic optimization. *arXiv preprint arXiv:1412.6980*.

Rita Kukafka, Michael E. Bales, Ann Burkhardt, and Carol Friedman. 2006. Human and Automated Coding of Rehabilitation Discharge Summaries According to the International Classification of Functioning, Disability, and Health. *Journal of the American Medical Informatics Association*, 13(5):508–515.

Guillaume Lample, Miguel Ballesteros, Sandeep Subramanian, Kazuya Kawakami, and Chris Dyer. 2016. Neural architectures for named entity recognition. *arXiv preprint arXiv:1603.01360*.

Zengjian Liu, Buzhou Tang, Xiaolong Wang, and Qingcai Chen. 2017. De-identification of clinical notes via recurrent neural network and conditional random field. *Journal of Biomedical Informatics*, 75:S34–S42.

Andrew McCallum and Wei Li. 2003. Early results for named entity recognition with conditional random fields, feature induction and web-enhanced lexicons. In *Proceedings of the Seventh Conference on Natural Language Learning at HLT-NAACL 2003-Volume 4*, pages 188–191. Association for Computational Linguistics.

Tomas Mikolov, Kai Chen, Greg Corrado, and Jeffrey Dean. 2013. Efficient Estimation of Word Representations in Vector Space. *arXiv preprint arXiv:1301.3781*, pages 1–12.

Serguei V S Pakhomov, Greg Finley, Reed McEwan, Yan Wang, and Genevieve B Melton. 2016. Corpus domain effects on distributional semantic modeling of medical terms. *Bioinformatics*, 32(August):btw529.

Guergana K Savova, James J Masanz, Philip V Ogren, Jiaping Zheng, Sunghwan Sohn, Karin C Kipper-Schuler, and Christopher G Chute. 2010. Mayo clinical Text Analysis and Knowledge Extraction System (cTAKES): architecture, component evaluation and applications. *Journal of the American Medical Informatics Association : JAMIA*, 17(5):507–513.

Burr Settles. 2004. Biomedical named entity recognition using conditional random fields and rich feature sets. In *Proceedings of the International Joint Workshop on Natural Language Processing in Biomedicine and its Applications*, pages 104–107. Association for Computational Linguistics.

Steven J Skube, Elizabeth A Lindemann, Elliot G Arsoniadis, Elizabeth C Wick, and Genevieve B Melton. 2018. Characterizing Functional Health Status of Surgical Patients in Clinical Notes. In *2018 AMIA Summit on Clinical Research Informatics*. American Medical Informatics Association.

Ergin Soysal, Jingqi Wang, Min Jiang, Yonghui Wu, Serguei Pakhomov, Hongfang Liu, and Hua Xu. 2018. CLAMP a toolkit for efficiently building customized clinical natural language processing pipelines. *Journal of the American Medical Informatics Association*, 25(3):331–336.

G Stucki, J Bickenbach, and J Melvin. 2017. Strengthening Rehabilitation in Health Systems Worldwide by Integrating Information on Functioning in National Health Information Systems. *Am J Phys Med Rehabil*, 96(9):677–681.

Gerold Stucki and Jerome Bickenbach. 2017. Functioning: the third health indicator in the health system and the key indicator for rehabilitation. *European Journal of Physical and Rehabilitation Medicine*, 53(1):134–138.

Thanh Thieu, Jonathan Camacho, Pei-Shu Ho, Diane Brandt, Julia Porcino, Denis Newman-Griffis, Ao Yuan, Min Ding, Lisa Nelson, Elizabeth Rasch, Chunxiao Zhou, Albert M Lai, and Leighton Chan. 2017. Inductive identification of functional status information and establishing a gold standard corpus A case study on the Mobility domain. *IEEE International Conference on Bioinformatics and Biomedicine (BIBM)*, pages 2300–2302.

Erik F Tjong Kim Sang and Fien De Meulder. 2003. Introduction to the CoNLL-2003 Shared Task: Language-Independent Named Entity Recognition. In *Proceedings of the Seventh Conference on Natural Language Learning at HLT-NAACL 2003*, pages 142–147.

Richard Tzong-Han Tsai, Cheng-Lung Sung, Hong-Jie Dai, Hsieh-Chuan Hung, Ting-Yi Sung, and Wen-Lian Hsu. 2006. Nerbio: using selected word conjunctions, term normalization, and global patterns to improve biomedical named entity recognition. In *BMC Bioinformatics*, volume 7, page S11. BioMed Central.

Özlem Uzuner, Brett R South, Shuying Shen, and Scott L DuVall. 2012. 2010 i2b2/VA challenge on concepts, assertions, and relations in clinical text. *Journal of the American Medical Informatics Association : JAMIA*, 18(5):552–6.

WHO. 2001. *International Classification of Functioning, Disability and Health: ICF.* World Health Organization.

Long Xia, G Alan Wang, and Weiguo Fan. 2017. A Deep Learning Based Named Entity Recognition Approach for Adverse Drug Events Identification and Extraction in Health Social Media. In *Smart Health*, pages 237–248, Cham. Springer International Publishing.

Wei Yang, Wei Lu, and Vincent Zheng. 2017. A simple regularization-based algorithm for learning cross-domain word embeddings. In *Proceedings of the 2017 Conference on Empirical Methods in Natural Language Processing*, pages 2898–2904.

Multi-task learning for interpretable cause-of-death classification using key phrase prediction

Serena Jeblee
Dept of Computer Science
University of Toronto
Toronto, Ontario, Canada
sjeblee@cs.toronto.edu

Mireille Gomes
St. Michael's Hospital
Toronto, Ontario, Canada
mireille.m.gomes@gmail.com

Graeme Hirst
Dept of Computer Science
University of Toronto
Toronto, Ontario, Canada
gh@cs.toronto.edu

Abstract

We introduce a multi-task learning model for cause-of-death classification of verbal autopsy narratives that jointly learns to output interpretable key phrases. Adding these key phrases outperforms the baseline model and topic modeling features.

1 Introduction

Verbal autopsies (VAs) are written records of the events leading up to a person's death, typically in situations where there was no physical autopsy and the cause of death (CoD) was not determined by a physician. As per World Health Organization recommendations, most VAs contain structured information from answers to a questionnaire, and may also contain a free-text narrative that captures additional information, such as the time and sequence of the subject's symptoms and details of their medical history (Nichols et al., 2018). VAs are used in countries such as India to gain a better idea of the most prevalent causes of death, which are not accurately captured by only the small number of well-documented deaths that occur in health facilities.

Typically, VAs are collected by non-medical surveyors who record the information on a form that is later reviewed by physicians who assign the record an International Classification of Diseases (ICD-10) code (World Health Organization, 2008). Automating some of this coding process would reduce the time and costs of VA surveys.

Previous work has shown that machine learning methods can be useful for medical text classification. However, many models do not provide interpretable explanations for their output, which are crucial in health care.

Multi-task learning methods use a shared architecture to learn several classification tasks, which has been shown to improve performance especially when the tasks are closely related. In this work we aim to use a multi-task learning model to classify VA narratives according to CoD and simultaneously provide automatically determined key phrases in order to increase the interpretability of the model.

2 Related work

Several automated methods for coding VAs are currently in use, including InterVA (Byass et al., 2012), InSilicoVA (McCormick et al., 2016), and the Tariff Method (Serina et al., 2015). However, these methods are largely based on the structured data (which varies depending on the particular VA survey form used) and on physician-curated conditional probabilities of symptoms and diseases, which are time-consuming to collect. The performance of these methods is typically less than .60 precision for 15 to 30 CoD categories (Desai et al., 2014).

Miasnikof et al. (2015) used a naïve Bayes classifier with structured data and achieved comparable or better results than the expert-driven models. Danso et al. (2013) used linguistic features to classify VA narratives of neonatal deaths into 16 CoD categories with a support vector machine (SVM), achieving .406 recall.

TextRank (Mihalcea and Tarau, 2004) is a popular method that uses document graphs to extract key phrases. However, unsupervised models such as TextRank can extract text only from the document itself, in which the physician-generated key phrases that we use in this work might or might not be explicitly present. Latent Dirichlet Allocation (LDA) (Blei et al., 2003) is a topic modeling framework that is often used for text classification. We will compare our key phrase clusters to LDA topics learned from the same narrative data.

Proceedings of the BioNLP 2018 workshop, pages 12–17
Melbourne, Australia, July 19, 2018. ©2018 Association for Computational Linguistics

3 Data

Our dataset consists of 12,045 records of adult deaths from the Million Death Study (MDS) (Westly, 2013; Aleksandrowicz et al., 2014; Gomes et al., 2017), which is a program to collect and code VAs from India. In the MDS coding process, two physicians separately assign an ICD-10 code to each record and disagreements are resolved by a third physician if necessary. Because there are hundreds of possible ICD-10 codes and our dataset is fairly small, the codes are grouped into 18 CoD categories, which are broader groupings of the WHO 2012 VA categories (World Health Organization, 2012).

The narratives, written by non-medical surveyors, range from a couple of sentences to a few paragraphs and describe the person's medical history and symptoms before death. In addition to the free-text narratives, the VA records from the MDS also contain key phrases assigned by the coding physicians. By highlighting important symptoms and events, these phrases are meant to explain the code assigned. They may be taken directly from the narrative or written in by the physician.

We represent the narrative text and key phrases with 100-dimensional word embeddings trained with the word2vec CBOW algorithm[1], which learns vector space representations for words based on their context (Mikolov et al., 2013). The key phrase representation for clustering is the average of the embeddings of each word in the phrase. The narrative representation is a matrix containing the embeddings for each word in order, padded with zero vectors to a maximum length of 200 words.

Because the dataset is rather small for training word2vec, we include Indian English text from the International Corpus of English[2] and 1.7M posts from MedHelp[3], an online medical advice forum that contains informal health-related language.

The text of both the narratives and the key phrases is lowercased, punctuation is removed, and spelling is corrected using PyEnchant's English dictionary (Kelly, 2015) and a 5-gram language model for scoring candidate replacements, using KenLM (Heafield et al., 2013). After preprocessing we remove duplicate key phrases.

---

[1]We used a context window of 5, min count of 1 (due to the small dataset), and no negative sampling.

[2]http://ice-corpora.net/ice/avail.htm

[3]http://www.medhelp.org

4 Model

The model used for both key phrase cluster prediction and CoD classification is a neural network that contains a gated recurrent unit layer (GRU) (Cho et al., 2014) with 0.1 dropout followed by a convolutional layer (CNN) with filters of size $1 \times d$ through $5 \times d$ where d is the word embedding size (100), followed by a max-pooling layer. The output of the pooling layer is then used as input to a dense softmax layer that outputs the classification. The models are implemented in Python using Keras (Chollet, 2015), with the Theano backend (Theano Development Team, 2016).

For CoD classification, the prediction layer outputs the probabilities over the 18 CoD categories, and we choose the one with the highest probability. For key phrase prediction, it outputs the probabilities over the key phrase clusters, and we take each cluster as a label if it has a probability of 0.1 or higher (since there can be any number of key phrases per record). A higher cutoff will result in slightly higher precision but lower recall. The loss functions are categorical cross-entropy for CoD classification and mean squared error for key phrase cluster prediction.

The multi-task learning model consists of a shared GRU/CNN model that generates a vector representation that is then used by two parallel prediction layers, one for the CoD category and one for the key phrase clusters. The key phrase loss function has a weight of 0.1 to emphasize the CoD coding task. All three of these models use only the narrative word embedding matrix as input.

5 Key phrase clustering

5.1 Unsupervised clustering

The key phrases from the training data are grouped into 100 clusters using the k-means algorithm with Euclidean distance from scikit-learn (Pedregosa et al., 2011).

We need a sufficient number of clusters to capture specific symptoms and event, but not so many that we cannot predict them accurately. In our case, we want to favor precision over recall because we would rather generate fewer, more-correct key phrases than more phrases that are less accurate. We chose 100 clusters based on early experiments to maximize precision and the number of clusters.

Label	Key phrases in cluster
cough	cough, cough with sputum, cough with phlegm, had sputum cough, . . .
rigours	fear, sudden chest pain one day and died in short while, h/o headache, epileptic, . . .
h/o chest pain	sudden chest pain, occasional chest pain, sudden pain in middle of chest, . . .
breathing difficulty	difficulty in eating, difficulty in urination, . . .

Table 1: Examples of key phrase clusters with generated labels ('h/o' means 'history of')

Model	CoD classification			Key phrase cluster prediction		
	Precision	Recall	F_1	Precision	Recall	F_1
Majority class	.027	.163	.046	.292	.070	.105
Key phrase only	-	-	-	**.498**	**.283**	**.317**
CoD only	.755	.746	.743	-	-	-
Multi-task	**.760**	**.753**	**.750**	.481	.276	.310

Table 2: Weighted average scores from 10-fold cross-validation using the GRU/CNN model

CoD classification			
Features	Precision	Recall	F_1
Majority class	.027	.163	.046
Embeddings	.757	.752	.747
Emb + LDA	.726	.703	.699
Emb + key phrases	**.779**	**.778**	**.774**

Table 3: Results using a CNN model with a parallel feed-forward network (inputs are word embeddings and key phrases or LDA topics respectively)

5.2 Cluster prediction

For new, uncoded records, we will have only the narrative and therefore will need to predict the key phrase clusters. For evaluation, because the clustering is unsupervised and we have no gold standard mapping of key phrases in the test data to clusters, we assign each test key phrase to a cluster using a k-nearest neighbor classifier ($k = 5$). We treat these clusters as the "true" labels for the key phrase prediction task.

5.3 Cluster interpretation

In order for these clusters to be useful to physicians, we need a text label for each. We could simply take the most frequent key phrase in each cluster as the label, but many key phrases are variations of the the same idea, or have extra details in them, so the most frequent phrase might not be the most representative. Therefore, to get a text label that is representative of the cluster, we choose the key phrase that is closest to the center of the cluster in vector space.

However, there are some key phrases which are much longer than average. Since the vector representation of each phrase is the average of the word embeddings, a phrase with many words is more likely to be closer to the center. Also, we want to favor shorter labels that are general enough to describe the members of the cluster. Therefore we introduce a length penalty: the score used for selecting the label phrase is the distance of the phrase embedding from the center of the cluster multiplied by the number of words in the phrase. This gives us cluster labels that are usually one or two words.

Table 1 shows some of the generated cluster labels and the associated key phrases.[4]

6 Results

Table 2 shows the results from 10-fold cross-validation for key phrase cluster prediction and CoD classification, using the multi-task learning model, as well as separate models. The majority-class baseline is the scores obtained by assigning every record to the most frequent class in the training set ('road traffic incidents').

Some key phrase clusters are much larger and more frequent than others, which can render them unhelpful if too many different key phrases are grouped together. For the key phrase majority

---

[4] All examples are from the first iteration of 10-fold cross-validation, since different clusters are generated for each training set.

Record CoD category	Physician-assigned key phrases	Nearest-neighbor clusters	Predicted clusters
Ischemic heart disease	stroke patient, fever, dizziness for days, severe abdominal pain, diggings's, sudden pain abd.	oliguria, fever, sometime, abdominal pain, oliguria, diahorrea	pain abdomen, fever
Chronic respiratory infections	cough, wheezing, breathlessness edema	cough, h/o cough, breathlessness, h/o edema	h/o cough, breathlessness
Liver and alcohol	heavy alcohol intake, less food, not having food at regular interval, excess consumption of alcohol	pesticide, pesticide, oliguria, pesticide	died in5 mts., oliguria, progressive

Table 4: Examples of predicted key phrase clusters and CoD categories from the test set. Nearest neighbor clusters are the clusters from the training set that are closest to the embeddings of the physician key phrases.

baseline, we assign the most frequent key phrase cluster from the training set to each record in the test set. Even though there are 100 possible clusters and multiple clusters per record, we get .292 precision from the most frequent cluster alone.

We also use the predicted key phrase clusters as features for CoD classification. We use the clusters predicted by the 'key phrase only' model as input to a CNN CoD classifier. The input to the CNN layer is the matrix of word embeddings from the narratives, as in the previous model, and key phrase clusters are represented as a binary array that is the input to a feed-forward layer of 100 nodes. The outputs of the CNN module and the feed-forward module are concatenated and used as input for the final softmax classification layer, which outputs the CoD category.

Table 3 shows the results of this model, compared to the same model architecture using 100 LDA topics as the second feature set. The model using predicted key phrase features performs much better than that using the LDA topics. It also outperforms both the CNN model using only the narrative embeddings (without the feed-forward layer), and the majority class baseline.

7 Discussion

Table 4 shows some examples of the key phrase clusters predicted by the multi-task model. As we can see from the first two examples, many of the predicted phrases capture the same type of information as the physician-generated key phrases, although not as thoroughly.

However, as seen in Table 1, the clustering doesn't always capture the type of similarity we're interested in, such as the 'breathing difficulty' cluster, which captures phrases containing 'difficulty', although these often represent different symptoms. In Table 4 we see that the cluster representing alcohol intake has been labeled as 'pesticide' (along with several other clusters), and the predicted clusters for the third record do not contain any useful information related to the CoD (alcohol consumption).

Despite the key phrase prediction accuracy being fairly low, adding these predicted clusters as features for CoD classification improves both the precision and recall of the model.

We suspect that topic modeling does not help in this case because the distribution of words is very similar between documents, since they all deal with symptoms leading up to death. In addition, the key phrases are generated by physicians, and can capture information that is not explicitly present in the narrative.

8 Conclusion

We have demonstrated that learning key phrases along with CoD categories can improve CoD classification accuracy for verbal autopsies. The text representation of the key phrase clusters also adds interpretability to the model. In future work, we will aim to improve the cluster prediction accuracy, and we will investigate unsupervised methods of extracting important information from VA narratives.

Acknowledgments

We thank Prabhat Jha of the Centre for Global Health Research for providing the dataset. Our work is supported by funding from the Natural Sciences and Engineering Research Council of Canada and from a Google Faculty Award.

References

Lukasz Aleksandrowicz, Varun Malhotra, Rajesh Dikshit, Rajesh Kumar Prakash C Gupta, Jay Sheth, Suresh Kumar Rathi, Wilson Suraweera, Pierre Miasnikof, Raju Jotkar, Dhirendra Sinha, Shally Awasthi, Prakash Bhatia, and Prabhat Jha. 2014. Performance criteria for verbal autopsy-based systems to estimate national causes of death: Development and application to the Indian Million Death Study. *BMC Medicine*, 12:21.

David M Blei, Andrew Y Ng, and Michael I Jordan. 2003. Latent Dirichlet allocation. *Journal of Machine Learning Research*, 3:993–1022.

Peter Byass, Daniel Chandramohan, Samuel Clark, Lucia D'Ambruoso, Edward Fottrell, Wendy Graham, Abraham Herbst, Abraham Hodgson, Sennen Hounton, Kathleen Kahn, Anand Krishnan, Jordana Leitao, Frank Odhiambo, Osman Sankoh, and Stephen Tollman. 2012. Strengthening standardised interpretation of verbal autopsy data: The new InterVA-4 tool. *Global Health Action*, 5:19281.

Kyunghyun Cho, Bart van Merriënboer, Caglar Gulcehre, Dzmitry Bahdanau, Fethi Bougares, Holger Schwenk, and Yoshua Bengio. 2014. Learning phrase representations using RNN encoder-decoder for statistical machine translation. In *Proceedings of the 2014 Conference on Empirical Methods in Natural Language Processing (EMNLP)*, pages 1724–1734.

François Chollet. 2015. Keras. https://github.com/fchollet/keras.

Samuel Danso, Eric Atwell, and Owen Johnson. 2013. Linguistic and statistically derived features for cause of death prediction from verbal autopsy text. In *Language Processing and Knowledge in the Web*, pages 47–60. Springer Berlin Heidelberg.

Nikita Desai, Lukasz Aleksandrowicz, Pierre Miasnikof, Ying Lu, Jordana Leitao, Peter Byass, Stephen Tollman, Paul Mee, Dewan Alam, Suresh Kumar Rathi, Abhishek Singh, Rajesh Kumar, Faujdar Ram, and Prabhat Jha. 2014. Performance of four computer-coded verbal autopsy methods for cause of death assignment compared with physician coding on 24,000 deaths in low- and middle-income countries. *BMC Medicine*, 12:20.

Mireille Gomes, Rehana Begum, Prabha Sati, Rajesh Dikshit, Prakash C Gupta, Rajesh Kumar, Jay Sheth, Asad Habib, and Prabhat Jha. 2017. Nationwide mortality studies to quantify causes of death: Relevant lessons from India's Million Death Study. *Health Affairs*, 36(11):1887–1895.

Kenneth Heafield, Ivan Pouzyrevsky, Jonathan H. Clark, and Philipp Koehn. 2013. Scalable modified Kneser-Ney language model estimation. In *Proceedings of the 51st Annual Meeting of the Association for Computational Linguistics*, pages 690–696, Sofia, Bulgaria.

Ryan Kelly. 2015. Pyenchant. http://pythonhosted.org/pyenchant/.

Tyler H McCormick, Zehang Richard Li, Clara Calvert, Amelia C Crampin, Kathleen Kahn, and Samuel Clark. 2016. Probabilistic cause-of-death assignment using verbal autopsies. *Journal of the American Statistical Association*, 111(15):1036–1049.

Pierre Miasnikof, Vasily Giannakeas, Mireille Gomes, Lukasz Aleksandrowicz, Alexander Y Shestopaloff, Dewan Alam, Stephen Tollman, Akram Samarikhalaj, and Prabhat Jha. 2015. Naïve Bayes classifiers for verbal autopsies: Comparison to physician-based classification for 21,000 child and adult deaths. *BMC Medicine*, 13(1):286–294.

Rada Mihalcea and Paul Tarau. 2004. TextRank: Bringing order into texts. In *Proceedings of the Conference on Empirical Methods in Natural Language Processing (EMNLP 2004)*.

Tomas Mikolov, Ilya Sutskever, Kai Chen, Greg S Corado, and Jeff Dean. 2013. Distributed representations of words and phrases and their compositionality. In *Advances in Neural Information Processing Systems 26*, pages 3111–3119.

Erin K. Nichols, Peter Byass, Daniel Chandramohan, Samuel J. Clark, Abraham D. Flaxman, Robert Jakob, Jordana Leitao, Nicolas Maire, Chalapati Rao, Ian Riley, and Philip W. Setel. 2018. The WHO 2016 verbal autopsy instrument: An international standard suitable for automated analysis by InterVA, InSilicoVA, and Tariff 2.0. *PLOS Medicine*, 15(1):e1002486.

Fabian Pedregosa, Gaël Varoquaux, Alexandre Gramfort, Vincent Michel, Bertrand Thirion, Olivier Grisel, Mathieu Blondel, Peter Prettenhofer, Ron Weiss, Vincent Dubourg, Jake Vanderplas, Alexandre Passos, David Cournapeau, Matthieu Brucher, Matthieu Perrot, and Édouard Duchesnay. 2011. Scikit-learn: Machine learning in Python. *Journal of Machine Learning Research*, 12(Oct):2825–2830.

Peter Serina, Ian Riley, Andrea Stewart, Spencer L James, Abraham D Flaxman, Rafael Lozano, et al. 2015. Improving performance of the Tariff method for assigning causes of death to verbal autopsies. *BMC Medicine*, 13(1):291.

Theano Development Team. 2016. Theano: A Python framework for fast computation of mathematical expressions. *arXiv e-prints*, abs/1605.02688. http://arxiv.org/abs/1605.02688.

Erica Westly. 2013. Global health: One million deaths. *Nature*, 504(7478):22–23.

World Health Organization. 2008. *International statistical classifications of diseases and related health problems. 10th rev*, volume 1. World Health Organization, Geneva, Switzerland.

World Health Organization. 2012. *The 2012 WHO Verbal Autopsy Instrument*. World Health Organization, Geneva, Switzerland.

Identifying Risk Factors For Heart Disease in Electronic Medical Records: A Deep Learning Approach

Thanat Chokwijitkul[1], Anthony Nguyen[2], Hamed Hassanzadeh[2], Siegfried Perez[3]

[1]School of Information Technology and Electrical Engineering, The University of Queensland
[2]The Australian e-Health Research Centre, CSIRO
[3]Emergency Department, Logan Hospital

t.chokwijitkul@uqconnect.edu.au
{Anthony.Nguyen, Hamed.Hassanzadeh}@csiro.au
SiegfriedRobert.Perez@health.qld.gov.au

Abstract

Automatic identification of heart disease risk factors in clinical narratives can expedite disease progression modelling and support clinical decisions. Existing practical solutions for cardiovascular risk detection are mostly hybrid systems entailing the integration of knowledge-driven and data-driven methods, relying on dictionaries, rules and machine learning methods that require a substantial amount of human effort. This paper proposes a comparative analysis on the applicability of *deep learning*, a re-emerged data-driven technique, in the context of clinical text classification. Various deep learning architectures were devised and evaluated for extracting heart disease risk factors from clinical documents. The data provided for the 2014 i2b2/UTHealth shared task focusing on identifying risk factors for heart disease was used for system development and evaluation. Results have shown that a relatively simple deep learning model can achieve a high micro-averaged F-measure of 0.9081, which is comparable to the best systems from the shared task. This is highly encouraging given the simplicity of the deep learning approach compared to the heavily feature-engineered hybrid approaches that were required to achieve state-of-the-art performances.

1 Introduction

Heart disease is a leading cause of morbidity and mortality worldwide (Benjamin et al., 2017). As failure to recognise atypical representations of such serious illness may lead to adverse outcomes, accurate diagnosis is crucial to ensure that patients are placed on the proper treatment pathway. Electronic medical records (EMR) can be used to improve the diagnosis ability along with measuring the quality of care. The rapid adoption of EMRs along with the necessity to enhance the quality of health care has incentivised the development of natural language processing (NLP) in the medical domain. An abundant amount of clinical information used for medical investigation is organised in unstructured narrative form, which is suitable for expressing medical concepts or events but challenging for analysis and decision support as gaining a full aspect of a patients medical history by reading through EMRs is significantly time-consuming, especially when only a specific piece of information is needed. The difficulty of this process increases in the case of heart disease due to its complex progression, which regularly involves various factors including lifestyle and social factors as well as specific medical conditions (Stubbs and Uzuner, 2015). Various methods have been proposed in the field of clinical concept extraction, ranging from simple pattern matching to systems based on symbolic or statistical data and machine learning (Meystre et al., 2008; Gonzalez-Hernandez et al., 2017). Those previously proposed approaches have shown promising results but it is very difficult to reach that point due to the assiduous process of defining rules and extracting features. This is where deep learning comes in as this intriguing re-emerged concept can alleviate heavily human dependent efforts required for knowledge-based approaches and the lack of the ability of many conventional machine learning algorithms to learn without the necessity of careful feature engineering with considerable domain expertise (LeCun et al., 2015).

This paper presents a comparative analysis of two widely used deep learning architectures, namely convolutional neural network (CNN) and

Proceedings of the BioNLP 2018 workshop, pages 18–27
Melbourne, Australia, July 19, 2018. ©2018 Association for Computational Linguistics

recurrent neural network (RNN) as well as three RNN variants, including long short-term memory (LSTM) (Hochreiter and Schmidhuber, 1997), bidirectional long short-term memory (BLSTM), and gated recurrent unit (GRU) (Cho et al., 2014), for extracting cardiac risk factors from EMRs. Using the data set from the i2b2/UTHealth shared task (Stubbs and Uzuner, 2015), the goal is to determine the risk factor indicators contained within each document along with the temporal attributes with respect to the document creation time (DCT).

2 Related Work

2.1 Deep Learning for Clinical Information Extraction

Many recent publications have focused on extracting relevant clinical information from EMRs using deep learning. One of the most fundamental tasks involves the extraction of medical concepts from unstructured clinical notes. This concept extraction problem can be treated as a sequence labelling problem where the goal is to assign a clinically relevant tag to each word in an EMR (Jagannatha and Yu, 2016). Jagannatha and Yu (2016) experimented with different deep learning architectures based on recurrent networks, including GRUs, LSTMs and BLSTMs. It turned out that all the RNN variants outperformed the conditional random field (CRF) baselines, which had previously been considered the state-of-the-art method for information extraction in general.

As patient EMRs evolve over time, the sequentiality of clinical events can be used for disease progression analysis and the prediction of impending disease conditions (Cheng et al., 2016). Its temporality induces the necessity of assigning notions of time to each extracted medical concept. Fries (2016) devised a solution for such more complex problems by using a standard RNN initialised with word2vec (Mikolov et al., 2013a) vectors along with utilising DeepDive (Shin et al., 2015) for forming relationships and predictions. Li and Huang (2016) and Chikka (2016) also employed word embedding vectors within their frameworks but used CNNs to extract the temporal attributes instead. While still not state-of-the-art, these approaches produced competitive results in the field of temporal event extraction but also required a separate model for each subtask (extracting concepts and temporal attributes) and slight manual engineering (Shickel et al., 2017; Bethard et al.,

2016). One thing to remark is that none of the existing systems has ever tried using a single, universal model that naturally learns the temporal characteristics of those concepts based on their contexts and incorporates them into the feature learning process, which can be used for extracting medical concepts and temporal attributes simultaneously. This work intends to explore this idea and prove that the aforementioned capability is well within the reach of deep learning.

2.2 i2b2/UTHealth Shared Task

In 2014, the Informatics for Integrating Biology and the Bedside (i2b2) issued an NLP shared task focusing on identifying risk factors for heart disease in clinical narratives. According to Stubbs et al. (2015), a total of 49 systems from 20 teams were submitted. The systems varied broadly, from rule-based systems to complex hybrid systems with a combination of machine learning techniques. Nevertheless, some similarities were found among the top systems including the use of preprocessing tools to obtain syntactic information and section headers for determining temporal labels. The results revealed that the top 10 systems achieved micro-averaged F1 scores over 0.87 while the top 6 systems were able to reach micro-averaged F1 scores over 0.90. The most successful system managed to achieve an F1 score of 0.928 (Roberts et al., 2015) while the averaged F1 score among all the systems was 0.815. While half of the top 10 teams used a combination of knowledge-driven methods, such as lexicon and rules, and machine learning algorithms, including CRF, support vector machine (SVM), Naïve Bayes classifier and Maximum Entropy, none of the participants attempted to integrate neural networks or deep learning into their systems. Furthermore, there has not existed any approaches that use deep learning to extract risk factor indicators from the shared task data since its inception in 2014, which is a research gap that this work intends to fill.

3 Methodology

3.1 Dataset

The dataset used in this work is the corpus provided for the 2014 i2b2/UTHealth shared task. The corpus consists of 1,304 medical records describing 296 diabetic patients for cardiovascular risk factors and time attributes with respect to the DCT. The dataset was split by the challenge

Risk Factor	Indicator	Training Instances	Testing Instances	Time Attribute
CAD	mention, event, test, symptom	1186	784	✓
Diabetes	mention, high A1c, high glucose	1695	1180	✓
Obesity	mention, high BMI	433	262	✓
Hyperlipidemia	mention, high cholesterol, high LDL	1062	751	✓
Hypertension	mention, high blood pressure	1926	1293	✓
Medication	ACE inhibitor, amylin, anti-diabetes, ARB, aspirin, beta blocker, calcium channel blocker, diuretic, DPP4 inhibitors, ezetimibe, fibrate, GLP1 agonist, insulin, Meglitinide, metformin, niacin, nitrate, obesity medications, statin, sulfonylurea, thiazolidinedione, thienopyridine	8638	5674	✓
Smoking	current, past, ever, never, unknown	771	512	n/a
Family history	present, not present	790	514	n/a

Table 1: The indicators associated with each cardiac risk factor and the number of training and testing instances at annotation level

Evidence Type	Example
Phrase-based	Significant PMH for **CAD**, **HTN**, **GERD**, and past cerebral embolism
Logic-based	Seen in Cardiac rehab locally last week and **BP 170/80**
Discourse-based	**Findings suggestive of an obstructive, coronary lesion in the left circumflex distribution**

Table 2: Three types of evidence

Risk Factor	Phrase-based	Logic-based	Discourse-based
CAD	mention	n/a	event, test, symptom
Diabetes	mention	high A1c, high glucose	n/a
Obesity	mention	BMI	n/a
Hyperlipidemia	mention	high cholesterol, high LDL	n/a
Hypertention	mention	high blood pressure	n/a
Medication	all types	n/a	n/a
Smoking	n/a	n/a	all statuses
Family history	n/a	n/a	all statuses
Percentage of training instances	85.33%	8.10%	6.57%

Table 3: Relationships between the indicators and evidence types and the percentage of training instances belonging to each type

provider. The training set consists of 60% of the entire dataset (790 records) and the test set contains the remaining 40% (514 records). The annotation guidelines describe a set of annotations to indicate the presence of diseases (*coronary artery disease (CAD)* and *diabetes*), relevant risk factors (*hyperlipidaemia, hypertension, obesity, smoking status* and *family history*) and associated medications. Each annotation for a risk factor also has an indicator value from its own set (see Table 1) as well as the time attribute (*before, during* or *after* the DCT). Figure 1 shows an example of annotations used for training and evaluation. The ultimate goal is to classify risk factors and time indicators at document level as per Gold Standard annotation.

The evidence of risk factor indicators can be categorised into three types according to the terminologies described by Chen et al. (2015), which include phrase-based, logic-based and discourse-based indicators as presented in Table 2. Phrase-based indicators are those that can be identified directly by locating relevant phrases or particular names. Logic-based indicators are indirect information that needs a comparison or further analysis after being identified. Finally, discourse-based indicators are those that appear in the form of sentences and may require a parsing process. The relationships between indicators and evidence types are listed in Table 3.

3.2 Problem Formation and Evaluation

The classification of risk factors and time indicators was posed as a document-level classification problem. This can be seen as a multilabel classification task where multiple labels are identified given an EMR. However, unique to the annotation guideline (Stubbs and Uzuner, 2015) and

Complete version (for training):
<DIABETES start="122" end="130" text="diabetes" time="before DCT" indicator="mention"/>
<DIABETES start="512" end="528" text="diabetes type II" time="before DCT" indicator="mention"/>
<DIABETES start="701" end="718" text="diabetes mellitus" time="before DCT" indicator="mention"/>

Gold standard version (for evaluation):
<DIABETES time="before DCT" indicator="mention"/>

Figure 1: Each complete annotation contains token-level information (risk factor tag, risk factor indicator, offset, text information, and time attribute) while each gold standard annotation contains document-level information (risk factor tag, risk factor indicator and time attribute) and cannot be duplicated.

the structure of the training data, which contains phrase-level risk factor and time indicator annotations (see Figure 1), *it seems appropriate to formulate the problem as an information extraction task instead.* This approach regards data as a sequence of tokens labelled using the Inside-Outside (IO) scheme: *I* represents named entity tokens and *O* indicates non-entity ones. As the main goal is to determine the risk factor indicators contained within the record along with the temporal categories of those indicators with respect to the DCT, each entity is tagged with a label using the following format:

I-risk_factor.indicator.time

Figure 2 shows a sample EMR (represented by a sequence of words) and associated labels. In this example, the word "coronary" with the label "I-cad.mention.before_dct" can be interpreted that as a mention of CAD which was present before the document creation time.

Words: he, has, coronary, artery, disease, and, diabetes

Labels: O, O, I-cad.mention.before_dct, I-cad.mention.before_dct, I-cad.mention.before_dct, O, I-diabetes.mention.before_dct

Figure 2: A sample phrase in an EMR and associated labels

Given an EMR as input, the output is a sequence of labels, with each label belonging to a given word. After removing duplicate labels, *the final output will be a set of unique labels identified for that record* (excluding the O label). For the example in Figure 2, the final output will be generated as a set of two unique labels, including "I-cad.mention.before_dct" and "I-diabetes.mention.before_dct". These labels will

be used to generate system annotations similar to the one presented in Figure 1 which will subsequently be evaluated against the gold standard annotations provided by the challenge provider using the micro-averaged recall, precision and F-measure as the primary evaluation metrics[1].

3.3 Deep Neural Network Models

3.3.1 Convolutional Neural Network

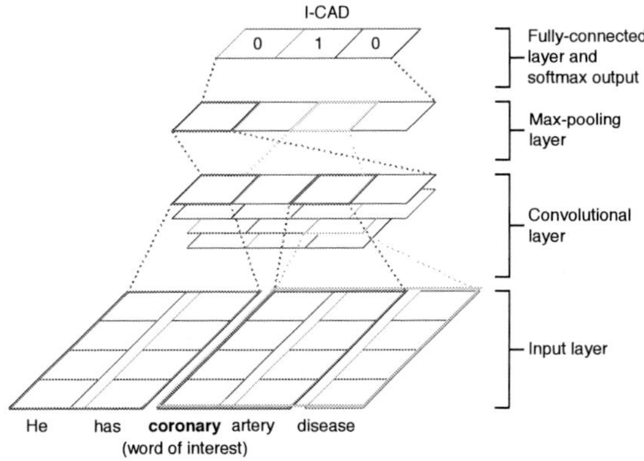

Figure 3: CNN architecture with multiple filter region sizes

The CNN model, as shown in Figure 3, is based on the CNN architecture of Kim (2014) but uses the window approach for NER, introduced by Collobert et al. (2011), to classify each individual word at a time instead of the entire sentence. This approach assumes the label of a word is dependent on its neighbouring words. Given a word to tag, a fixed size window of n words around the target word where n is odd is taken into account. A window of n words is represented as a matrix $\mathbf{S} \in \mathbb{R}^{d \times n}$:

$$\mathbf{S} = \begin{bmatrix} \mathbf{w}_1 & \cdots & \mathbf{w}_{n-\frac{(n-1)}{2}} & \cdots & \mathbf{w}_n \end{bmatrix} \quad (1)$$

[1]The official evaluation script is available at https://github.com/kotfic/i2b2_evaluation_scripts

where $\mathbf{w}_i \in \mathbb{R}^d$ is the d-dimensional word vector representing the ith word in $\mathbf{S}$ and $\mathbf{w}_{n-\frac{(n-1)}{2}}$ is the target word. Let $\mathbf{w}_{i:i+j}$ be the concatenation of words $\mathbf{w}_i, \mathbf{w}_{i+1}, ..., \mathbf{w}_{i+j}$. A convolution operation involves applying a *filter* $\mathbf{k} \in \mathbb{R}^{d \times h}$ to a window of h words, where $h < n$, to generate a new feature. For instance, a feature x_i is computed by

$$x_i = f(\mathbf{k} \cdot \mathbf{w}_{i:i+h-1} + b) \qquad (2)$$

where f is an activation function and $b \in \mathbb{R}$ is a bias. Note that this CNN architecture can employ multiple filter region sizes for extracting multiple features. This operation is applied to every possible window of words in the sequence $\{\mathbf{w}_{1:h}, \mathbf{w}_{2:h+1}, ..., \mathbf{w}_{n-h+1:n}\}$ to generate a *feature map* $\mathbf{x} = (x_1, x_2, ..., x_{n-h+1})$ where $\mathbf{x} \in \mathbb{R}^{n-h+1}$. The pooling layer then applies the max-pooling operation to down-sample each feature map by taking the maximum value $\hat{x} = \max(\mathbf{x})$ which represents the most important feature. Finally, multiple down-sampled feature maps form a fully-connected layer, which is used as inputs to the softmax distribution over all classes. The subsampled feature maps provide a sequence representation for softmax to map to an appropriate class.

3.3.2 Recurrent Neural Network

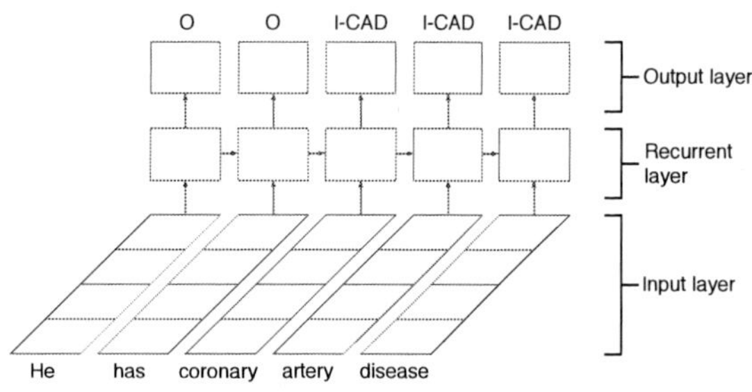

Figure 4: Basic structure of an RNN

A recurrent neural network is a class of neural networks specialised for processing sequential data. Unlike the CNN, the RNN uses a recurrent layer to learn the representation of clinical text, as shown in Figure 4. The input to an RNN is a word sequence of length l representing the *entire document*, denoted by a matrix $\mathbf{S} \in \mathbb{R}^{d \times l}$:

$$\mathbf{S} = \begin{bmatrix} \mathbf{w}_1 & \mathbf{w}_2 & ... & \mathbf{w}_l \end{bmatrix} \qquad (3)$$

where $\mathbf{w}_i \in \mathbb{R}^d$ is the d-dimensional word vector representing the ith word in $\mathbf{S}$. In an *Elman-type*

network (Elman, 1990), a hidden state output $\mathbf{h}_i$ is a result of nonlinear transformation of an input vector $\mathbf{w}_i$ and the previous hidden state $\mathbf{h}_{i-1}$:

$$\mathbf{h}_i = f(\mathbf{h}_{i-1}, \mathbf{w}_i) \qquad (4)$$

where f is a recurrent unit, such as a standard recurrent unit, LSTM and GRU. Finally, the hidden state $\mathbf{h}_i$ is then used as an input to softmax for identifying a risk factor in the IO format.

Bidirectionality. A bidirectional recurrent neural network (Schuster and Paliwal, 1997) consists of two separated recurrent layers for computing the forward hidden states $(\overrightarrow{\mathbf{h}_1}, \overrightarrow{\mathbf{h}_2}, ..., \overrightarrow{\mathbf{h}_l})$ and the backward hidden states $(\overleftarrow{\mathbf{h}_1}, \overleftarrow{\mathbf{h}_2}, ..., \overleftarrow{\mathbf{h}_l})$. In this settings, $\overrightarrow{\mathbf{h}_i}$ and $\overleftarrow{\mathbf{h}_i}$ can be regarded as preserved information from the past and the future respectively. By using the hidden states from both directions combined, the network has complete past and future context for every point in the input sequence.

3.4 Pre-trained Word Embeddings

Due to the incapability of neural networks to process text input, each word is fed to the network as an index taken from a finite dictionary. As this simple representation does not contain much semantic information, the first layer of each network maps each index into its vector representation using pre-trained word embeddings. The pre-trained vectors were trained on the 2014 i2b2 dataset. The number of embedding dimensions was determined empirically. Given a small vocabulary (36,663 words) and a range of embedding dimensions from 20 to 300, an embedding dimension of 20 yielded best results. Each vector was trained via the word2vec's continuous bag-of-words (CBOW) model (Mikolov et al., 2013b) similar to that used by Kim (2014).

3.5 Hyperparameters and Training

The CNN model used 5-gram of each EMR as input since a window of 5 words has shown to be effective for many NLP tasks (Collobert et al., 2011). Based on the hyperparameters described by Kim (2014) and Zhang and Wallace (2015), the convolutional layer uses multiple filter region sizes $\{2, 3, 4\}$, each of which has 32 filters, and a rectifier (ReLU) as the activation function. For the RNN approach, experiments were performed on the standard RNN as well as its variants: LSTM,

BLSTM and GRU. All the recurrent networks use the hyperbolic tangent as activation functions as it was considered one of the most common choices for RNN-type networks (Graves, 2012).

The hyperparameters apart from the above mentioned were tuned on the validation set (20% of the training set) using the hyperparameter tuning library within the framework of Bayesian optimisation, namely *Hyperopt* (Bergstra et al., 2013). Based on the hyperparameter optimisation results, all the networks were trained with mini-batch stochastic gradient descent using *Nadam* (Adam RMSprop with Nesterov momentum) (Dozat, 2016) with a batch size of 32. Dropout regularisation was also applied to the penultimate layer of each network for overfitting prevention. The resulting optimal values of other hyperparameters, including the number of hidden units (hidden), learning rate (lr), dropout rate and the number of epochs are listed in Table 4.

	CNN	RNN	GRU	LSTM	BLSTM
hidden	256[*]	256	256	512	256
lr	0.001	0.002	0.002	0.002	0.004
dropout	0.2	0.3	0.1	0.3	0.5
epochs	15	40	50	45	40

[*] The number of units in the fully connected layer

Table 4: Hyperparameters estimated by Hyperopt

4 Results and Discussion

4.1 Overall Performance

Results for each deep learning model's best run against state-of-the-art models from the 2014 i2b2/UTHealth shared task are listed in Table 5. Among the deep learning approaches, RNN-type networks outperformed CNN in the context of clinical text classification. Although the CNN model achieved the highest recall, its precision is far from being competitive, which results in a relatively low F-measure. A comparison between the RNN-type models shows that BLSTM achieved the highest micro-averaged F-measure (0.9081) on the test data, followed closely by GRU and LSTM. A two-tailed unpaired t-test was also performed to determine the significance of the difference in F-measure between the two best-performing networks. Over 50 independent training and testing sessions with different weight initialisation (drawn from the uniform distribution), the test yielded a statistically significant difference between the performance of BLSTM ($\mu = 0.903$, $\sigma = 0.002$) and GRU ($\mu = 0.899$, $\sigma = 0.002$) with $p < 0.05$, which implies that the improvement in performance of the BLSTM model is also statistically significant compared with that of other remaining models.

In comparison with the top performing systems from the previous work, the results reveal that the BLSTM model without employing any knowledge-driven approaches ranked in the top 6 systems, and was substantially better than the overall average (0.815) of all the participating systems in the shared task. As a universal classifier, the performance of the BLSTM model is auspicious since it produced only 0.0195 loss in F-measure when comparing against the first-ranked system (Roberts et al., 2015) which involves the use of a series of SVMs along with a rule-based classifier and additional annotations. Besides the best-performing model, the LSTM and GRU models ranked in the top 7 systems while the CNN and standard RNN models performed well within the top 10 systems from the shared task. This outcome concludes that simple deep learning models still can rank within the top 10 heavily feature-engineered best-performing systems from the shared task.

Model	Recall	Precision	F-score
BLSTM	0.9180	0.8983	**0.9081**
GRU	0.9091	**0.9002**	0.9046
LSTM	0.9191	0.8836	0.9010
RNN	0.8956	0.8844	0.8900
CNN	**0.9245**	0.8383	0.8793
Roberts et al. (2015)[*]	**0.9625**	0.8951	**0.9276**
Chen et al. (2015)[*]	0.9436	**0.9106**	0.9268
Torii et al. (2014)[*]	0.9409	0.8972	0.9185
Cormack et al. (2015)[†]	0.9375	0.8975	0.9171
Yang and Garibaldi (2014)[*]	0.9488	0.8847	0.9156
Shivade et al. (2015)[†]	0.9261	0.8907	0.9081
Chang et al. (2015)[*]	0.9387	0.8594	0.8973
NCU[‡]	0.9256	0.8586	0.8909
Karystianis et al. (2015)[†]	0.9007	0.8557	0.8776
Khalifa and Meystre (2015)[†]	0.8951	0.8552	0.8747

[*] A combination of knowledge- and data-driven approaches (hybrid)
[†] Knowledge-driven approaches only e.g. lexicon and rules
[‡] Unknown (National Central University did not submit a paper)

Table 5: Experimental results and state-of-the-art systems from 2014 i2b2/UTHealth shared task

4.2 Performance on Individual Risk Factors

Table 6 shows the performance of the deep learning models on individual risk factors. All five architectures achieved micro-averaged F-measures

	CNN	RNN	GRU	LSTM	BLSTM
CAD	0.6553	0.7966	0.7972	0.8010	**0.8074**
Diabetes	0.9133	0.9227	0.9177	**0.9272**	0.9171
Obesity	0.8717	0.8739	0.8819	**0.8880**	**0.8880**
Hyperlipidemia	0.9154	0.9209	0.9100	0.9243	**0.9323**
Hypertension	0.8839	0.9093	0.9102	0.9043	**0.9187**
Medication	0.9075	0.8901	**0.9192**	0.9090	0.9171
Smoking	0.8350	0.8077	0.8146	0.8152	**0.8409**
Family history	0.9397	**0.9630**	0.9572	0.9591	**0.9630**
Overall	0.8798	0.8900	0.9046	0.9010	**0.9081**

Table 6: Micro-averaged F-measure for individual risk factor categories (best runs); highest F-measures for each category are bolded

over 0.87. These deep networks performed best on the family history category, achieved F-measures above 0.90 for the hyperlipidemia and diabetes risk factors, and maintained F-measures over 0.87 for the hypertension and obesity risk factors along with relevant medications. The worst classification performance of all the models was obtained for the CAD risk factor, followed by the smoking status.

Among the deep learning models, highest micro-averaged F-measures for most of the risk factor categories were achieved by the BLSTM network while the top performance for the diabetes and medication categories were obtained by the LSTM and GRU networks respectively. Lowest classification scores for most of the risk factor categories were achieved by the CNN model, which implies its inferiority in comparison with the RNN-type models for extracting cardiac risk factor information from EMRs. The overall outcome also reveals that even though the neural network architectures with the integration of recurrent units can be potentially applied to this particular task with higher success rate, the capability of the standard RNN is far from being highly efficient and thus using the gating mechanism as well and introducing bidirectionality can substantially increase the chance of achieving better performances.

4.3 Performance on Individual Risk Factor Indicators

The results in Table 7 reveals that phrase-based indicators have comparatively high F-measures in all models. As the deep learning approach for clinical concept extraction can be posed as a standard the named entity recognition task, specific keywords play a significant role in identifying named entities and an increase in the predictive performance

is simply due to a tremendous amount of sample instances in the training data.

In contrast, the logic-based and discourse-based indicators have substantially lower F-measure. As both types of indicators infrequently appear in the training data (see Table 3), the primary cause of poor performance is likely due to the sparsity and imbalance of training instances.

	CNN	RNN	GRU	LSTM	BLSTM
Phrase-based	0.7679	0.6818	0.7810	0.7342	0.7808
Logic-based	0.3643	0.1857	0.2185	0.2114	0.2640
Discourse-based	0.5341	0.4983	0.5425	0.5328	0.5721

Table 7: The average of F-measure performances across all risk factor indicators for each evidence type

4.4 Error Analysis

4.4.1 Complex Textual Evidence

Even though phrase-based evidence may vary (e.g. CAD can appear as "heart disease" or "CAD"), these phrases along with a sufficiently large amount of samples are generally enough for deep neural networks to achieve high classification accuracy. However, the context of discourse-based evidence may appear to be as complex as "probable inferior and old anteroseptal myocardial infarction" or "Cath (5/88): 3v disease: RCA 90%, LAD 30% mid, 80% distal, D1 70%, D2 40% and 60%, LCx 30%, OM2 80%". The difficulty of learning the patterns and identifying these indicators implies the need for a higher amount of training instances and perhaps amended semantic matching of medical terms to medical terminology resources such as the UMLS Metathesaurus

(Bodenreider, 2004) or Systematised Nomenclature of Medicine – Clinical Terms (SNOMED CT) (Stearns et al., 2001), such that information in EMRs can be more accurately extracted using deep learning.

4.4.2 Conditional Textual Evidence

Although deep learning requires less human effort and time than dictionary-based and rule-based approaches as it can automatically learn the patterns in data which results in more flexible predictive power, the experimental results demonstrate the limitation of such data-driven approach as it is infeasible to accurately identify logic-based indicators in the test set without having seen the numbers and their contexts in the training set. For example, it is unlikely for deep learning models to classify the evidence "glucose 420" as the diabetes.glucose indicator without learning that particular pattern during training as it is unable to perform comparison during classification whether 420 is greater than 126 (the glucose level greater than 126 is considered a risk factor (Stubbs and Uzuner, 2015)). A decrease in classification accuracy is primarily due to a massive amount of unforeseen evidence in the test data i.e. many numbers that imply the risk of heart disease never appear in the training set. In this case, utilising dictionaries and rules based on the domain knowledge would be more optimal than collecting more data in which every possible pattern, which may include every number that is considered a risk factor as well as its context, is required.

4.4.3 Data Sparsity and Class Imbalance

Figure 5 illustrates the relationship between classification performance of the BLSTM network[2] and the number of training instances in terms of risk factor indicators. When the number of samples is low (less than approximately 200 samples), each network's performance significantly varies depending on risk factor indicator. However, the prediction capability raises and tends to be more stable as the number of training instance increases. As many of the machine learning algorithms greatly suffer from insufficient and imbalanced data where the classes are not equally presented, it is not surprising if deep learning is

---

[2]The relationship between classification performance of the BLSTM network and the number of training instances is selected as it is the best-performing model from the experiment and the patterns found among other deep learning architectures are very similar.

severely impacted by the same problem. Inadequate training samples typically result in failure of pattern recognition while imbalanced classes in the training set tend to bias the trained models towards more common classes. These non-trivial issues likely explain the relatively poor classification results for various risk factor indicators, especially those that belong to the logic-based and discourse-based types, due to misclassification of either indicators or time attributes or both. Regarding the report from the 2014 i2b2/UTHealth risk factor challenge (Stubbs et al., 2015), all the participating systems also produced similar sets of results due to these problems.

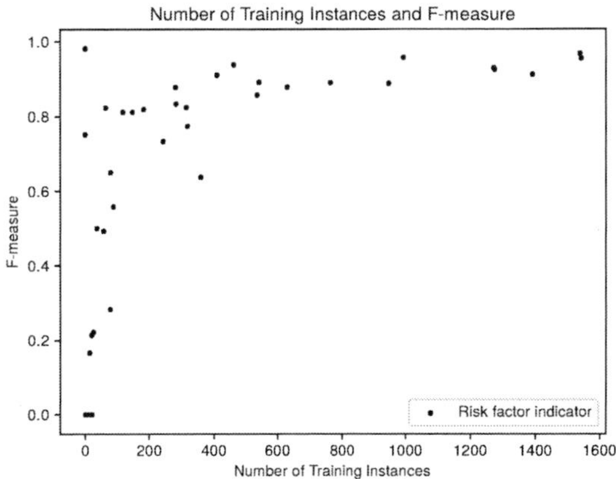

Figure 5: Effect of training-sample size illustrated by the relationship between classification performance of the BLSTM network and the number of training instances (risk factor indicator-level)

5 Conclusion

This work empirically evaluated the performance of different deep learning architectures for identifying risk factors for heart disease in clinical text. The experimental results showed that the deep learning approaches were not only comparable to highly feature-engineered hybrid systems but most importantly achieved relatively high performances without the help of any knowledge-driven methods. The findings leads to an anticipation that leveraging knowledge-based approaches with the BLSTM model could potentially provide significant performance improvements over best systems for extracting key cardiac risk factors from EMRs.

References

Emelia J Benjamin, Michael J Blaha, Stephanie E Chiuve, Mary Cushman, Sandeep R Das, Rajat Deo, Sarah D de Ferranti, James Floyd, Myriam Fornage, Cathleen Gillespie, et al. 2017. Heart disease and stroke statistics 2017 update: a report from the american heart association. *Circulation*, 135(10):e146–e603.

James Bergstra, Dan Yamins, and David D Cox. 2013. Hyperopt: A python library for optimizing the hyperparameters of machine learning algorithms. In *Proceedings of the 12th Python in Science Conference*, pages 13–20.

Steven Bethard, Guergana Savova, Wei-Te Chen, Leon Derczynski, James Pustejovsky, and Marc Verhagen. 2016. SemEval-2016 task 12: Clinical TempEval. In *Proceedings of the 10th International Workshop on Semantic Evaluation (SemEval-2016)*, pages 1052–1062.

Olivier Bodenreider. 2004. The unified medical language system (UMLS): integrating biomedical terminology. *Nucleic Acids Research*, 32(1):D267–D270.

Nai-Wen Chang, Hong-Jie Dai, Jitendra Jonnagaddala, Chih-Wei Chen, Richard Tzong-Han Tsai, and Wen-Lian Hsu. 2015. A context-aware approach for progression tracking of medical concepts in electronic medical records. *Journal of Biomedical Informatics*, 58:S150–S157.

Qingcai Chen, Haodi Li, Buzhou Tang, Xiaolong Wang, Xin Liu, Zengjian Liu, Shu Liu, Weida Wang, Qiwen Deng, Suisong Zhu, et al. 2015. An automatic system to identify heart disease risk factors in clinical texts over time. *Journal of Biomedical Informatics*, 58:S158–S163.

Yu Cheng, Fei Wang, Ping Zhang, and Jianying Hu. 2016. Risk prediction with electronic health records: A deep learning approach. In *Proceedings of the 2016 SIAM International Conference on Data Mining*, pages 432–440. SIAM.

Veera Raghavendra Chikka. 2016. CDE-IIITH at SemEval-2016 task 12: Extraction of temporal information from clinical documents using machine learning techniques. In *Proceedings of the 10th International Workshop on Semantic Evaluation (SemEval-2016)*, pages 1237–1240.

Kyunghyun Cho, Bart Van Merriënboer, Dzmitry Bahdanau, and Yoshua Bengio. 2014. On the properties of neural machine translation: Encoder-decoder approaches. *arXiv preprint arXiv:1409.1259*.

Ronan Collobert, Jason Weston, Léon Bottou, Michael Karlen, Koray Kavukcuoglu, and Pavel Kuksa. 2011. Natural language processing (almost) from scratch. *Journal of Machine Learning Research*, 12:2493–2537.

James Cormack, Chinmoy Nath, David Milward, Kalpana Raja, and Siddhartha R Jonnalagadda. 2015. Agile text mining for the 2014 i2b2/UTHealth cardiac risk factors challenge. *Journal of Biomedical Informatics*, 58:S120–S127.

Timothy Dozat. 2016. Incorporating nesterov momentum into adam. *4th International Conference on Learning Representations (ICLR 2016)*.

Jeffrey L Elman. 1990. Finding structure in time. *Cognitive science*, 14(2):179–211.

Jason Alan Fries. 2016. Brundlefly at SemEval-2016 task 12: Recurrent neural networks vs. joint inference for clinical temporal information extraction. *arXiv preprint arXiv:1606.01433*.

G Gonzalez-Hernandez, A Sarker, K O'Connor, and G Savova. 2017. Capturing the patients perspective: a review of advances in natural language processing of health-related text. *Yearbook of Medical Informatics*, 26(01):214–227.

Alex Graves. 2012. *Supervised sequence labelling with recurrent neural networks*. Springer, Berlin, Heidelberg.

Sepp Hochreiter and Jürgen Schmidhuber. 1997. Long short-term memory. *Neural computation*, 9(8):1735–1780.

Abhyuday N Jagannatha and Hong Yu. 2016. Structured prediction models for rnn based sequence labeling in clinical text. In *Proceedings of the Conference on Empirical Methods in Natural Language Processing. Conference on Empirical Methods in Natural Language Processing*, volume 2016, page 856.

George Karystianis, Azad Dehghan, Aleksandar Kovacevic, John A Keane, and Goran Nenadic. 2015. Using local lexicalized rules to identify heart disease risk factors in clinical notes. *Journal of Biomedical Informatics*, 58:S183–S188.

Abdulrahman Khalifa and Stéphane Meystre. 2015. Adapting existing natural language processing resources for cardiovascular risk factors identification in clinical notes. *Journal of Biomedical Informatics*, 58:S128–S132.

Yoon Kim. 2014. Convolutional neural networks for sentence classification. *Proceedings of the 2014 Conference on Empirical Methods in Natural Language Processing (EMNLP)*, pages 1746–1751.

Yann LeCun, Yoshua Bengio, and Geoffrey Hinton. 2015. Deep learning. *Nature*, 521(7553):436–444.

Peng Li and Heng Huang. 2016. University of Texas at Arlington (UTA) with deep learning based natural language processing (DLNLP) at SemEval-2016 task 12: deep learning based natural language processing system for clinical information identification from clinical notes and pathology reports. In

Proceedings of the 10th International Workshop on Semantic Evaluation (SemEval-2016), pages 1268–1273.

Stéphane M Meystre, Guergana K Savova, Karin C Kipper-Schuler, and John F Hurdle. 2008. Extracting information from textual documents in the electronic health record: a review of recent research. *IMIA Yearbook of Medical Informatics 2008*, 35(128):128–144.

Tomas Mikolov, Kai Chen, Greg Corrado, and Jeffrey Dean. 2013a. Efficient estimation of word representations in vector space. *Proceedings of the International Conference on Learning Representations (ICLR 2013)*, pages 1–12.

Tomas Mikolov, Ilya Sutskever, Kai Chen, Greg S Corrado, and Jeff Dean. 2013b. Distributed representations of words and phrases and their compositionality. In *Advances in Neural Information Processing Systems*, pages 3111–3119.

Kirk Roberts, Sonya E Shooshan, Laritza Rodriguez, Swapna Abhyankar, Halil Kilicoglu, and Dina Demner-Fushman. 2015. The role of fine-grained annotations in supervised recognition of risk factors for heart disease from EHRs. *Journal of Biomedical Informatics*, 58:S111–S119.

Mike Schuster and Kuldip K Paliwal. 1997. Bidirectional recurrent neural networks. *IEEE Transactions on Signal Processing*, 45(11):2673–2681.

Benjamin Shickel, Patrick James Tighe, Azra Bihorac, and Parisa Rashidi. 2017. Deep EHR: A survey of recent advances in deep learning techniques for electronic health record (EHR) analysis. *IEEE Journal of Biomedical and Health Informatics*.

Jaeho Shin, Sen Wu, Feiran Wang, Christopher De Sa, Ce Zhang, and Christopher Ré. 2015. Incremental knowledge base construction using DeepDive. *Proceedings of the VLDB Endowment*, 8(11):1310–1321.

Chaitanya Shivade, Pranav Malewadkar, Eric Fosler-Lussier, and Albert M Lai. 2015. Comparison of UMLS terminologies to identify risk of heart disease using clinical notes. *Journal of Biomedical Informatics*, 58:S103–S110.

Michael Q Stearns, Colin Price, Kent A Spackman, and Amy Y Wang. 2001. SNOMED clinical terms: overview of the development process and project status. In *Proceedings of the AMIA Symposium*, page 662. American Medical Informatics Association.

Amber Stubbs, Christopher Kotfila, Hua Xu, and Özlem Uzuner. 2015. Identifying risk factors for heart disease over time: Overview of 2014 i2b2/UTHealth shared task track 2. *Journal of Biomedical Informatics*, 58:S67–S77.

Amber Stubbs and Özlem Uzuner. 2015. Annotating risk factors for heart disease in clinical narratives for diabetic patients. *Journal of Biomedical Informatics*, 58:S78–S91.

Manabu Torii, Jung-wei Fan, Wei-li Yang, Theodore Lee, Matthew T Wiley, Daniel Zisook, and Yang Huang. 2014. De-identification and risk factor detection in medical records. In *Seventh i2b2 Shared Task and Workshop: Challenges in Natural Language Processing for Clinical Data*.

Hui Yang and Jonathan Garibaldi. 2014. Automatic extraction of risk factors for heart disease in clinical texts. *Proceeding of the i2b2/UTHealth NLP Challenge*.

Ye Zhang and Byron Wallace. 2015. A sensitivity analysis of (and practitioners' guide to) convolutional neural networks for sentence classification. *arXiv preprint arXiv:1510.03820*.

Keyphrases Extraction from User-Generated Contents in Healthcare Domain Using Long Short-Term Memory Networks

Ilham Fathy Saputra, Rahmad Mahendra, Alfan Farizki Wicaksono
Faculty of Computer Science, Universitas Indonesia
Depok 16424, West Java, Indonesia
`ilham.fathy@ui.ac.id, {rahmad.mahendra, alfan}@cs.ui.ac.id`

Abstract

We propose keyphrases extraction technique to extract important terms from the healthcare user-generated contents. We employ deep learning architecture, i.e. Long Short-Term Memory, and leverage word embeddings, medical concepts from a knowledge base, and linguistic components as our features. The proposed model achieves 61.37% F-1 score. Experimental results indicate that our proposed approach outperforms the baseline methods, i.e. RAKE and CRF, on the task of extracting keyphrases from Indonesian health forum posts.

1 Introduction

The growth of Internet access facilitates users to share and obtain contents related to healthcare topic. There has been growing interest in using Internet to find information related to healthcare concerns, including symptoms management, medication side effects, alternative treatment, and fitness plan. The tremendous amounts of user-generated health contents are available on the web pages, online forums, blogs, and social networks. These user-generated contents are actually potential sources for enriching medical-related knowledge. It is highly desirable if the knowledge contained in the user generated contents can be extracted and reused for Natural Language Processing and text mining application.

Keyphrases, which is a concise representation of document, describe important information contained in that document. The number of text processing tasks can take advantage of keyphrases, e.g. document indexing, text summarization, text classification, topic detection and tracking, information visualization.

We believe that extracting keyphrases from documents in healthcare domain can be beneficial. A medical question answering system is expected to provide concise answers in response to clinical questions. The keyphrases extracted from the question can be used to formulate a query to retrieve the answer passage from a collection of medical documents. On the other hand, the health-related web forums usually contain very large archives of forum threads and posts. To make use of those archives, it is critical to have functionality facilitating users to search previous forum contents. Keyphrases identification is an important step to tackle this kind of document retrieval problem.

Most previous works on keyphrases extraction task focused on long documents, e.g. scientific articles and web pages, while few works attempt to identify keyphrases from user-generated contents, e.g. e-mail messages (Dredze et al., 2008), chats (Kim and Baldwin, 2012; Habibi and Popescu-Belis, 2013), and tweets (Li et al., 2010; Zhao et al., 2011; Zhang et al., 2016). Extracting keyphrases from an online forum is simply not a trivial task, since the contents are written in free text format (i.e. unstructured format), and often prone to grammatical and typographical glitches.

In this paper, we address the task of keyphrases extraction from user-generated posts in online healthcare forums. We present the technique that treats keyphrase extraction as a sequence labeling task. In our experiment, we employ and combine deep learning architectures, i.e. bi-directional Long Short-Term Memory networks, to exploit high level features between neighboring word positions. To improve the quality of our model, we leverage several new hand-crafted features that can handle our keyphrase extraction problems in medical user-generated contents.

Proceedings of the BioNLP 2018 workshop, pages 28–34
Melbourne, Australia, July 19, 2018. ©2018 Association for Computational Linguistics

2   Related Work

Keyphrases are usually selected phrases or clauses that can capture the main topic of a given document (Turney, 2000). Keyphrases are expected to provide readers with highly valuable and representative information, such that looking at keyphrases is sufficient to understand the whole body of a document.

In general, keyphrases extraction methods can be categorized into unsupervised and supervised approach (Hasan and Ng, 2014). For the unsupervised line of research, keyphrases extraction can be formulated as a ranking problem, in which each candidate keyphrase is assigned the score. RAKE (Rose et al., 2010) uses ratio of word degree and frequency to rank terms. Mihalcea and Tarau (2004) studies the graph-based approach that treats words as vertices and constructs edge between words using co-occurrence.

On the other hand, supervised machine learning approach requires training data that contain a collection of documents with their labeled keywords which are often very difficult to obtain. Using this approach, keyphrase extraction is formulated as a classification or sequence labeling task in the level of words or phrases. The supervised learning approach starts from generating candidate keyphrases from a particular document. Then, each candidate is classified as either a keyphrase or non-keyphrase. The well-known method for this approach is KEA (Witten et al., 1999), which applied machine-learning (i.e. Naive Bayes) for classifying candidate keyphrases. Sarkar et al. (2010) utilizes neural network algorithm in classifying candidate phrase as a keyphrase.

Other supervised approach for keyphrases extraction is based on sequence labeling problem (Zhang, 2008; Cao et al., 2010; Zhang et al., 2016). The assumption behind this model is that the decision on whether a particular word serves as a keyword is affected by the information from its neighboring word positions. Zhang (2008) apply Conditional Random Fields (CRF) algorithm to find keyphrases from the Chinese documents. Zhang et al. (2016) proposes a joint-layer recurrent neural network model to extract keyphrases from tweets, which is an application of deep neural networks in the context of keyphrase extraction.

As far as our knowledge, there are limited works regarding the task of keyphrase extraction from user-generated contents, especially for healthcare domain. Sarkar (2013) applies hybrid statistical and knowledge-base approach to extract keyphrases from medical articles. Stopwords are used to split candidate keyphrases, then candidates were ranked based on two aspects: Phrase Frequency * Inverse Document Frequency (PF-IDF) and domain knowledge which is extracted from medical articles. Cao et al. (2010) uses CRF for extracting keywords from medical questions in online health forum. They harness information about word location and length as features in their experiments.

3   Methodology

In this work, we see the problem of keyphrase extraction as a sequence labeling, in which each word w_i is associated with a hidden label $y_i \in \{\text{keyword, non-keyword}\}$. Formally, given a medical forum posts containing N words $W = (w_1, w_2, ..., w_N)$, we want to find the best sequence of labels $Y = (y_1, y_2, ..., y_N)$, in which each label is determined using probabilities $P(y_i | w_{i-l}, ..., w_{i+l}, y_{i-l}, ..., y_{i+l})$; and l is a small integer.

3.1   Proposed Model

To cope with our problem, we employ a deep neural network-based approach specially designed for sequence labeling problem, such as Long Short-Term Memory (LSTM) Networks and its variants, to extract high-level sequential features, before they are fed into the last layer that determines the most probable label for each word or timestep. Since we employ neural networks, we can also view our model as a function $F : R^{N \times M} \to R^{N \times 2}$:

$$[z_1, z_2, ..., z_N] = F([x_1, x_2, ..., x_N])$$

where, M is the size of input vector in each timestep, N is the number of timestep, $x_i \in R^M$ is the vector representation of word w_i, and $z_i \in R^2$ is the output vector in each timestep ($\sum_j z_{i,j} = 1$). Furthermore, the vector representation of word can be obtained using state-of-the-art technique, such as the one proposed by (Mikolov et al., 2013). Finally, y_i can be determined as follows.

$$y_i = \begin{cases} \text{keyphrase} & \text{if } z_{i,0} > z_{i,1} \\ \text{non-keyphrase} & \text{otherwise} \end{cases}$$

In our work, the operational definition of keyphrase is actually extractive, in the sense that keyphrase is explicitly extracted from a sequence of words found in the document. For example, from the following sentence: *"Doc, I have a frequent back pain. What happen?"*, we can extract *"frequent back pain"* as our keyphrase.

In order to know that *"back"* and *"frequent"* are part of the keyphrase along with *"pain"*, we need to consider such phrasal structure information, i.e. the word *"back"* that serves as the modifier of symptom *"pain"*, as well the word *"frequent"* that informs its intensity level. Therefore, we argue that sequential-based neural networks, such as Long Short-Term Memory (LSTM) networks and their variants can better fit our problem since they can naturally leverage neighboring information.

To get a better inference process, the information from the past and future of current position in the sequence can be integrated. This approach has been proven effective in the number of sequence labeling tasks, such as semantic role labeling (Zhou and Xu, 2015) and named-entity recognition (Ma and Hovy, 2016). Based on those previous studies, we utilize a bi-directional LSTM (B-LSTM) in our work to extract structural knowledge by doing forward and backward processing in the sequence. In order to do that, we build two LSTMs with different parameters and subsequently concatenate the outputs from both LSTMs. Moreover, we build our layers for up to two layers of B-LSTM. Finally, the locally normalized distribution over output labels is computed via a softmax layer. In the other scenario, we also employ Conditional Random Fields (CRFs) (Lafferty et al., 2001) to produce label predictions in the last layer. Following Rei et al. (2016), we used Viterbi algorithm to efficiently find the sequence of labels $[y_1, y_2, ..., y_M]$ with the largest score $s(Y)$. As can be seen in the following equation, $s(Y)$ computes CRF score of a sequence, which means the likelihood of the output labels.

$$s(Y) = \sum_{t=1}^{M} A_{t,y_t} + \sum_{t=0}^{M} B_{y_t,y_{t+1}}$$

where A_{t,y_t} shows the confident score of of label y_t at timestep t, $B_{y_t,y_{t+1}}$ show the likelihood of transitioning from label y_t to label y_{t+1}. It is worth to note that all these parameters are trainable.

The spirit of deep learning is basically to au-tomatically extract features from the input without the need of expensive feature engineering. However, this idea works well when we have a significant amount of training samples, which is not applicable in our case since the size of our data is small enough. As a result, to cope with this problem, we combine deep learning technique with several feature engineering steps. The idea is that several hand-crafted features are leveraged to help deep learning architectures understand the main characteristics of the data before they actually learn more high-level features from those hand-crafted features. Suppose, $F_{i,j}$ represents a type of feature vector extracted from an input x_i in one timestep and K is the number of feature types. A feature vector for x_i is defined as follows.

$$x_i = concatenate(F_{i,1}, F_{i,2}, ..., F_{i,K})$$

The detail of our proposed feature types is explained in the next subsection. Moreover, we also argue that each feature has different contribution to the model. Instead of concatenating all features into one vector, we try to assign weights to every feature type before we pass it to the next layer. In order to do so, we create a new layer underneath our model to do the weighting scenario. Suppose, $W_i \in \mathbb{R}^{a \times b_i}$ is a trainable weight for feature vector F_i, where a is the size of input size in each timestep and b_i is the size of feature vector F_i. The following equation presents our idea for weighting the feature vectors.

$$x_i = tanh(W_1.F_{i,1} + W_2.F_{i,2} + .. + W_K.F_{i,K})$$

3.2 Proposed Features

We perform automatic representation learning in the input layer, in which vector representation of a particular word is automatically learned. However, we argue that the end-to-end learning approach alone will have not effectively worked in our case since the tiny size of dataset. So, we leverage nine hand-crafted features that can help our model to characterize the sequence of keyphrases.

WORD EMBEDDING. We use pre-trained word embedding that fills several slots in our feature vector. A skip-gram model from Mikolov et al. (2013) was used to generate a 128-dimensional vector of a particular word. The word embedding we used in this work is trained using documents that are collected from online forums, medical articles, and medical question answering forums. By using embedding feature, slank words and lexical variants can be naturally handled since all variants should have similar vector representation.

MEDICAL DICTIONARY. We also devise feature vector using the knowledge derived from a dictionary containing medical terminologies. Technically, we generate a one-hot feature vector for each word in the sentence by checking whether the word is listed in the dictionary.

WORD LENGTH. This feature represents the length of each word (i.e., the number of characters in every word) in the sentence. The rationale behind proposing this kind of feature is that the medical domain-specific words (e.g. "influenza", "tuberculosis") tend to be lengthy compared to general words. Cao et al. (2010) found that there is a correlation between the length of a word and its informativeness value (inverse document frequency value).

WORD POSITION. The important term most likely appears in either beginning or end of document. The first sentence of document typically contains phrasal topic, while a few last sentences usually emphasize the content discussed in the document. In a medical consultation forum, a user often starts her post with a statement explaining the problem, then gives more explanation in form of several narrative sentences. In the end, she asks one or two questions. Keyphrases are potentially extracted from a problem statement and the questions in the forum posts.

POS-TAG. Part-of-Speech category of word can be also exploited as a feature, since it may feed our model with grammatical information and a better understanding of ambiguous words. Based on our observation, many keyphrases have a common POS pattern, e.g. a verb followed by the sequence of nouns. The POS-tag feature is represented as a one-hot-vector, whose length is the number of tags.

MEDICAL ENTITY. We extract four types of medical entity from the text, i.e. *drug*, *treatment*, *symptom*, and *disease*. The medical entity often become part of a keyphrase of the sentence or document. Furthermore, this feature complements the Medical Dictionary feature. While a drug or disease name is not available in the training data or a medical dictionary, it is still possible to learn it using medical entity recognizer.

ABBREVIATION AND ACRONYM. We also identify whether a word is an abbreviation or acronym. We compile an acronym dictionary and then check whether a word in the forum post is found in the dictionary. We have observed that important words are rarely shortened by the users.

WORD CENTRALITY. The role of this feature is to rank words in a document by their importance. To extract this feature, we adapt TextRank algorithm (Mihalcea and Tarau, 2004). We build undirected graph, in which the word is represented as the vertice and the distance between words as the edge. We use word similarity score as weights for the edge. Pre-trained word embedding model is used to calculate the cosine similarity between two word vectors. In our work, two words (vertices) are adjacent (having an edge between them) if their similarity are not negative. Moreover, we use a modified PageRank algorithm (Page et al., 1998) that consider weight of edge in calculation.

Formally, let $G = (V, E)$ be an undirected graph with the set of vertices V and set of edges E, where E is a subset of $V \times V$. For a given vertex V_i, let $In(V_i)$ be the set of vertices that points to it (predecessors) and $Out(V_i)$ be the set of vertices that vertex V_i points to (successors), the modified PageRank equation proposed by (?) can be seen in the following formula.

$$WS(V_i) = (1-d) + d * \sum_{V_i \in In(V_i)} \frac{w_{ji}}{\sum_{V_i \in Out(V_j)}} w_{jk}$$

WORD STICKINESS. There are typical noises found in the user-generated contents, such as lack of proper punctuation usage. For example, in the sentence *"I have a pain on forehead stomachache and blurred vision"*, there is no comma between the words *"forehead"* and *"stomachache"*, as well between *"stomachache"* and *"and"*; while it should be required. The model may mistakenly select the sequence *"forehead stomachache"* as a single keyphrase.

To address this problem, we propose a feature that is able to capture how likely a given word is occurred together with the preceding and succeeding words. We compute Pointwise Mutual Information (PMI) of all bigrams to capture such information using documents from health-related online forum.

Table 1: Statistical data of Indonesian healthcare user-generated posts dataset

Number of posts	416
Total number of words	26,747
Number of keyphrases	1,861
Avg number of words per posts	64
Avg number of keyphrases per posts	4

The PMI formula can be seen as follows.

$$PMI(x,y) = log(\frac{P(x,y)}{P(x).P(y)})$$

where $p(x)$ is the occurrence probability of word x, $p(y)$ is the occurrence probability of word y, and $p(x,y)$ is the probability of word x and y co-occur together.

The feature function is formally described as $f_s(w) = [x,y]$, where w is a word in a particular document, x is the stickiness value between w and its preceding word, and y is the stickiness value between w and its succeeding word. For example, given the word "cancer" in the sentence "How to prevent cancer doc?", the feature value is $f_s(cancer) = [0.56, 0.1]$, where f_s is feature function for stickiness value. It is worth to note that the word "cancer" rarely co-occur with the word "doc". It is reflected that the stickiness value of the word "cancer" relative to the word "doc" is smaller than the stickiness value to the word "prevent".

4   Evaluation Result

4.1   Data and Resources

The data for the experiment is taken from the collection of consumer-health questions crawled through Indonesian healthcare consultation forums (Hakim et al., 2017). Due to resources limitation, we only manually annotate 416 sample of user-generated posts. The description of the dataset for experiment can be seen in Table 1

We use the dictionary from The Medical Council of Indonesia[1] to extract MEDICAL DICTIONARY feature. On the other hand, POS-Tag feature is learned by model that is trained using data from Dinakaramani et al. (2014).

---

[1] (www.kki.go.id/assets/data/arsip/ SKDI_Perkonsil,_11_maret_13.pdf)

4.2   Experiment

There are two main scenarios for the experiment. First, feature ablation study aims to determine feature's contribution to the model performance. Second, model selection finds the model that outputs best result. The performance of a model is measured by precision, recall, and F1 metric. Precision is the number of keyphrases that are correctly extracted, divided by the total number of keyphrases labeled by our system. Recall is the number of keyphrases that are correctly extracted, divided by the total number of keyphrases in the gold-standard.

4.2.1   Feature Ablation Study

Ablation study is done by systematically removing feature sets to identify the most important features. We adopt leave one out (LOO) technique for feature ablation study. First, the model that uses all proposed features is evaluated. After that, 9 other different models are constructed, each of which uses combination of 8 features (another one feature is ablated in each model). The difference of F1-score between original model using all features and model with one missing feature indicate the contribution of (missing) feature to model performance. For ablation study, we split the data into 80% training set and 20% testing set. In this work, feature ablation is conducted using LSTM model.

Negative delta percentage score, as shown in Table 2, means that our proposed features contribute positively to improve model performance. The WORD EMBEDDING, which is most basic feature in our model, contributes the most. By removing word embedding feature, precision and recall decrease by 13.40% and 23.48% respectively. WORD STICKINESS is the second most important feature, indicated by change of 8.12% F-1 score. Based on this ablation study, WORD POSITION is not part of best feature combination.

We re-evaluate ablation study result using partial match score. In this scheme, suppose that the expected keyphrase consists of two words or more and the predicted contains only one word of it, partial match will still count it as a true positive. We find that removing WORD POSITION feature causes partial-match precision of model drops. So, our decision is to include WORD POSITION feature along with all other features in the final model.

32

Table 2: Feature Ablation Evaluation (in %)

Removed Features	Precision	Recall	F-1
Word Embedding	-13.40	-23.48	-18.91
Medical Dictionary	-3.44	-5.28	-4.30
Word Length	-4.92	-5.78	-5.33
Word Position	+2.91	+0.52	+1.70
POS-Tag	-5.41	-6.25	-5.80
Medical Entity	-6.15	-6.73	-6.40
Abbreviation and Acronym	-4.85	-5.99	-5.35
Word Centrality	-3.92	-5.44	-4.61
Word Stickiness	-7.35	-9.02	-8.12

Table 3: Model Evaluation (in %)

Models	Precision	Recall	F-1
RAKE	11.60	12.54	12.05
CRF	20.98	18.60	19.23
LSTMs	52.03	55.04	53.43
B-LSTM	55.12	58.07	56.48
Stacked-B-LSTMs	56.19	59.84	57.93
Stacked-B-LSTMs-CRF	59.12	60.11	59.22
Weight-Stacked-B-LSTMs	**60.06**	**63.08**	**61.37**

4.2.2 Model Selection

We test our model using 10-cross-validation scenario. As the baselines, we implement RAKE (Rose et al., 2010) and CRF (Cao et al., 2010). For CRF model, we apply the similar features with used in LSTM. The summary of various model evaluation is presented in Table 3.

RAKE performs the worst on extracting keyphrases from user-generated healthcare forum posts, since it is actually devoted for formal text. Performance of CRF model is also not good. Based on our observation from the predicted output by CRF, this method fails to predict the long sequences as the keyphrases. LSTM outperforms the baselines by achieving 53.43% F-1 score, 35% higher than CRF.

Using bidirectional concept, LSTM is able to integrate information from previous and after timestep, so that B-LSTM deliver better result compared to LSTM. Stacked-B-LSTMs using two layers performs better than B-LSTM for this task. Moreover, the weighting layer, which learns the weight for each feature, improves model performance. Hence, the best result was obtained by Weight-Stacked-B-LSTMs, whose Precision, Recall, and F-1 are respectively 60.06%, 63.06%, and 61.37%. It indicates that the feature weighting process worked well and, on some degree, demonstrate the reliability of our model in keyphrases extraction for user-generated contents in healthcare domain.

5 Conclusion

We proposed the model to address the task of keyphrases extraction from user-generated contents in medical domain. Extracting information about health-related concerns from user-generated forum post is not a trivial task, due to the fact that the content is usually short and written in an unstructured format, as opposed to formal text. Our model is based on sequence labeling task that employs deep learning approach using Long Short-Term Memory networks. Furthermore, several handcrafted features are proposed, including word centrality to detect important word in a document and word stickiness to obtain complete sequence of words as a keyphrase. We also propose a new layer in the neural network architecture for weighting the features. Our model successfully outperforms baseline methods for keyphrase extraction.

References

Yong-gang Cao, James J Cimino, John Ely, and Hong Yu. 2010. Automatically extracting information needs from complex clinical questions. *Journal of biomedical informatics*, 43(6):962–971.

Arawinda Dinakaramani, Fam Rashel, Andry Luthfi, and Ruli Manurung. 2014. Designing an indonesian part of speech tagset and manually tagged indonesian corpus. In *IALP*, pages 66–69.

Mark Dredze, Hanna M. Wallach, Danny Puller, and Fernando Pereira. 2008. Generating summary keywords for emails using topics. In *Proceedings of the 13th International Conference on Intelligent User Interfaces*, IUI '08, pages 199–206, New York, NY, USA. ACM.

Maryam Habibi and Andrei Popescu-Belis. 2013. Diverse keyword extraction from conversations. In *Proceedings of the ACL 2013 (51th Annual Meeting of the Association for Computational Linguistics)*, *Short Papers*, pages 651–657. ACL.

Abid Nurul Hakim, Rahmad Mahendra, Mirna Adriani, and Adrianus Saga Ekakristi. 2017. Corpus development for indonesian consumer-health question answering system. In *Proceedings of the 2017 International Conference on Advanced Computer Science and Information Systems (ICACSIS)*, pages 221–226.

Kazi Saidul Hasan and Vincent Ng. 2014. Automatic keyphrase extraction: A survey of the state of the art. In *Proceedings of the 52nd Annual Meeting of the Association for Computational Linguistics (Volume 1: Long Papers)*, pages 1262–1273, Baltimore, Maryland. Association for Computational Linguistics.

Su Nam Kim and Timothy Baldwin. 2012. Extracting keywords from multi-party live chats. In *Proceedings of the 26th Pacific Asia Conference on Language, Information and Computation, PACLIC 26, Bali, Indonesia, December 16-18, 2012*, pages 199–208.

John D. Lafferty, Andrew McCallum, and Fernando C. N. Pereira. 2001. Conditional random fields: Probabilistic models for segmenting and labeling sequence data. In *Proceedings of the Eighteenth International Conference on Machine Learning*, ICML '01, pages 282–289, San Francisco, CA, USA. Morgan Kaufmann Publishers Inc.

Zhenhui Li, Ding Zhou, Yun-Fang Juan, and Jiawei Han. 2010. Keyword extraction for social snippets. In *Proceedings of the 19th International Conference on World Wide Web*, WWW '10, pages 1143–1144, New York, NY, USA. ACM.

Xuezhe Ma and Eduard Hovy. 2016. End-to-end sequence labeling via bi-directional lstm-cnns-crf. In *Proceedings of the 54th Annual Meeting of the Association for Computational Linguistics*, pages 1064–1074. ACM.

Rada Mihalcea and Paul Tarau. 2004. TextRank: Bringing order into texts. In *Proceedings of EMNLP-04and the 2004 Conference on Empirical Methods in Natural Language Processing*.

Tomas Mikolov, Ilya Sutskever, Kai Chen, Greg S Corrado, and Jeff Dean. 2013. Distributed representations of words and phrases and their compositionality. In *Advances in neural information processing systems*, pages 3111–3119.

Lawrence Page, Sergey Brin, Rajeev Motwani, and Terry Winograd. 1998. The pagerank citation ranking: Bringing order to the web. In *Proceedings of the 7th International World Wide Web Conference*, pages 161–172, Brisbane, Australia.

Marek Rei, Gamal K. O. Crichton, and Sampo Pyysalo. 2016. Attending to characters in neural sequence labeling models. *CoRR*, abs/1611.04361.

Stuart Rose, Dave Engel, Nick Cramer, and Wendy Cowley. 2010. Automatic keyword extraction from individual documents. *Text Mining*, pages 1–20.

Kamal Sarkar. 2013. A hybrid approach to extract keyphrases from medical documents. *International Journal of Computer Applications*, 63:14–19.

Kamal Sarkar, Mita Nasipuri, and Suranjan Ghose. 2010. A new approach to keyphrase extraction using neural networks. *arXiv preprint arXiv:1004.3274*.

Peter D Turney. 2000. Learning algorithms for keyphrase extraction. *Information retrieval*, 2(4):303–336.

Ian H Witten, Gordon W Paynter, Eibe Frank, Carl Gutwin, and Craig G Nevill-Manning. 1999. Kea: Practical automatic keyphrase extraction. In *Proceedings of the fourth ACM conference on Digital libraries*, pages 254–255. ACM.

Chengzhi Zhang. 2008. Automatic keyword extraction from documents using conditional random fields. *Journal of Computational Information Systems*, 4(3):1169–1180.

Qi Zhang, Yang Wang, Yeyun Gong, and Xuanjing Huang. 2016. Keyphrase extraction using deep recurrent neural networks on twitter. In *Proceedings of the 2016 Conference on Empirical Methods in Natural Language Processing*, pages 836–845.

Wayne Xin Zhao, Jing Jiang, Jing He, Yang Song, Palakorn Achananuparp, Ee-Peng Lim, and Xiaoming Li. 2011. Topical keyphrase extraction from twitter. In *Proceedings of the 49th Annual Meeting of the Association for Computational Linguistics: Human Language Technologies - Volume 1*, HLT '11, pages 379–388, Stroudsburg, PA, USA. Association for Computational Linguistics.

Jie Zhou and Wei Xu. 2015. End-to-end learning of semantic role labeling using recurrent neural networks. In *ACL (1)*, pages 1127–1137.

Identifying Key Sentences for Precision Oncology Using Semi-Supervised Learning

Jurica Ševa, Martin Wackerbauer and Ulf leser
Knowledge Managment in Bioinformatics
Humboldt Universität zu Berlin
Berlin, Germany
{seva,wackerbm,leser}@informatik.hu berlin.de

Abstract

We present a machine learning pipeline that identifies key sentences in abstracts of oncological articles to aid evidence-based medicine. This problem is characterized by the lack of gold standard datasets, data imbalance and thematic differences between available silver standard corpora. Additionally, available training and target data differs with regard to their domain (professional summaries vs. sentences in abstracts). This makes supervised machine learning inapplicable. We propose the use of two semi-supervised machine learning approaches: To mitigate difficulties arising from heterogeneous data sources, overcome data imbalance and create reliable training data we propose using transductive learning from positive and unlabelled data (PU Learning). For obtaining a realistic classification model, we propose the use of abstracts summarised in relevant sentences as unlabelled examples through Self-Training. The best model achieves 84% accuracy and 0.84 F1 score on our dataset.

1 Introduction

The ever-growing amount of biomedical literature accessible online is a valuable source of information for clinical decisions. The PubMed database (National Library of Medicine, 1946-2018), for instance, lists approximately 30 million articles' abstracts. As a consequence, machine learning (ML) based text mining (TM) is increasingly employed to support evidence-based medicine by finding, condensing and analysing relevant information (Kim et al., 2011). Practitioners in this field search for clinically relevant articles and findings, and are typically not interested in the bulk of search results which are devoted to basic research. However, defining clinical relevance in a given abstract is not a trivial task. On top, although abstracts provide a very brief summary of their corresponding articles' content, practitioners determine abstracts' clinical relevance based on only a few key sentences (McKnight and Srinivasan, 2003). To optimally support such users, it is thus necessary to first retrieve only clinically relevant articles and next to identify the sentences in those articles which express their clinical relevance.

Any survival benefit of dMMR was lost in N2 tumors. *Mutations in BRAF(V600E) (HR, 1.37; 95% CI, 1.08 to 1.70; P = .009) or KRAS (HR, 1.44; 95% CI, 1.21 to 1.70; P ¡ .001) were independently associated with worse DFS.* *The observed MMR by tumor site interaction was validated in an independent cohort of stage III colon cancers (P(interaction) = .037).*

Example 1: Snippet of highlighted clinically relevant (or key; yellow background color) and irrelevant (no background color) sentences in a precision oncology setting. Source document with PMID 24019539.

In this work, we present an ML pipeline to identify key (clinically relevant) sentences, in a precision oncology setting, in abstracts of oncological articles to aid evidence-based medicine. This setting is implied throughout the text when referring to clinical relevance or key (clinically relevant) sentences. An example of relevant and irrelevant sentences is shown in Example 1. For solving this problem no gold standard corpora is available. Additionally, clinical relevance has

Proceedings of the BioNLP 2018 workshop, pages 35–46
Melbourne, Australia, July 19, 2018. ©2018 Association for Computational Linguistics

only a vague definition and is a subjective measure. As manually labelling text is expensive, semi-supervised learning offers the possibility to utilize related annotated corpora. We focus on Self-Training (Wang et al., 2008), which mostly relies on supervised classifiers trained on labelled data and use of unlabelled examples to improve the decision boundary. Several corpora can be used to mitigate the issues arising from the lack of gold standard data set and data imbalance. These corpora implicitly define characteristics of key sentences, but cannot be considered as gold standards. In the following, we call them "silver standard" corpora - collections of sentences close to the intended semantic but with large amounts of noise. Specifically, we employ *Clinical Interpretations of Variants in Cancer* (CIViC) (Griffith et al., 2017) for implicit notion of clinical relevance and positive data points, i.e. sentences or abstracts which have clinical relevance. Unfortunately, negative data points, i.e. sentences or abstracts which do not have clinical relevance, are not present in this data set. *PubMed abstracts*, referenced by CIViC, are used as unlabelled data. Since we consider all sentences in CIViC to be positive examples and the corresponding abstracts are initially unlabelled, additional data for negative examples is required. We utilize the *Hallmarks of Cancer Corpus* (HoC) (Baker et al., 2016) as an auxiliary source of noisy labelled data. To expand on our set of labelled data points we propose transductive learning from positive and unlabelled data (PU Learning) to identify noise within HoC, with CIViC as a guide set for determining the relevance of sentences from HoC. This gives us additional, both positive and negative data points, used as an initialization for Self-Training. The pipeline is available at https://github.com/nachne/semisuper.

2   Related Work

Sentence classification is a special case of text categorisation. It has been used in a wide range of fields, like sentiment analysis (Yu and Hatzivassiloglou, 2003; Go et al., 2009; Vosoughi et al., 2015), rhetorical annotation, and automated summarisation (Kupiec et al., 1995; Teufel and Moens, 2002). Between the two, feature engineering has been reported as the major difference. For instance, common stop words like "but", "was", and "has" are often among the top features for sentence classification, and verb tense is useful to

determine a sentence's precise meaning (Agarwal and Yu, 2009; Khoo et al., 2006). Additional features beyond the pure language level have also been proposed. For sentiment analysis, Yu and Hatzivassiloglou (2003) use a dictionary of semantically meaningful seed words to estimate the likely positive or negative polarity of co-occurring words, from which in turn a sentences' polarity is determined. Teufel and Moens (2002) focus on identifying rhetorical roles of sentences for automatic summarisation of scientific articles. They use sentence length and location, the presence of citations and of words included in headlines, labels of preceding sentences, and predefined cue words and formulaic expressions accompanying Bag of Words (BOW).

Text represented as high dimensional BOW vectors has been reported to be often linearly separable, making Support Vector Machines (Joachims, 1998) (SVM) a popular choice for classifiers. Conditional Random Fields (CRF) have been used to predict sequences of labels rather than labelling sentences one by one (Kim et al., 2011). In recent years, Neural Networks (NN) and Deep Learning (DL) has increasingly been used, e.g. using Convolutional Neural Networks (CNN) (Kim, 2014; Rios and Kavuluru, 2015; Zhang et al., 2016; Conneau et al., 2017). Other authors employ various versions of Recurrent Neural Networks (RNN): LSTM (Hassan and Mahmood, 2017), bi-directional LSTM (Dernoncourt et al., 2017; Zhou et al., 2016) or convolutional LSTM (Zhou et al., 2015). The use of DL has also popularised the use of pre-trained word embedding vectors. Habibi et al. (2017) show that the use of word embeddings, in general, increases the quality of biomedical named entity recognition pipelines.

Specific to the biomedical domain, sentence classification has been used to determine the rhetorical role sentences play in an article or abstract. Ruch et al. (2007) propose using the "Conclusion" section of abstracts as examples for key sentences. McKnight and Srinivasan (2003) have classified sentences in abstracts of randomised control trials as belonging to the categories "Introduction", "Methods", "Results", and "Discussion" (IMRaD), using section headlines as soft labels for training data in addition to a smaller hand annotated corpus. They also report that adding sentence location as a feature improved performance on the "Introduction" and "Discussion" categories. Kim et al. (2011) used a CRF for sequential classification, trained on the hand-annotated Population, Intervention, Background, Outcome, Study Design of evidence-based medicine, or Other (PIBOSO) corpus, with sentences annotated with one of the aforementioned categories. Unfortunately, since sentences in our primary source of positive data are from a different context than the abstracts to be classified, section headings, preceding sentences, and location are not available for our task.

2.1 Semi-Supervised Learning

Semi-supervised learning has the potential to match the performance of supervised learning while requiring considerably less labelled data (Wang et al., 2008; Thomas et al., 2012; Liu et al., 2013). Soft labelling (e.g. aforementioned heuristics for using section headlines as labels) is sometimes subsumed under semi-supervised learning as Distant Supervision (Go et al., 2009; Vosoughi et al., 2015; Wallace et al., 2016). Label Propagation (Zhu and Ghahramani, 2002) and Label Spreading (Zhou et al., 2003) can be seen as largely unsupervised classification, using labelled data to initialise and control clustering. Likewise, Nigam et al. (2011) propose Naive Bayes (NB) based variants of the unsupervised Expectation-Maximisation (EM) algorithm for utilising unlabelled data in semi-supervised text classification.

2.1.1 PU Learning

PU Learning is a special case of semi-supervised learning where examples in the unlabelled set U are to be classified as positive (label 1) or negative (label 0), with only positive labelled data P initially available. Therefore, the PU Learning problem can be approximated by learning to discriminate P from U (Mordelet and Vert, 2014). For that, learning should favour false positive errors over false negatives, e.g. by using class-specific weights for error penalisation. Approaches include *one-class SVMs*, which approximate the support of the positive class and treat negative examples as outliers; *ranking methods*, which rank unlabelled examples by their decreasing similarity to the mean positive example; and *two-step heuristics*, which try to identify reliable negative examples in the unlabelled data to initialise semi-supervised learning. We consider the aforementioned heuristics useful for outlier detection to reduce noise in our auxiliary data, and use variations of PU Learning algorithms in semi-

supervised learning, as our problem of finding summary-like sentences without explicitly defined negative sentences is closely related to PU Learning. An overview is available in (Liu et al., 2003). Additional information can be found in (Elkan and Noto, 2008; Plessis et al., 2014, 2015). An example of the use of PU Learning, for spotting online fake reviews, is available in (Li et al., 2014).

Without known negative examples, measuring classification performance using accuracy or F1-score in PU Learning is not possible. Lee and Liu (2003) suggest an alternative score, called PU-score, defined as $Pr[f(X) = 1|Y = 1]^2/Pr[f(X) = 1]$, for comparing PU Learning classifiers that can be derived from positive and unlabelled data alone. The authors show theoretically that maximising the PU-score is equivalent to maximising the F1-score and can be used to compare different models classifying the same data. Nonetheless, it should be noted that this metric is not bounded, making it viable only for comparing classifiers trained and tested on the same data; it is not an indicator for an individual classifier's performance.

2.1.2 Self-Training

Self-Training, used in this work, starts from an initial classifier trained on the labelled data. Previously unlabelled examples that were labelled with high confidence are added to the training data. This procedure repeats iteratively, retraining the classifier until a terminating condition is met. NB is a popular classifier for Self-Training because the probabilities it produces provide a confidence ranking, but any other algorithm may be used as long as confidence scores can be derived (Wang et al., 2008).

3 Methods

We present the data sources we use, the preprocessing pipeline and describe in detail the experiments performed with both PU Learning and Self-Training.

3.1 Used Corpora

CIViC is a database of clinical evidence summaries. Entries consist of evidence statements about gene variants, such as their association with diseases and the outcome of drug response trials. Additional information includes the names of the respective genes, variants, drugs, diseases, and a variant summary. Each entry contains the PubMed

ID of the respective publication the information is taken from. Evidence statements are prototypes of high-quality information in condensed form. However, they are not themselves contained in the abstracts they are based on, as they try to summarize the entire article. At time of writing, CIViC contains about 2,300 evidence statements consisting of 6,600 sentences (5,300 without duplicates). They make up our initial corpus of positive sentences (P).

PubMed abstracts referenced in CIViC. We extract about 12,700 sentences from 1,300 abstracts referenced in CIViC, and use them as the unlabelled corpus (U). We use CIViC summaries and the PubMed abstract corpus to estimate the acceptable range for the ratio of key sentences in an abstract. We use the ratio of overall sentence counts in the two corpora (CIViC summaries and PubMed abstracts) as an upper bound of ≈ 0.4 (5,300/12,700). As a rough estimate for the lower bound, based on an informed guess that half of the sentences could be redundant, since one abstract may correspond to multiple CIViC entries for different drug/variant combinations. This results in a lower bound ≈ 0.2. Although this is a simplifying assumption and disregards e.g. any differences in information density in our data sources' sentences, it provides a rough guideline for the ratio of key sentences in U a classifier should find.

Hallmarks of Cancer (HoC) describe common traits of all forms of cancer. We use it as a silver standard corpus consisting of about 13,000 sentences from 1,580 PubMed abstracts. Sentences not relevant to any of the hallmarks are left unlabelled. We assume unlabelled sentences are less likely to be clinically relevant than sentences with one or more labels, aggregating them in the likely negative set HoC_n (about 8,900 sentences) and the likely positive set HoC_p (about 4,300 sentences). In order to improve generalisation, as well as to be able to validate our classifier, which requires positive as well as negative labelled data, we use HoC as auxiliary data. To utilise HoC_p and HoC_n as sources of realistic positive and negative sentences for training and test data, but avoiding propagation of misclassification errors resulting from our simplifying assumption, they must be filtered for noise (Section 3.3).

3.2   Text Preprocessing and Feature Selection

As features, we use word n-grams, character n-grams, and sentence length, concatenating them to form a mixed feature space. All tokens are converted to lower-case. Biomedical scientific text exhibits some particularities that have to be taken into consideration during text preprocessing. To normalise all text, before sentence splitting with the PunktSentenceTokenizer of the Python Natural Language Toolkit (NLTK) (Bird et al., 2009), we use regular expressions: we substitute spaces for full stops in common abbreviations followed by a space and lower-case letter or digit (e.g. "ca._5" → "ca_5"). As the pattern "patient no. V[123]" is quite frequent in *CIViC*, we introduce a special rule for not splitting it despite the upper-case. All whitespace characters are replaced by spaces to avoid splitting on newlines. Afterwards, to avoid character encoding-related problems and to reduce alphabet size, we normalize all text to ASCII before tokenization.

For word-level tokenization, we use NLTK's TreebankWordTokenizer and split the resulting tokens at characters in {"-", "/", ".", ",", "_", "|", "¿"}. Sentences below a minimum character count of 8 are denoted by a special "_empty_sentence_" token. To prepare word n-grams, we replace tokens representing numbers or ordinals and their spelled out versions by a special "_num_" token and do the equivalent for e.g. ranges, inequalities, percentages, measurement units, and years. Tokens with suffixes common for drugs but not found in common speech, such as "-inib", are replaced by "_chemical_", and sequences that start with a letter, but contain digits, are replaced by "_abbrev_" in the hope of catching identifiers of biomedical entities. We evaluated the use of word n-grams with n bounded from (1,1) (the bag-of-words case) up to (1,4), thereby retaining information about the order of words in a sentence.

We evaluated the use of character n-gram with n in ranges from (2,3) to (2,6). To reduce alphabet size and avoid overfitting, all sequences of non-word characters except those in {"-", "%", "="}, which may carry semantic information, are replaced by single spaces, and all digits are replaced by 1.

In feature vectors, word and character n-grams are weighted by their tf-idf score (Aizawa, 2003) and sentence length is represented as inverse character count. Character n-grams proved to be more expressive than word n-grams, yielding better accuracy scores when we tested each of them in isolation. However, a combination of character and word level n-grams and text length performed best.

3.3 Noise reduction with PU Learning

Using PU Learning, we filter HoC_n and HoC_p for sentences that are likely to be useful in our classification task. We explored several approaches to PU Learning and subsequent noise reduction.

3.3.1 PU Learning

First, we explored several **Two-Step** techniques, which (1) identify a set of reliable negative examples (RN) from U and (2) train a classifier on P and RN and retrain the classifier using the predicted labels until a stopping criterion is met. We present them next. ***i-EM*** is a variation of the Expectation-Maximisation algorithm that relies on a NB classifier, with predicted probabilities of labels in range $[0, 1]$. ***s-EM*** is an extension to i-EM. Initially, a subset $S \subset P$ is added to U as spy documents. After training an initial classifier on $P \setminus S$ vs. $U \cup S$, the predicted probabilistic labels for the known hidden positives in S are used to determine a threshold t; all $u \in U$ with probabilistic labels $p(y_u) < t$ are moved to RN. In Step 2, starting from a classifier trained to discriminate P from RN, the EM algorithm is iterated as in i-EM until it converges or the estimated classification error is deteriorating. ***Roc-SVM*** uses the Rocchio algorithm for Step 1: Firstly, prototype vectors $\bar{p}$ for P and $\bar{u}$ for U are computed as a weighted differences between the two sets' respective average examples. Using these vectors, RN is defined as $\{u \in U : \cos(u, \bar{u}) < \cos(u, \bar{p})\}$, i.e. all unlabelled sentences that are more similar to the prototype of the unlabelled set than the positive sets. Step 2 uses SVMs to expand RN. Initially, an SVM is trained to discriminate P from RN. Afterwards, all $u \in U \setminus RN$ with predicted label 0 are added to RN for iteratively retraining the classifier as long as RN changes. This iteration may go wrong and result in poor recall on the positive class; as a fallback strategy, if the classifier at convergence misclassifies too large a portion of P, the initial classifier is returned instead. ***CR-SVM*** is a minor extension to Roc-SVM. P and U are each ranked by decreasing cosine similarity to the mean positive example; a probably negative set PN is built from the $u \in U$ with a lower

score than a given ratio of least typical examples in P. The negative prototype vector is then computed using PN rather than U. Step 2 is the same as in Roc-SVM. Additionally, we explored **Biased SVM**, a soft-margin SVM that uses class-specific weights for positive and negative errors. Weight parameters are selected in a grid search manner to find a combination that optimises the PU-score; this effectively assumes U to contain only negligible amounts of hidden positive examples.

3.3.2 Noise reduction

We experiment with two heuristics for noise reduction in HoC. For both of them, let $\mathrm{clf}(P, U)(x)$ be the label for x predicted by classifier clf trained on P and U. Appendix B (Figure 1) summarises corpora used for this task.

Strict mode: Remove *CIViC*-like sentences, i.e. likely hidden positives, from HoC_n for the reliable negative set HoC_n'. Keep only *CIViC*-like sentences in HoC_p for a reliable positive set HoC_p'. This implies rather pessimistic assumptions about HoC_p's relevance, considering only outliers as key sentences.

$$HoC_n' := HoC_n \setminus \{x \in HoC_n : \mathrm{clf}(CIViC, HoC_n)(x) = 1\}$$
$$HoC_p' := \{x \in HoC_p : \mathrm{clf}(CIViC, HoC_p)(x) = 1\}$$

Tolerant mode: Remove *CIViC*-like sentences from HoC_n as before. But rather than requiring sentences from HoC_p to be *CIViC*-like, remove those sentences from HoC_p that are similar to reliable negative sentences, i.e. the purified HoC_n'. In doing so, HoC_p is assumed to be largely relevant, contaminated with non-key sentences.

$$HoC_n' := HoC_n \setminus \{x \in HoC_n : \mathrm{clf}(CIViC, HoC_n)(x) = 1\}$$
$$HoC_p' := HoC_p \setminus \{x \in HoC_p : \mathrm{clf}(HoC_n', HoC_p)(x) = 1\}$$

3.4 Semi-supervised learning with Self-Training

In the following, let the labelled set be $L := P \cup N$, with positive labelled set $P := CIViC \cup HoC_p'$ and negative labelled set $N := HoC_n'$. The unlabelled set of original abstracts is denoted by U. The purified sets HoC_p' and HoC_n' are obtained using either of the above heuristics. Appendix B (Figure 2) summarises corpora used for this task. We use: ***Standard Self-Training (ST)*** with a confidence threshold. Having experimented with different values, we use a threshold

Method	Reliable negatives: $P := CIViC$, $U := HoC_n$		Strict mode: $P := CIViC$, $U := HoC_p$		Tolerant mode: $P := HoC'_n$ $U := HoC_p$	
	PU-score	pos. ratio in U_{test}	PU-score	pos. ratio in U_{test}	PU-score	pos. ratio in U_{test}
i-EM	2.06	0.11	1.61	0.06	0.94	0.36
s-EM	2.06	0.11	1.61	0.06	0.94	0.40
Roc-SVM	**2.19**	**0.07**	**1.67**	**0.06**	**1.07**	**0.31**
CR-SVM	2.19	0.08	1.67	0.06	1.04	0.57
Biased-SVM	2.28	0.03	1.70	0.05	1.13	0.31

Table 1: Removing noise from HoC_n and HoC_p. Results for different PU Learning techniques, averaged over 10 runs, on 20% reserved test sets $P_{test} \subset P$ and $U_{test} \subset U$. To generate HoC'_n as required for tolerant mode, Roc-SVM (highlighted in bold) was used in the previous step.

Algorithm 1 Self-Training

1: **procedure** SELF-TRAINING(training data L with labels, unlabelled data U)
2:     **while** U is not empty **do**
3:         train classifier clf on L
4:         predict labels for U with clf
5:         move examples with most confidently predicted labels from U to L
6:     **end while**
7:     **return** clf
8: **end procedure**

---

of 0.75 for classifiers producing class probabilities, and 0.5 for the absolute values of SVM's decision function; *"Negative" Self-Training (NST)*: Rather than using a confidence criterion, all unlabelled examples classified as negative are added to the training data for the next iteration. This is analogous to the iterative SVM step of Roc-SVM, except for the predefined rather than heuristically estimated initial negative set, and has shown to help avoid an unrestricted propagation of positive labels; A variant of the *Expectation-Maximisation (EM)* algorithm as used in i-EM. Starting with P and N as initial fixed-label examples, iterate a NB classifier until convergence, using the class probabilities predicted for U as labels for the next training iteration; *Label Propagation and Label Spreading*: These algorithms propagate labels through high-density regions using a graph representation of the data. Both are implemented in Scikit-learn with Radial Basis Function (RBF) and k-Nearest-Neighbour (kNN) kernels available. We were unable to obtain competitive results with these techniques. In the Self-Training

algorithm (shown in Algorithm 1), we use Scikit-learn's implementations of SVM, NB, and Logistic Regression (LR) as underlying classifiers.

4 Results

Section 4.1 describes the effects of noise reduction heuristics using PU Learning. The performances of different semi-supervised approaches for training a classifier, with both strict and tolerant noise reduction scenarios, are shown in Section 4.2.

4.1 PU Learning for Noisy Data

Table 1 summarises the PU-scores and ratio of examples in U classified as positive for different algorithms for reducing noise in HoC_n and HoC_p using the strict vs. tolerant heuristics.

Cleaning up HoC_n removes some 2 to 7% of examples, depending on the classifier. Additional manual inspection of a subset of the sentences removed confirms them as true negatives with respect to key sentences.

Regarding HoC_p, the strict heuristics keeps only 5.5%, some 250 sentences, of positive examples. We suspect this is due to the different thematic foci of HoC and the articles summarised in $CIViC$, as well as the summaries' different writing style. This leaves us with $N := 8,300$ sentences, $P := 5,600$ sentences and $U := 12,700$ sentences. As our experiments show, choosing this very selective approach drastically improves the nominal accuracy of subsequent steps; however, it leaves a lack of real-world data in the positive training set and harbours the risk of overfitting.

On the other hand, using the tolerant strategy, roughly 25% of HoC_p are removed due to being very similar to HoC'_n. This results in a 50%

Method	parameters	acc	P_{test}:			N_{test}:			U:
			p	r	F1	p	r	F1	pos. ratio
NST(SVM)	$C = 0.3$	0.94	0.95	0.90	0.92	0.93	0.97	0.95	0.33
NST(LR)	$C = 6.0$	0.89	0.99	0.75	0.84	0.86	0.99	0.92	0.13
NST(NB)	$\alpha = 0.1$	0.90	0.97	0.78	0.86	0.87	0.98	0.92	0.31
ST(SVM)	$C = 0.4$	0.96	0.97	0.93	0.95	0.95	0.98	0.97	0.62
ST(LR)	$C = 4.0$	0.96	0.96	0.94	0.95	0.96	0.98	0.97	0.62
ST(NB)	$\alpha = 0.1$	0.92	0.93	0.88	0.90	0.92	0.96	0.94	0.60
EM	$\alpha = 0.1$	0.91	0.92	0.85	0.88	0.91	0.95	0.93	0.62
Label Propagation	RBF kernel	0.83	0.91	0.63	0.74	0.79	0.96	0.87	0.35
Label Propagation	kNN kernel	0.69	0.96	0.22	0.36	0.66	0.99	0.79	0.03
Label Spreading	RBF kernel	0.85	0.93	0.68	0.79	0.82	0.97	0.89	0.50
Label Spreading	kNN kernel	0.79	0.92	0.54	0.68	0.76	0.96	0.85	0.32

Table 2: Performance of different semi-supervised approaches trained on P, N, and U after strict noise filtering. ST = Self-Training. NST = "Negative" Self-Training. Results averaged over 10 runs with randomised 20% validation sets from P and N; min-df threshold = 0.002, 25% of most relevant features selected with χ^2.

Method	parameters	acc	P_{test}:			N_{test}:			U:
			p	r	F1	p	r	F1	pos. ratio
NST(SVM)	**$C = 0.3$**	**0.84**	**0.84**	**0.84**	**0.84**	**0.84**	**0.83**	**0.83**	**0.32**
NST(LR)	$C = 6.0$	0.81	0.88	0.72	0.79	0.76	0.90	0.82	0.17
NST(NB)	$\alpha = 0.1$	0.76	0.85	0.64	0.73	0.70	0.88	0.78	0.30
ST(SVM)	$C = 0.4$	0.85	0.90	0.81	0.85	0.83	0.89	0.86	0.62
ST(LR)	$C = 6.0$	0.86	0.87	0.85	0.85	0.84	0.86	0.85	0.66
ST(NB)	$\alpha = 0.1$	0.76	0.88	0.62	0.72	0.69	0.91	0.79	0.70
EM	$\alpha = 0.1$	0.74	0.88	0.58	0.70	0.68	0.91	0.78	0.70
Label Propagation	RBF kernel	0.72	0.88	0.50	0.64	0.64	0.92	0.76	0.36
Label Propagation	kNN kernel	0.58	0.90	0.20	0.32	0.54	0.98	0.70	0.02
Label Spreading	RBF kernel	0.74	0.88	0.56	0.68	0.67	0.92	0.77	0.56
Label Spreading	kNN kernel	0.68	0.91	0.43	0.58	0.62	0.96	0.77	0.34

Table 3: Performance of different semi-supervised approaches trained on P, N, and U after tolerant noise filtering. Results averaged over 10 runs with randomised 20% validation sets from P and N; min-df threshold = 0.002, 25% of most relevant features selected with χ^2. The model we consider most suitable for identifying key sentences is highlighted in bold.

larger and less homogenous positive labelled set compared to strict noise filtering, which we expect to provide greater generality and robustness to our classifier. This leaves us with $N := 8,300$ sentences, $P := 8,600$ sentences and $U := 12,700$ sentences. This is enough to assume a noticeable reduction of noise and easier distinction between HoC'_n and HoC'_p, but it still contributes a considerable amount of data to the positive set and is not suspect to overfitting. Typical topics of sentences removed as irrelevant include biochemical research hypotheses and non-human study subjects; however, as this heuristic is indirectly defined, its decisions are not quite as clearly correct as those directly linked to *CIViC*.

Our results confirm Biased-SVM nominally performs best among the PU Learning techniques described above; this is simply because the PU-score is maximised by minimising the amount of positive examples found in U, which Biased-SVM does by regarding U as negative and performing supervised classification. However, we do not find this to be useful for our purpose of noise detection, or for finding hidden positive data in unlabelled data in general. The EM-based techniques tend to go the opposite direction and consider comparably large ratios of U as positive, were more sensitive to distributions, and misclassified positive labelled data. Roc-SVM, on the other hand, had stable performance in our tests and scores close to those of Biased-SVM, which is why we use this approach to filter HoC for the subsequent steps. Our results also suggest the iterative second step is more crucial than the exact heuristics for choosing a reliable negative set from the unknown data.

4.2 Semi-Supervised classification of key sentences with Self-Training

We report accuracy (acc), precision (p), recall (r), F1-score ($F1$) and the ratio of key sentences found in U ($pos.ratio$) of different semi-supervised learning methods for strict (Table 2) and tolerant (Table 3) noise filtering scenarios. We consider classification to have gone wrong if the ratio of positive sentences in U significantly deviates from the acceptable range $[0.2, 0.4]$ (as defined in Section 3.1). Additionally, results of supervised ML pipeline on data sets generated after noise filtering are available in Appendix A.

Our experiments show that strict noise filtering leads to greatly improved classification accuracy; however, it may be fallacious to judge this approach only by the scores it produces. Given $CIViC$'s deviations from typical language in scientific articles, the different thematic foci of $CIViC$ and HoC, and the negligible amount of realistic positive sentences added in this scenario (Table 1), we suspect classifiers may overfit to superficial and incidental differences rather than learning to generalise to correctly identify key sentences in unseen abstracts. In order to avoid this, we discard strict noise filtering.

On the other hand, tolerant filtering of HoC_n and HoC_p still allows for reasonable classification accuracy considering the data's heterogeneity. We expect additional positive sentences to provide improvements to generalisation that outweigh the lower nominal performance scores and possible errors propagated due to remaining noise. Although HoC's notion of relevant sentences is not identical to that implied by $CIViC$, our experiments show that removing only the least suitable sentences is enough to use HoC_p' as meaningful training data.

Standard Self-Training yields performance results very similar to supervised classification, analogous to what can be observed in strict mode, but

a larger ratio of positive predictions for U. The linear classifiers SVM and Logistic Regression perform much better than NB, the latter modelling an inaccurate probability distribution. In both strict and tolerant mode, methods with an emphasis on unsupervised clustering (EM, Label Propagation, and Label Spreading) underperform, with a strong bias towards the negative class. Label Propagation with k-Nearest-Neighbours kernel performs particularly poorly, failing to find any positive examples in the unlabelled set. In contrast, NST with base classifiers leads to positive ratios in U close to our preliminary estimate, as well as acceptable classification accuracy. SVM performs better than Logistic Regression and has balanced precision and recall for both classes, appearing the more robust choice.

5 Conclusion

We have developed a pipeline for identifying the most informative key sentences in oncological abstracts, judging sentences' clinical relevance implicitly by their similarity to clinical evidence summaries in the $CIViC$ database. To account for deviations from typical content between professional summaries and sentences appearing in abstracts, we use the abstracts corresponding to these summaries as unlabelled data in an semi-supervised learning setting. An auxiliary silver standard corpus is used for more realistic training and validation data. To mitigate introducing errors due to miscategorised examples in partitions of the auxiliary data, we propose using PU Learning techniques in a noise detection preprocessing step.

We evaluate different heuristics for semi-supervised learning and measure their performance with heterogenous data. While methods with an emphasis on unsupervised clustering perform poorly, (which we attribute to the data violating smoothness assumptions) Self-Training with linear classifiers proved robust to unfavourably distributed data, reaching performance scores similar to those of supervised classifiers trained without the unlabelled data. By adapting Self-Training with SVMs to iteratively expand only the negative training set as in PU Learning, we were able to restrict the amount of hidden positive examples found in unlabelled data while maintaining good accuracy scores. Our best model using this method reaches 84% accuracy and 0.84 F1-score.

As a byproduct of the proposed pipeline, we obtain a silver standard corpus consisting of approximately 12,700 sentences from our unlabelled set, annotated with sentences' estimated clinical relevance, which may be useful for future classification tasks. Our final pipeline can be used to help clinicians quickly assess which articles are relevant to their work, e.g. by incorporating it into workflows for the retrieval of cancer-related literature. As such, it has been integrated in to Variant Information Search Tool[1] (VIST), a query-based document retrieval system which ranks scientific abstracts according to the clinical relevance of their content given a (set of) variations and/or genes.

We encountered various difficulties resulting from using a gold standard with atypical and solely positive examples and the heterogeneity of different training corpora. Although our problem of finding key sentences is a standard PU Learning task, the methods described in the PU Learning literature cannot be used in a verifiable way on real-world data without negative validation data. Even for semi-supervised learning with positive as well as negative labelled data, standard metrics alone are not enough to judge a classifier's adequacy, since the amount of noise in automatically gathered training data is never completely certain and the way unlabelled data is handled by a classifier is not represented in performance scores. By using heuristics for noise filtering and adapting self-training to incorporate unlabelled data in a way suitable to our goal, we alleviate these difficulties.

References

Shashank Agarwal and Hong Yu. 2009. Automatically classifying sentences in full-text biomedical articles into introduction, methods, results and discussion. *Bioinformatics*, 25(23):3174–3180.

Akiko Aizawa. 2003. An information-theoretic perspective of tf–idf measures. *Information Processing & Management*, 39(1):45–65.

Simon Baker, Ilona Silins, Yufan Guo, Imran Ali, Johan Högberg, Ulla Stenius, and Anna Korhonen. 2016. Automatic semantic classification of scientific literature according to the hallmarks of cancer. *Bioinformatics*, 32(3):432–440.

Steven Bird, Ewan Klein, and Edward Loper. 2009. *Natural language processing with Python: analyzing text with the natural language toolkit*. O'Reilly.

Alexis Conneau, Holger Schwenk, Loïc Barrault, and Yann Lecun. 2017. Very deep convolutional networks for text classification. In *Proceedings of the 15th Conference of the European Chapter of the Association for Computational Linguistics: Volume 1, Long Papers*, volume 1, pages 1107–1116.

Franck Dernoncourt, Ji Young Lee, and Peter Szolovits. 2017. Neural networks for joint sentence classification in medical paper abstracts. In *Proceedings of the 15th Conference of the European Chapter of the Association for Computational Linguistics: Volume 2, Short Papers*, pages 694–700. Association for Computational Linguistics.

Charles Elkan and Keith Noto. 2008. Learning classifiers from only positive and unlabeled data. In *Proceedings of the 14th ACM SIGKDD international conference on Knowledge discovery and data mining*, pages 213–220. ACM.

Alec Go, Richa Bhayani, and Lei Huang. 2009. Twitter sentiment classification using distant supervision. *CS224N Project Report, Stanford*, 1(12).

Malachi Griffith, Nicholas C Spies, Kilannin Krysiak, Joshua F McMichael, Adam C Coffman, Arpad M Danos, Benjamin J Ainscough, Cody A Ramirez, Damian T Rieke, Lynzey Kujan, et al. 2017. Civic is a community knowledgebase for expert crowdsourcing the clinical interpretation of variants in cancer. *Nature genetics*, 49(2):170.

Maryam Habibi, Leon Weber, Mariana Neves, David Luis Wiegandt, and Ulf Leser. 2017. Deep learning with word embeddings improves biomedical named entity recognition. *Bioinformatics*, 33(14):i37–i48.

A. Hassan and A. Mahmood. 2017. Deep learning for sentence classification. In *2017 IEEE Long Island Systems, Applications and Technology Conference (LISAT)*, pages 1–5.

Thorsten Joachims. 1998. Text categorization with support vector machines: Learning with many relevant features. In *Machine Learning: ECML-98, Chemnitz, Germany*, volume 1398 of *Lecture Notes in Computer Science*, pages 137–142. Springer.

Anthony Khoo, Yuval Marom, and David Albrecht. 2006. Experiments with sentence classification. In *Proceedings of the Australasian Language Technology Workshop 2006*, pages 18–25.

Su Kim, David Martínez, Lawrence Cavedon, and Lars Yencken. 2011. Automatic classification of sentences to support evidence based medicine. *BMC Bioinformatics*, 12(S-2):S5.

Yoon Kim. 2014. Convolutional neural networks for sentence classification. In *Proceedings of the 2014 Conference on Empirical Methods in Natural Language Processing (EMNLP)*, pages 1746–1751. Association for Computational Linguistics.

---

[1] https://triage.informatik.hu-berlin.de:8080/

Julian Kupiec, Jan O. Pedersen, and Francine Chen. 1995. A trainable document summarizer. In *SIGIR'95, Seattle, Washington, USA*, pages 68–73. ACM Press.

Wee Sun Lee and Bing Liu. 2003. Learning with positive and unlabeled examples using weighted logistic regression. In *Machine Learning, Proceedings of the Twentieth International Conference (ICML 2003), August 21-24, 2003, Washington, DC, USA*, pages 448–455. AAAI Press.

Huayi Li, Zhiyuan Chen, Bing Liu, Xiaokai Wei, and Jidong Shao. 2014. Spotting fake reviews via collective positive-unlabeled learning. In *Proceedings of the 2014 IEEE International Conference on Data Mining*, ICDM '14, pages 899–904, Washington, DC, USA. IEEE Computer Society.

Bing Liu, Yang Dai, Xiaoli Li, Wee Sun Lee, and Philip S. Yu. 2003. Building text classifiers using positive and unlabeled examples. In *Proceedings of the 3rd IEEE International Conference on Data Mining (ICDM 2003), Melbourne, Florida, USA*, pages 179–188. IEEE Computer Society.

Zhiguang Liu, Xishuang Dong, Yi Guan, and Jinfeng Yang. 2013. Reserved self-training: A semi-supervised sentiment classification method for chinese microblogs. In *Sixth International Joint Conference on Natural Language Processing, Nagoya, Japan*, pages 455–462. Asian Federation of Natural Language Processing / ACL.

Larry McKnight and Padmini Srinivasan. 2003. Categorization of sentence types in medical abstracts. In *American Medical Informatics Association Annual Symposium, Washington, DC, USA*. AMIA.

Fantine Mordelet and Jean-Philippe Vert. 2014. A bagging SVM to learn from positive and unlabeled examples. *Pattern Recognition Letters*, 37:201–209.

National Library of Medicine. 1946-2018. Pubmed. `https://www.ncbi.nlm.nih.gov/pubmed`. Accessed: 2018-02-01.

Bhawna Nigam, Poorvi Ahirwal, Sonal Salve, and Swati Vamney. 2011. Document classification using expectation maximization with semi supervised learning. *CoRR*, abs/1112.2028.

Marthinus Du Plessis, Gang Niu, and Masashi Sugiyama. 2014. Analysis of learning from positive and unlabeled data. In *Proceedings of the 27th International Conference on Neural Information Processing Systems - Volume 1*, NIPS'14, pages 703–711, Cambridge, MA, USA. MIT Press.

Marthinus Du Plessis, Gang Niu, and Masashi Sugiyama. 2015. Convex formulation for learning from positive and unlabeled data. In *Proceedings of the 32nd International Conference on Machine Learning*, volume 37 of *Proceedings of Machine Learning Research*, pages 1386–1394, Lille, France. PMLR.

Anthony Rios and Ramakanth Kavuluru. 2015. Convolutional neural networks for biomedical text classification: Application in indexing biomedical articles. In *Proceedings of the 6th ACM Conference on Bioinformatics, Computational Biology and Health Informatics*, BCB '15, pages 258–267, New York, NY, USA. ACM.

Patrick Ruch, Célia Boyer, Christine Chichester, Imad Tbahriti, Antoine Geissbühler, Paul Fabry, Julien Gobeill, Violaine Pillet, Dietrich Rebholz-Schuhmann, Christian Lovis, and Anne-Lise Veuthey. 2007. Using argumentation to extract key sentences from biomedical abstracts. *I. J. Medical Informatics*, 76(2-3):195–200.

Simone Teufel and Marc Moens. 2002. Summarizing scientific articles: Experiments with relevance and rhetorical status. *Computational Linguistics*, 28(4):409–445.

Philippe Thomas, Tamara Bobić, Ulf Leser, Martin Hofmann-Apitius, and Roman Klinger. 2012. Weakly labeled corpora as silver standard for drug-drug and protein-protein interaction. In *Proceedings of the Workshop on Building and Evaluating Resources for Biomedical Text Mining (BioTxtM) on Language Resources and Evaluation Conference (LREC)*.

Soroush Vosoughi, Helen Zhou, and Deb Roy. 2015. Enhanced Twitter Sentiment Classification Using Contextual Information. In *Proceedings of the 6th Workshop on Computational Approaches to Subjectivity, Sentiment and Social Media Analysis*, pages 16–24, Stroudsburg, PA, USA. Association for Computational Linguistics.

Byron C. Wallace, Joël Kuiper, Aakash Sharma, Mingxi (Brian) Zhu, and Iain James Marshall. 2016. Extracting PICO sentences from clinical trial reports using supervised distant supervision. *Journal of Machine Learning Research*, 17:132:1–132:25.

Bin Wang, Bruce Spencer, Charles X. Ling, and Harry Zhang. 2008. Semi-supervised self-training for sentence subjectivity classification. In *Advances in Artificial Intelligence , 21st Conference of the Canadian Society for Computational Studies of Intelligence Windsor, Canada*, pages 344–355. Springer.

Hong Yu and Vasileios Hatzivassiloglou. 2003. Towards answering opinion questions: Separating facts from opinions and identifying the polarity of opinion sentences. In *Proceedings of the 2003 Conference on Empirical Methods in Natural Language Processing*, EMNLP '03, pages 129–136, Stroudsburg, PA, USA. Association for Computational Linguistics.

Ye Zhang, Stephen Roller, and Byron C. Wallace. 2016. MGNC-CNN: A simple approach to exploiting multiple word embeddings for sentence classification. In *NAACL HLT 2016, San Diego California, USA*, pages 1522–1527. The Association for Computational Linguistics.

Chunting Zhou, Chonglin Sun, Zhiyuan Liu, and Francis C. M. Lau. 2015. A c-lstm neural network for text classification. *CoRR*, abs/1511.08630.

Dengyong Zhou, Olivier Bousquet, Thomas Navin Lal, Jason Weston, and Bernhard Schölkopf. 2003. Learning with local and global consistency. In *Advances in Neural Information Processing Systems 16 [Neural Information Processing Systems, NIPS 2003, December 8-13, 2003, Vancouver and Whistler, British Columbia, Canada]*, pages 321–328. MIT Press.

Peng Zhou, Zhenyu Qi, Suncong Zheng, Jiaming Xu, Hongyun Bao, and Bo Xu. 2016. Text classification improved by integrating bidirectional lstm with two-dimensional max pooling. pages 3485–3495.

Xiaojin Zhu and Zoubin Ghahramani. 2002. Learning from labeled and unlabeled data with label propagation.

A Semi-supervised ML on filtered datasets

Table 4 shows results of supervised baseline classifiers trained on P and N after strict filtering. Performance is very good for all classifiers tested, which is not surprising as *CIViC* and HoC_n are easy to separate even without filtering. The ratio of the unlabelled set U classified as positive, however, is outside of the acceptable range [0.2, 0.4] for selecting key sentences, probably due to the more similar contents of *CIViC* and the corresponding abstracts compared to HoC.

Table 5 shows the results of supervised classifiers trained on only P and N after tolerant filtering. Accuracies and F1-scores are about 10 percent points lower compared to results in the strict filtering scenario, which can be explained by HoC_p and HoC_n being comparably difficult to separate. However, performance is better compared to distinguishing $CIViC \cup HoC_p$ vs. HoC_n without any noise filtering.

B PU Learning and Self-Training: used corpora

| | | P_{test}: | | | | N_{test}: | | | U: |
Method	parameters	acc	p	r	F1	p	r	F1	pos. ratio
SVM	$C = 3.0$	0.96	0.97	0.94	0.95	0.96	0.98	0.97	0.63
LR	$C = 6.0$	0.96	0.96	0.94	0.95	0.96	0.97	0.97	0.64
NB	$\alpha = 0.1$	0.94	0.95	0.91	0.93	0.94	0.97	0.95	0.61

Table 4: Supervised classifiers trained on P and N after strict noise filtering. Results averaged over 10 runs with randomised 20% reserved test sets; min-df threshold = 0.002, 25% of most relevant features selected with χ^2.

| | | P_{test}: | | | | N_{test}: | | | U: |
Method	parameters	acc	p	r	F1	p	r	F1	pos. ratio
SVM	$C = 3.0$	0.86	0.88	0.84	0.86	0.84	0.88	0.86	0.63
LR	$C = 6.0$	0.86	0.88	0.85	0.86	0.85	0.87	0.86	0.63
NB	$\alpha = 0.1$	0.79	0.89	0.67	0.77	0.72	0.91	0.81	0.70

Table 5: Supervised classifiers trained on P and N after tolerant noise filtering. Results averaged over 10 runs with randomised 20% validation sets; min-df threshold = 0.002, 25% of most relevant features selected with χ^2.

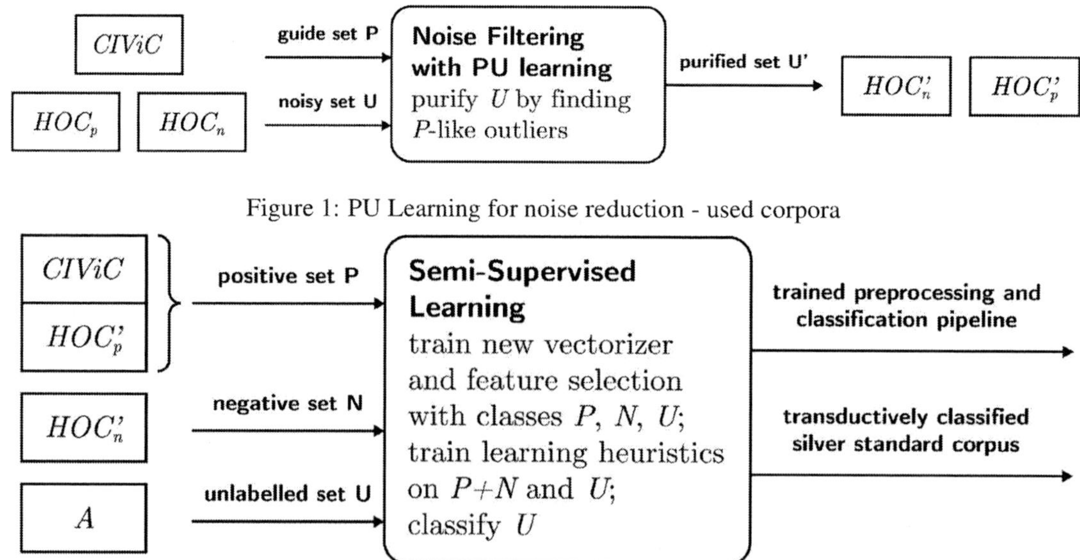

Figure 1: PU Learning for noise reduction - used corpora

Figure 2: Semi-supervised training with Self-Training - used corpora

Ontology Alignment in the Biomedical Domain Using Entity Definitions and Context

**Lucy Lu Wang[†], Chandra Bhagavatula, Mark Neumann,
Kyle Lo, Chris Wilhelm, and Waleed Ammar**

Allen Institute for Artificial Intelligence
[†]Department of Biomedical Informatics and Medical Education, University of Washington
Seattle, Washington, USA
`lucylw@uw.edu`

Abstract

Ontology alignment is the task of identifying semantically equivalent entities from two given ontologies. Different ontologies have different representations of the same entity, resulting in a need to de-duplicate entities when merging ontologies. We propose a method for enriching entities in an ontology with external definition and context information, and use this additional information for ontology alignment. We develop a neural architecture capable of encoding the additional information when available, and show that the addition of external data results in an F1-score of 0.69 on the Ontology Alignment Evaluation Initiative (OAEI) largebio SNOMED-NCI subtask, comparable with the entity-level matchers in a SOTA system.

1 Introduction

Ontologies are used to ground lexical items in various NLP tasks including entity linking, question answering, semantic parsing and information retrieval.[1] In biomedicine, an abundance of ontologies (e.g., MeSH, Gene Ontology) has been developed for different purposes. Each ontology describes a large number of concepts in healthcare, public health or biology, enabling the use of ontology-based NLP methods in biomedical applications. However, since these ontologies are typically curated independently by different groups, many important concepts are represented inconsistently across ontologies (e.g., "Myoclonic Epilepsies, Progressive" in MeSH is a broader concept that includes "Dentatorubral-pallidoluysian atrophy" from OMIM).

This poses a challenge for bioNLP applications where multiple ontologies are needed for grounding, but each concept must be represented by only one entity. For instance, in `www.semanticscholar.org`, scientific publications related to carpal tunnel syndrome are linked to one of multiple entities derived from UMLS terminologies representing the same concept,[2] making it hard to find all relevant papers on this topic. To address this challenge, we need to automatically map semantically equivalent entities from one ontology to another. This task is referred to as ontology alignment or ontology matching.

Several methods have been applied to ontology alignment, including rule-based and statistical matchers. Existing matchers rely on entity features such as names, synonyms, as well as relationships to other entities (Shvaiko and Euzenat, 2013; Otero-Cerdeira et al., 2015). However, it is unclear how to leverage the natural language text associated with entities to improve predictions. We address this limitation by incorporating two types of natural language information (definitions and textual contexts) in a supervised learning framework for ontology alignment. Since the definition and textual contexts of an entity often provide complementary information about the entity's meaning, we hypothesize that incorporating them will improve model predictions. We also discuss how to automatically derive labeled data for training the model by leveraging existing resources. In particular, we make the following contributions:

- We propose a novel neural architecture for ontology alignment and show how to effectively

---

[1]Ontological resources include ontologies, knowledgebases, terminologies, and controlled vocabularies. In the rest of this paper, we refer to all of these with the term 'ontology' for consistency.

[2]See `https://www.semanticscholar.org/topic/Carpal-tunnel-syndrome/248228` and `https://www.semanticscholar.org/topic/Carpal-Tunnel-Syndrome/3076`

47

Proceedings of the BioNLP 2018 workshop, pages 47–55
Melbourne, Australia, July 19, 2018. ©2018 Association for Computational Linguistics

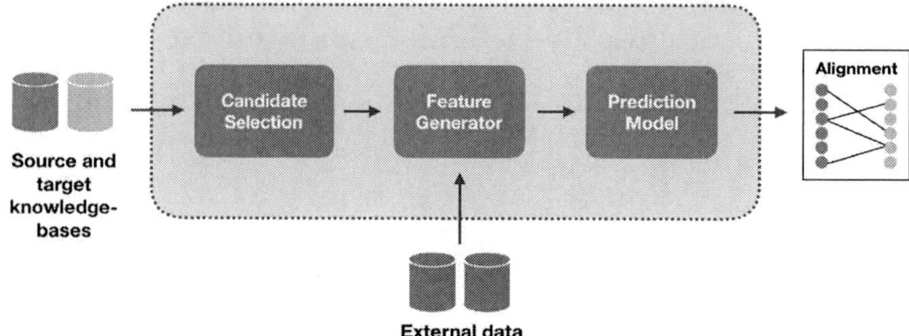

Figure 1: OntoEmma consists of three modules: a) candidate selection (see §2.2 for details), b) feature generation (see §2.2 for details), and c) prediction (see §2.3 for deatils). OntoEmma accepts two ontologies (a source and a target) as inputs, and outputs a list of alignments between their entities. When using a neural network, the feature generation and prediction model are combined together in the network.

integrate natural language inputs such as definitions and contexts in this architecture (see §2 for details).[3]

- We use the UMLS Metathesaurus to extract large amounts of labeled data for supervised training of ontology alignment models (see §3.1). We release our data set to help future research in ontology alignment.[3]

- We use external resources such as Wikipedia and scientific articles to find entity definitions and contexts (see §3.2 for details).

2 OntoEmma

In this section, we describe OntoEmma, our proposed method for ontology matching, which consists of three stages: candidate selection, feature generation and prediction (see Fig. 1 for an overview).

2.1 Problem definition and notation

We start by defining the ontology matching problem: Given a source ontology O^s and a target ontology O^t, each consisting of a set of entities, find all semantically equivalent entity pairs, i.e., $\{(\mathbf{e}^s, \mathbf{e}^t) \in O^s \times O^t : \mathbf{e}^s \equiv \mathbf{e}^t\}$, where $\equiv$ indicates semantic equivalence. For consistency, we preprocess entities from different ontologies to have the same set of attributes: a canonical name ($\mathbf{e}_{\mathrm{name}}$), a list of aliases ($\mathbf{e}_{\mathrm{aliases}}$), a textual definition ($\mathbf{e}_{\mathrm{def}}$),

and a list of usage contexts ($\mathbf{e}_{\mathrm{contexts}}$).[4]

2.2 Candidate selection and feature generation

Many ontologies are large, which makes it computationally expensive to consider all possible pairs of source and target entities for alignment. For example, the number of all possible entity pairs in our training ontologies is on the order of 10^{11}. In order to reduce the number of candidates, we use an inexpensive low-precision, high-recall candidate selection method using the inverse document frequency (*idf*) of word tokens appearing in entity names and definitions. For each source entity, we first retrieve all target entities that share a token with the source entity. Given the set of shared word tokens $\mathbf{w_{s+t}}$ between a source and target entity, we sum the *idf* of each token over the set, yielding $idf_{total} = \sum_{i \in \mathbf{w_{s+t}}} idf(i)$. Tokens with higher *idf* values appear less frequently overall in the ontology and presumably contribute more to the meaning of a specific entity. We compute the *idf* sum for each target entity and output the $K = 50$ target entities with the highest value for each source entity, yielding $|O^s| \times K$ candidate pairs.

For each candidate pair $(\mathbf{e}^s, \mathbf{e}^t)$, we precompute a set of 32 features commonly used in the ontology matching literature including the token Jaccard distance, stemmed token Jaccard distance, character n-gram Jaccard distance, root word equivalence, and other boolean and probability values

---

[3]Implementation and data available at `https://www.github.com/allenai/ontoemma/`

[4]Some of these attributes may be missing or have low coverage. See §3.2 for coverage details.

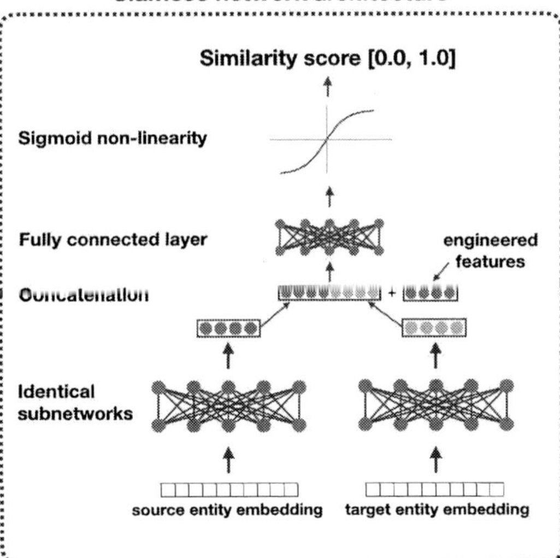

Figure 2: Siamese network architecture for computing entity embeddings for each source and target entity in a candidate entity pair.

over the entity name, aliases, and definition.[5]

2.3 Prediction

Given a candidate pair (e^s, e^t) and the precomputed features $f(e^s, e^t)$, we train a model to predict the probability that the two entities are semantically equivalent. Figure 2 illustrates the architecture of our neural model for estimating this probability which resembles a siamese network (Bromley et al., 1993). At a high level, we first encode each of the source and target entities, then concatenate their representations and feed it into a multilayer perceptron ending with a sigmoid function for estimating the probability of a match. Next, we describe this architecture in more detail.

Entity embedding. As shown in Fig. 2 (*left*), we encode the attributes of each entity as follows:

- A canonical name e_{name} is a sequence of tokens, each encoded using pretrained `word2vec` embeddings concatenated with a character-level convolutional neural network (CNN). The token vectors feed into a bi-directional long short-term memory network (LSTM) and the hidden layers at both ends of the bi-directional LSTM

are concatenated and used as the name vector v_{name}.

- Each alias in $e_{aliases}$ is independently embedded using the same encoder used for canonical names (with shared parameters), yielding a set of alias vectors $v_{alias-i}$ for $i = 1, \ldots, |e_{aliases}|$.

- An entity definition e_{def} is a sequence of tokens, each encoded using pretrained embeddings then fed into a bi-directional LSTM. The definition vector v_{def} is the concatenation of the final hidden states in the forward and backward LSTMs.

- Each context in $e_{contexts}$ is independently embedded using the same encoder used for definitions (with shared parameters), then averaged yielding the context vector $v_{contexts}$.

The name, alias, definition, and context vectors are appended together to create the entity embedding, e.g., the source entity embedding e^s is: $v^s = [v^s_{name}; v^s_{alias-i*}; v^s_{def}; v^s_{contexts}]$. In order to find representative aliases for a given pair of entities, we pick the source and target aliases with the smallest Euclidean distance, i.e., $i^*, j^* = \arg\min_{i,j} \|v^s_{alias-i} - v^t_{alias-j}\|_2$

Siamese network. After the source and target entity embeddings are computed, they are fed into two subnetworks with shared parameters followed by a parameterized function for estimating similarity. Each subnetwork is a two layer feedforward

---

[5]Even though neural models may obviate the need for feature engineering, feeding highly discriminative features into the neural model improves the inductive bias of the model and reduces the amount of labeled data needed for training.

network with ReLU non-linearities and dropout (Srivastava et al., 2014). The outputs of the two subnetworks are then concatenated together with the engineered features and fed into another feedforward network with a ReLU layer followed by a sigmoid output layer. We train the model to minimize the binary cross entropy loss for gold labels.

To summarize, the network estimates the probability of equivalence between $\mathbf{e}^s$ and $\mathbf{e}^t$ as follows:

$$\mathbf{h}^s = \textsc{Relu}(\textsc{Relu}(\mathbf{v}^s; \theta_1); \theta_2)$$
$$\mathbf{h}^t = \textsc{Relu}(\textsc{Relu}(\mathbf{v}^t; \theta_1); \theta_2)$$
$$P(\mathbf{e}^s \equiv \mathbf{e}^t) = \textsc{Sigmoid}(\textsc{Relu}([\mathbf{h}^s; \mathbf{h}^t]; \theta_3); \theta_4)$$

3 Deriving and enriching labeled data

In this section, we discuss how to derive a large amount of labeled data for training, and how to augment entity attributes with definitions and contexts from external resources.

3.1 Deriving training data from UMLS

The Unified Medical Language System (UMLS) Metathesaurus, which integrates more than 150 source ontologies, illustrates the breadth of coverage of biomedical ontologies (Bodenreider, 2004). Also exemplified by the UMLS Metathesaurus is the high degree of overlap between the content of some of these ontological resources, whose terms have been (semi-)manually aligned. Significant time and effort has gone into cross-referencing semantically equivalent entities across the ontologies, and new terms and alignments continue to be added as the field develops. These manual alignments are high quality, but considered to be incomplete (Morrey et al., 2011; Mougin and Grabar, 2014).

To enable supervised learning for our models, training data was derived from the UMLS Metathesaurus. By exposing our models to labeled data from the diverse subdomains covered in the UMLS Metathesaurus, we hope to learn a variety of patterns indicating equivalence between a pair of entities which can generalize to new ontologies not included in the training data.

We identified the following set of ontologies within UMLS to use as the source of our labeled data, such that they cover a variety of domains without overlapping with the test ontologies used for evaluation in the OAEI: Current Procedural Terminology (CPT), Gene Ontology (GO), Hugo Nomenclature (HGNC), Human Phenotype Ontology (HPO), Medical Subject Headings (MeSH),

Online Mendelian Inheritance in Man (OMIM), and RxNorm.

Our labeled data take the form $(\mathbf{e}^s, \mathbf{e}^t, l \in \{0, 1\})$, where $l = 1$ indicates positive examples where $\mathbf{e}^s \equiv \mathbf{e}^t$. For each pair of ontologies, we first derive all the positive mappings from UMLS. We retain the positive mappings for which there are no name equivalences. Then, for each positive example $(\mathbf{e}^s, \mathbf{e}^t_+, 1)$, we sample negative mappings $(\mathbf{e}^s, \mathbf{e}^t_-, 0)$ from the other entities in the target ontology. One "easy" negative and one "hard" negative are selected for each positive alignment, where easy negatives consist of entities with little overlap in lexical features while hard negatives have high overlap. Easy negatives are obtained by randomly sampling entities from the target ontology, for each source entity. Hard negatives are obtained using the same candidate selection method described in §2. In both easy and hard examples, we exclude all target entities which appear in a positive example.[6]

Over all seven ontologies, 50,523 positive alignments were extracted from UMLS. Figure 3 reports the number of positive alignments extracted from each ontology pair. For these positives, 98,948 hard and easy negatives alignments were selected. These positive and negative labeled examples were pooled and randomly split into a 64% training set, a 16% development set, and a 20% test set.

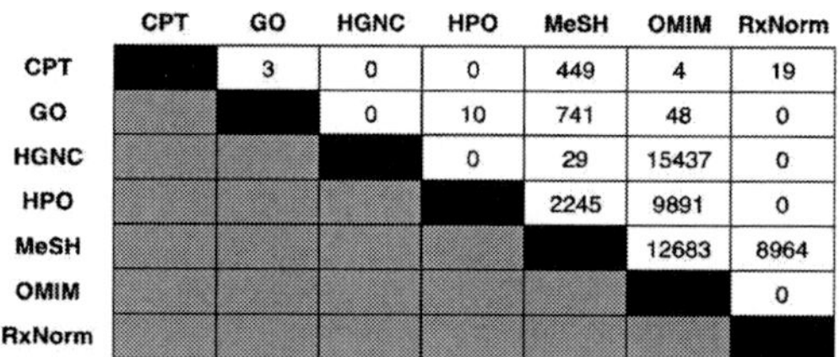

	CPT	GO	HGNC	HPO	MeSH	OMIM	RxNorm
CPT		3	0	0	449	4	19
GO			0	10	741	48	0
HGNC				0	29	15437	0
HPO					2245	9891	0
MeSH						12683	8964
OMIM							0
RxNorm							

Figure 3: Number of positive alignments extracted from each pair of ontologies from UMLS.

3.2 Deriving definitions and mention contexts

Many ontologies do not provide entity definitions (Table 1). In fact, only a few (GO, HPO, MeSH) of the ontologies we included have any definitions at all.

---

[6]Although the negative examples we collect may be noisy due to the incompleteness of manual alignments in UMLS, this noise is also present in widely adopted evaluation of knowledge base completion problems and relation extraction with distant supervision (e.g., Li et al., 2016; Mintz et al., 2009).

Table 1: Entities with definitions and contexts for each of the training ontologies.

Ont.	# of entities	% w/ def.	% w/ con.
CPT	13,786	0.0	97.9
GO	44,684	100.0	30.5
HGNC	39,816	0.0	0.8
HPO	11,939	72.5	17.9
MeSH	268,162	10.5	35.1
OMIM	98,515	0.0	2.8
RxNorm	205,858	0.0	5.1
Total	682,760	11.9	20.1

We can turn to external sources of entity definitions in such cases. Many biomedical and healthcare concepts are represented in Wikipedia, a general purpose crowd-sourced encyclopedia. The Wikipedia API can be used to search for and extract article content. The first paragraph in each Wikipedia article offers a description of the concept, and can be used as a substitute for a definition. For each entity in the labeled dataset described in the previous section, we query Wikipedia using the entity's canonical name. The first sentence from the top Wikipedia article match is extracted and used to populate the attribute e_{def} when undefined in the ontology. For example, a query for OMIM:125370, "Dentatorubral-pallidoluysian atrophy," yields the following summary sentence from Wikipedia: *"Dentatorubral-pallidoluysian atrophy (DRPLA) is an autosomal dominant spinocerebellar degeneration caused by an expansion of a CAG repeat encoding a polyglutamine tract in the atrophin-1 protein."* Based on a human-annotated sample, the accuracy of our externally-derived definitions is 75.5%, based on a random sample of 200 definitions and two annotators with Cohen's kappa coefficient of $\kappa = 0.88$.[7]

Usage contexts are derived from scientific papers in Medline, leveraging entity annotations available via the Semantic Scholar project (Ammar et al., 2018). In order to obtain the annotations, an entity linking model was used to find mentions of UMLS entities in the abstracts of Medline papers. The sentences in which a UMLS entity were mentioned are added to the $e_{contexts}$ attribute of that entity. For UMLS entity C0751781, "Dentatorubral-Pallidoluysian At-

rophy," an example context: *"Dentatorubral-pallidoluysian atrophy (DRPLA) is an autosomal dominant neurodegenerative disease clinically characterized by the presence of cerebellar ataxia in combination with variable neurological symptoms,"* is extracted from Yoon et al (2012) (Yoon et al., 2012). This context sentence was scored highly by the linking model, and provides additional information about this entity, for example, its acronym (*DRPLA*), the type of disease (*autosomal dominant neurodegenerative*), and some of its symptoms (*cerebellar ataxia*). Because there are often numerous linked contexts for each entity, we sample up to 20 contexts per entity when available. The number of entities with context in our labeled data is given in Table 1. The accuracy of usage contexts extracted using this approach is 92.5%, based on human evaluation of 200 contexts with Cohen's kappa coefficient of $\kappa = 1$.[7]

4 Experiments

In this section, we experiment with several variants of OntoEmma: In the first variant (OntoEmma:NN), we only encode native attributes obtained from the source and target ontologies: canonical name, aliases, and native definitions. In the second variant (OntoEmma:NN+f), we also add the manually engineered features as described in §2.2. In the third variant (OntoEmma:NN+f+w), we incorporate external definitions from Wikipedia, as discussed in §3.2. In the fourth variant (OntoEmma:NN+f+w+c), we also encode the usage contexts we derived from Medline, also discussed in §3.2.

Data. We use the training section of the UMLS-derived labeled data to train the model and use the development section to tune the model hyperparameters. For evaluation, we use the test portion of our UMLS-derived data as well as the OAEI largebio subtrack SNOMED-NCI task, the largest task in OAEI largebio. The UMLS test set includes 29,859 positive and negative mappings. The OAEI reference alignments included 17,210 equivalent mappings and 1,623 uncertain mappings between the SNOMED and NCI ontologies.

Baselines. Our main baseline is a logistic regression model (OntoEmma:LR) using the same engineered features described in §2.2. To illustrate how our proposed method performs compared to previous work on ontology matching, we compare

---

[7]Annotations are available at `https://github.com/allenai/ontoemma#human-annotations`

Table 2: Model performance on UMLS test dataset

Model	Prec.	Recall	F1
OntoEmma:LR	0.98	0.92	0.95
OntoEmma:NN	0.87	0.85	0.86
OntoEmma:NN+f	0.93	0.96	0.95
OntoEmma:NN+f+w	0.93	0.97	0.95
OntoEmma:NN+f+w+c	0.94	0.97	0.96

Table 3: Model performance on OAEI largebio SNOMED-NCI task

Model	Prec.	Recall	F1
AML:entity	0.81	0.62	0.70
OntoEmma:LR	0.75	0.56	0.65
OntoEmma:NN+f+w+c	0.80	0.61	0.69

to AgreementMakerLight (AML) which has consistently been a top performer in the OAEI subtasks related to biomedicine (Faria et al., 2013). For a fair comparison to OntoEmma, we only use the entity-level matchers in AML; i.e., relation and structural matchers in AML are turned off.[8]

Implementation and configuration details. We provide an open source, modular, Python implementation of OntoEmma where different candidate selectors, feature generators, and prediction modules can be swapped in and out with ease.[3] We implement the neural model using `PyTorch` and `AllenNLP`[9] libraries, and implement the logistic regression model using `scikit-learn`. Our 100-dimensional pretrained embeddings are learned using the default settings of `word2vec` based on the Medline corpus. The character-level CNN encoder uses 50 filters of size 4 and 5, and outputs a token embedding of size 100 with dropout probability of 0.2. The LSTMs have output size 100, and have dropout probability of 0.2.

Results. The performance of the models is reported in terms of precision, recall and F1 score on the held-out UMLS test set and the OAEI largebio SNOMED-NCI task in Tables 2 and 3, respectively.

Table 2 illustrates how different variants of OntoEmma perform on the held-out UMLS test set. We note that the bare-bones neural network model (OntoEmma:NN) does not match the performance of the baseline logistic regression model (OntoEmma:LR), suggesting that the representations learned by the neural network are not sufficient. Indeed, adding the engineered features to the neural model in (OntoEmma:NN+f) provides substantial improvements, matching the F1 score of the baseline model. Adding definitions and usage context in (OntoEmma:NN+f+w+c) further improves the F1 score by one absolute point, outperforming the logistic regression model.

While the UMLS-based test set in Table 2 represents the realistic scenario of aligning new entities in partially-aligned ontologies, we also wanted to evaluate the performance of our method on the more challenging scenario where no labeled data is available in the source and target ontologies. This is more challenging because the patterns learned from ontologies used in training may not transfer to the test ontologies. Table 3 illustrates how our method performs in this scenario using SNOMED-NCI as test ontologies. For matching of the SNOMED and NCI ontologies, we enriched the entities first using Wikipedia queries. At test time, we also identified and aligned pairs of entities with exact string matches, using the OntoEmma matcher only for those entities without an exact string match. Unsurprisingly, the performance of OntoEmma on unseen ontologies (in Table 3) is much lower than its performance on seen ontologies (in Table 2). With unseen ontologies, we gain a large F1 improvement of 4 absolute points by using the fully-featured neural model (OntoEmma:NN+f+w+c) instead of the logistic regression variant (OntoEmma:LR), suggesting that the neural model may transfer better to different domains. We note, however, that the OntoEmma:NN+f+w+c matcher performs slightly worse than the AML entity matcher. This is to be expected since AML incorporates many matchers which we did not implement in our model, e.g., using background knowledge, acronyms, and other features.

5   Discussion

Through building and training a logistic regression model and several neural network models, we evaluated the possibility of training a supervised machine learning model for ontology alignment based on existing alignment data, and evalu-

---

[8]The performance of the full AML system on the SNOMED-NCI subtask for OAEI 2017 is: precision: 0.90, recall: 0.67, F1: 0.77.

[9]https://allennlp.org/

ated the efficacy of including definitions and usage context to improve entity matching. For the first question, we saw some success with both the logistic regression and neural network models. The logistic regression model performed better than the simple neural network model without engineered features. Hand-engineered features encode human knowledge, and are less noisy than features trained from a neural network. The NN model required more training data to learn the same sparse information encoded by pre-defined features.

To bolster performance, hand-engineered features and extensive querying of third-party resources were used to increase knowledge about each entity. Definitions and usage contexts had rarely been used by previous ontology matchers, and we sought to exploit the value of these additional pieces of information. Definitions especially, often offer information about an entity's relations and attributes, which may not be explicitly defined in the ontology. The ontologies used for training contained inconsistent information – some had definitions for all entities, some none; some were well-represented in our context linking model, some were not. To take advantage of such information, therefore, we had to turn to external sources of definitions and contexts, which are understandably more noisy than information provided in the ontology itself.

Using Wikipedia and the Medline corpus, we derived definitions and contexts for many of the entities in the UMLS training corpus. Adding definitions improved the performance of our neural network model. However, high quality definitions native to each terminology would likely have improved results further, since we could not ensure that externally derived definitions were always relevant to the entity of interest.

Limitations. Our ontology matcher did not implement any structural matchers, and did not take advantage of relationship data where it existed. In ontologies with well-defined hierarchy or relationships, the structural component provides orthogonal and extremely relevant information for matching. By choosing to focus on entity alignment, we were unable to be competitive on global ontology matching.

Of all the entities in our UMLS training, development, and test datasets, only 11.9% of entities had definitions from their source ontology (Table 1). Similarly, we were only able to derive contexts for 20.1% of the training entities from the Semantic Scholar entity linking model (Table 1). We were hoping for better coverage of the overall dataset. We were, however, able to use Wikipedia to increase the overall definition coverage of the entities in our data set to 82.1%.

Although Wikipedia is a dense resource containing curated articles on many concepts, it is by no means exhaustive. Many of the entities in our training and test data set did not correspond directly to entities in Wikipedia. We also could not review each query to ensure a correct match between the Wikipedia article and the entity. The data is therefore noisy and can introduce error in some cases. Although the overall performance improved upon querying Wikipedia for additional definitions, we believe that dedicated definitions from the source terminologies would perform better where available.

Future work. We are exploring other ways to derive high-quality definitions from external resources, for example, by deriving definitions from synonymous entities in other ontologies, or by generating textual definitions using the logical definitions given in an ontology. Similarly, we can incorporate usage context from other sources. For example, the Semantic MEDLINE Database (SemMedDB) is a database of semantic relationship predictions from PubMed articles (Kilicoglu et al., 2012). The entity-relation triples in this database can be used to retrieve PubMed article context mapped to UMLS terms.

Continuing on, we aim to develop a more flexible ontology matching system that takes into account all of the information available about an entity. Flexible entity embeddings would represent critical information for proper entity alignment, while accounting for a variety of data types, such as list-like and graph-like data. We would also like to incorporate ontology structure and relations in matching. Hierarchical structure is provided by most biomedical terminologies, and provides essential information for a matching system. One possibility is ensembling OntoEmma with other matcher systems that incorporate or focus on using structural features in matching.

Related work The OAEI has been driving ontology matching research in the biomedical domain since 2005. It provides evaluation data supporting several tracks such as the anatomy, largebio, and more recently introduced phenotype tracks (Faria et al., 2016). Participating matchers implement a variety of matching techniques including rule-based and statistical methods (Faria et al., 2016; Gross et al., 2016; Otero-Cerdeira et al., 2015; Shvaiko and Euzenat, 2013). Features used by matchers can be element-level (extracted from each individual entity), or structure-level (based on the topology of the ontology and its relationships). Content features can be based on terminology (i.e., names of entities), structure (i.e., how entities are connected), annotations (i.e., annotations made to entities), or reasoning output. Some features can also be derived from external sources, such as cross-references to other ontologies, or cross-annotations in other datasets, such as term coincidence in publications, or co-annotation of experiments with terms from different ontologies (Husein et al., 2016).

Notable general purpose matchers that have excelled in biomedical domain matching tasks include AgreementMakerLight (AML), YAM++, and LogMap. AML has consistently been a top performer in the OAEI subtasks related to biomedicine. It uses a combination of different matchers, such as the lexical matcher (looking for complete string matches between the names of entities), mediating matcher (performing the function of the lexical matcher through a third background ontology), word-based string similarity matcher (matching entities with minimal string edit distances), and others. AML then combines these various similarity scores to generate a global alignment between the two input ontologies (Faria et al., 2013). YAM++, another successful matcher, implemented a decision tree learning model over many string similarity metrics, but leaves the challenges of finding suitable training data to the user, defaulting to information retrieval-based similarity metrics for its decision-making when no training data is provided (Ngo and Bellahsene, 2016). LogMap is a matcher specifically designed to efficiently align large ontologies, generating logical output alignments. The matcher uses high-probability matches as anchors from which to deploy its lexical and structural matchers (Jiménez-Ruiz and Cuenca Grau, 2011).

Our system uses neural networks to learn entity representations and features for matching. Several published works discuss using neural networks to learn weights over similarity metrics pre-defined by the user or developer of the matching system (Djeddi and Khadir, 2013; Peng, 2010; Huang et al., 2008; Hariri et al., 2006). These systems do not use neural networks to generate and learn the features most appropriate for entity matching. Qiu et al. (2017) proposes and tests an autoencoder network for unsupervised entity representation learning over a bag of words vector that treats all descriptive elements of each entity (its name, definitions etc.) equally. We are interested in investigating how these various descriptive elements contribute to entity matching, how sparsity of specific descriptive fields can be offset by deriving information from external resources, and also whether we can use domain-specific training data to optimize a model for the biomedical domain.

Conclusion In this paper, we propose using natural language text associated with entities to improve ontology alignment. We describe a novel neural architecture for ontology alignment which can encode a variety of information, and derive large amounts of labeled data for training the model. To address the limited coverage of definitions and usage contexts describing entities, we turn to external resources to supplement the available information about entities in the test ontologies. Our empirical results illustrate that externally-derived definitions and contexts can effectively be used to improve the performance of ontology matching systems.

6   Acknowledgements

We would like to thank the anonymous reviewers for their helpful comments. We also thank John Gennari, Oren Etzioni, Joanna Power as well as the rest of the Semantic Scholar team at the Allen Institute for Artificial Intelligence for helpful comments and insights.

References

Waleed Ammar, Dirk Groeneveld, Chandra Bhagavatula, Iz Beltagy, Miles Crawford, Doug Downey, Jason Dunkelberger, Ahmed Elgohary, Sergey Feldman, Vu Ha, Rodney Kinney, Sebastian Kohlmeier, Kyle Lo, Tyler Murray, Hsu-Han Ooi, Matthew Peters, Joanna Power, Sam Skjonsberg, Lucy Lu Wang, Chris Wilhelm, Zheng Yuan, Madeleine van Zuylen, and Oren Etzioni. 2018. Construction of the literature graph in semantic scholar. In *NAACL (industry track)*.

Olivier   Bodenreider.   2004.   The   Unified

Medical Language System (UMLS): integrating biomedical terminology. *Nucleic Acids Res* 32(Database issue):D267–D270. https://doi.org/10.1093/nar/gkh061.

Jane Bromley, Isabelle Guyon, Yann LeCun, Eduard Säckinger, and Roopak Shah. 1993. Signature verification using a siamese time delay neural network. In *NIPS*.

Warith Eddine Djeddi and Mohamed Tarek Khadir. 2013. Ontology Alignment Using Artificial Neural Network for Large-scale Ontologies. *Int. J. Metadata Semant. Ontologies* 8(1):75–92. https://doi.org/10.1504/IJMSO.2013.054180.

Daniel Faria, Catia Pesquita, Booma S. Balasubramani, Catarina Martins, João Cardoso, Hugo Curado, Francisco M. Couto, and Isabel F. Cruz. 2016. OAEI 2016 results of AML. volume 1766, pages 138–145.

Daniel Faria, Catia Pesquita, Emanuel Santos, Matteo Palmonari, Isabel F. Cruz, and Francisco M. Couto. 2013. The AgreementMakerLight Ontology Matching System. In *On the Move to Meaningful Internet Systems: OTM 2013 Conferences*. Springer, Berlin, Heidelberg, Lecture Notes in Computer Science, pages 527–541. https://doi.org/10.1007/978-3-642-41030-7_38.

Anika Gross, Cedric Pruski, and Erhard Rahm. 2016. Evolution of biomedical ontologies and mappings: Overview of recent approaches. *Comput Struct Biotechnol J* 14:333–340. https://doi.org/10.1016/j.csbj.2016.08.002.

Babak Bagheri Hariri, Hassan Abolhassani, and Hassan Sayyadi. 2006. A Neural-Networks-Based Approach for Ontology Alignment. Japan Society for Fuzzy Theory and Intelligent Informatics, pages 1248–1252. https://doi.org/10.14864/softscis.2006.0.1248.0.

Jingshan Huang, Jiangbo Dang, Michael N. Huhns, and W. Jim Zheng. 2008. Use artificial neural network to align biological ontologies. *BMC Genomics* 9 Suppl 2:S16. https://doi.org/10.1186/1471-2164-9-S2-S16.

Inne Gartina Husein, Saiful Akbar, Benhard Sitohang, and Fazat Nur Azizah. 2016. Review of ontology matching with background knowledge. In *2016 International Conference on Data and Software Engineering (ICoDSE)*. pages 1–6. https://doi.org/10.1109/ICODSE.2016.7936159.

Ernesto Jiménez-Ruiz and Bernardo Cuenca Grau. 2011. Logmap: Logic-based and scalable ontology matching. In Lora Aroyo, Chris Welty, Harith Alani, Jamie Taylor, Abraham Bernstein, Lalana Kagal, Natasha Noy, and Eva Blomqvist, editors, *The Semantic Web – ISWC 2011*. Springer Berlin Heidelberg, Berlin, Heidelberg, pages 273–288.

Halil Kilicoglu, Dongwook Shin, Marcelo Fiszman, Graciela Rosemblat, and Thomas C. Rindflesch. 2012. Semmeddb: a pubmed-scale repository of biomedical semantic predications. *Bioinformatics* 28:3158–60. https://doi.org/10.1093/bioinformatics/bts591.

Xiang Li, Aynaz Taheri, Lifu Tu, and Kevin Gimpel. 2016. Commonsense knowledge base completion. In *ACL*.

Mike Mintz, Steven Bills, Rion Snow, and Daniel Jurafsky. 2009. Distant supervision for relation extraction without labeled data. In *ACL*.

Charles Paul Morrey, Ling Chen, Michael Halper, and Yehoshua Perl. 2011. Resolution of redundant semantic type assignments for organic chemicals in the UMLS. *Artif Intell Med* 52(3):141–151. https://doi.org/10.1016/j.artmed.2011.05.003.

Fleur Mougin and Natalia Grabar. 2014. Auditing the multiply-related concepts within the UMLS. *J Am Med Inform Assoc* 21(e2):e185–193. https://doi.org/10.1136/amiajnl-2013-002227.

DuyHoa Ngo and Zohra Bellahsene. 2016. Overview of YAM++(not) Yet Another Matcher for ontology alignment task. *Web Semantics: Science, Services and Agents on the World Wide Web* 41:30–49. https://doi.org/10.1016/j.websem.2016.09.002.

Lorena Otero-Cerdeira, Francisco J. Rodriguez-Martinez, and Alma Gomez-Rodriguez. 2015. Ontology matching: A literature review. *Expert Systems with Applications* 42(2):949–971. https://doi.org/10.1016/j.eswa.2014.08.032.

Yefei Peng. 2010. Ontology Mapping Neural Network: An Approach to Learning and Inferring Correspondences Among Ontologies.

Lirong Qiu, Jia Yu, Qiumei Pu, and Chuncheng Xiang. 2017. Knowledge entity learning and representation for ontology matching based on deep neural networks. *Cluster Comput* 20(2):969–977. https://doi.org/10.1007/s10586-017-0844-1.

Pavel Shvaiko and Jérôme Euzenat. 2013. Ontology Matching: State of the Art and Future Challenges. *IEEE Transactions on Knowledge and Data Engineering* 25(1):158–176. https://doi.org/10.1109/TKDE.2011.253.

Nitish Srivastava, Geoffrey E. Hinton, Alex Krizhevsky, Ilya Sutskever, and Ruslan Salakhutdinov. 2014. Dropout: a simple way to prevent neural networks from overfitting. *Journal of Machine Learning Research* 15:1929–1958.

Won Tae Yoon, Jinyoung Youn, and Jin Whan Cho. 2012. Is cerebral white matter involvement helpful in the diagnosis of dentatorubralpallidoluysian atrophy? *J Neurol* 259:1694–7. https://doi.org/10.1007/s00415-011-6401-6.

Sub-word information in pre-trained biomedical word representations: evaluation and hyper-parameter optimization

Dieter Galea   **Ivan Laponogov**   **Kirill Veselkov**
Department of Surgery and Cancer
Imperial College London
{d.galea14 | i.laponogov | kirill.veselkov04}@imperial.ac.uk

Abstract

Word2vec embeddings are limited to computing vectors for in-vocabulary terms and do not take into account sub-word information. Character-based representations, such as fastText, mitigate such limitations. We optimize and compare these representations for the biomedical domain. fastText was found to consistently outperform word2vec in named entity recognition tasks for entities such as chemicals and genes. This is likely due to gained information from computed out-of-vocabulary term vectors, as well as the word compositionality of such entities. Contrastingly, performance varied on intrinsic datasets. Optimal hyper-parameters were intrinsic dataset-dependent, likely due to differences in term types distributions. This indicates embeddings should be chosen based on the task at hand. We therefore provide a number of optimized hyper-parameter sets and pre-trained word2vec and fastText models, available   on https://github.com/dterg/bionlp-embed.

1   Introduction

word2vec (Mikolov et al., 2013) and GloVe (Pennington et al., 2014) models are a popular choice for word embeddings, representing words by vectors for downstream natural language processing. Optimization of word2vec has been thoroughly investigated by Chiu et al. (2016a) for biomedical text. However, word2vec has two main limitations: i) out-of-vocabulary (OOV) terms cannot be represented, losing potentially useful information; and ii) training is based on co-occurrence of terms, not taking into account sub-word information. With new entities such as genetic variants, pathogens, chemicals and drugs, these limitations can be critical in biomedical NLP.

Sub-word information has played a critical role in improving NLP task performances and has predominantly depended on feature-engineering. More recently, character-based neural networks for tasks such as named entity recognition have been developed and evaluated on biomedical literature (Gridach, 2017). This has achieved state-of-the-art performances but is limited by the quantity of supervised training data.

Character-based representation models such as fastText (Bojanowski et al., 2017; Mikolov et al., 2018) and MIMICK (Pinter et al., 2017) exploit word compositionality to learn distributional embeddings, allowing to compute vectors for OOV words. Briefly, fastText uses a feed-forward architecture to learn n-gram and word embeddings, whereas MIMICK uses a Bi-LSTM architecture to learn character-based embeddings in the same space of another pre-trained embeddings, such as word-based word2vec.

Here we evaluate and optimize pre-trained character-based word representations with the fastText implementation for biomedical terms. To compare with word2vec models, we also optimize word2vec by extending the work by Chiu et al. (2016a). We report that fastText outperforms word2vec in all named entity recognition tasks of feature-rich entities such as chemicals and genes. However, in intrinsic evaluation, results and optimal hyper-parameters vary. This is likely due to different entity type distributions within the intrinsic standards. This indicates representations should be selected and optimized based on the task at hand and the entities of interest. We evaluate and provide optimized generalized fastText and word2vec models and models optimized on

56

Proceedings of the BioNLP 2018 workshop, pages 56–66
Melbourne, Australia, July 19, 2018. ©2018 Association for Computational Linguistics

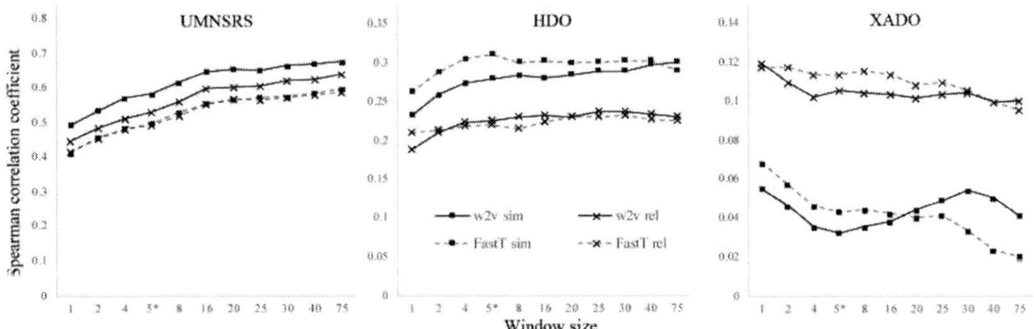

Figure 1: Intrinsic evaluation of window size in word2vec (w2v) and fastText (FastT) models on UMNSRS, HDO, and XADO datasets (Supp. Table 4).

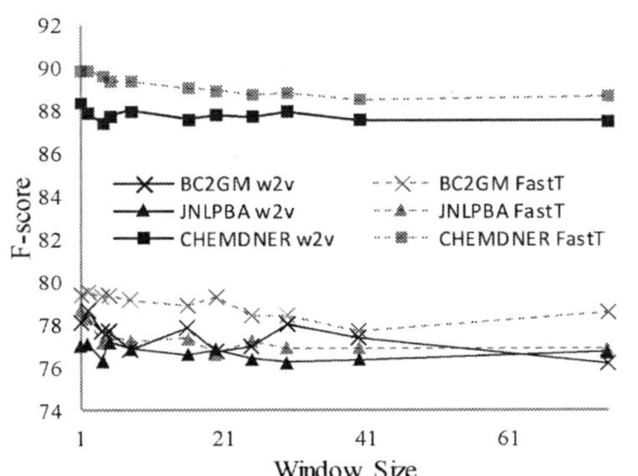

Figure 2: Extrinsic evaluation of window size in word2vec (w2v) and fastText (FastT) models on BC2GM, JNLPBA and CHEMDNER datasets (Supp. Table 5).

individual datasets, outperforming a number of current state-of-the-art embeddings.

2 Materials and Methods

2.1 Data and pre-processing

PubMed 2018 baseline abstracts and titles were parsed using PubMed parser (Achakulvisut and Acuna, 2016), each article was represented as a single line, and any new line characters within an article were replaced by a whitespace. Pre-processing was performed using the NLPre module (He and Chen, 2018). All upper-case sentences were lowered, de-dashed, parenthetical phrases identified, acronyms replaced with full term phrases (e.g. "Chronic obstructive pulmonary disease (COPD)" was changed to "Chronic pulmonary disease (Chronic_pulmonary_disease)), URLs removed, and single character tokens removed. Tokenization was carried out on whitespace. Punctuation was retained. This resulted in a training dataset of 3.4 billion tokens and a

vocabulary of up to 19 million terms (Supp. Table 1).

2.2 Embeddings and hyper-parameters

Word embeddings were trained on the pre-processed PubMed articles using Skip-Gram word2vec and fastText implementations in gensim (Řehůřek and Sojka, 2010). As in Chiu et al. (2016a), we tested the effect of hyper-parameter selection on embedding performance for each hyper-parameter: negative sample size, sub-sampling rate, minimum word count, learning rate (alpha), dimensionality, and window size. Extended parameter ranges were tested for some hyper-parameters, such as window size. Additionally, we test the range of character n-grams for the fastText models, as originally performed for language models (Bojanowski et al., 2017). Due to the computational cost, especially since fastText models can be up to 7.2x slower to train compared to word2vec (Supp. Figure 1), we modify one hyper-parameter at a time, while keeping all other hyper-parameters constant. Performance was measured both intrinsically and extrinsically on a number of datasets.

2.3 Intrinsic Evaluation

Intrinsic evaluation of word embeddings is commonly performed by correlating the cosine similarity between term pairs, as determined by the trained embeddings, and a reference list. We use the manually curated UMNSRS covering disorders, symptoms, and drugs (Pakhomov et al., 2016), and compute graph-based similarity and relatedness using the human disease ontology graph (Schriml et al., 2012) (HDO) and the *Xenopus* anatomy and development ontology graph (Segerdell et al., 2008) (XADO). 1 million pairwise combinations of entities and ontologies were

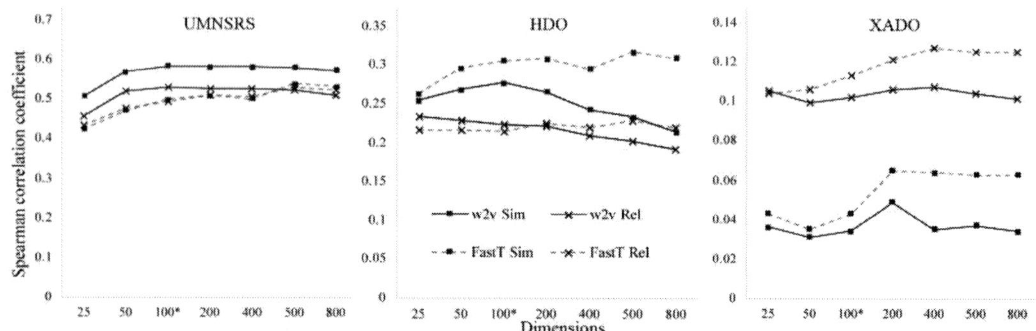

Figure 3: Intrinsic evaluation of dimension size in word2vec (w2v) and fastText (FastT) models on UMNSRS, HDO, and XADO datasets (Supp. Table 6).

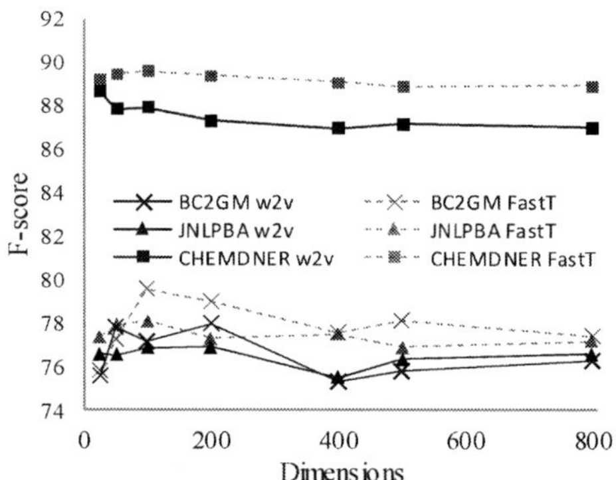

Figure 4: Extrinsic evaluation of dimension size in word2vec (w2v) and fastText (FastT) models on BC2GM, JNLPBA and CHEMDNER datasets (Supp. Table 7).

randomly computed from each graph and entities which did not map to the ontology map or were multi-token were not considered. Similarity between a pair of terms was computed using the Wu and Palmer (1994) similarity metric, and relatedness was determined by a simplified Lesk algorithm (1986). In the latter, token intersection (excluding stopwords) was calculated between definitions and normalized by the maximum definition length. Pairs which did not have definition statements for any of the terms were excluded.

As with UMNSRS, the computed similarity and relatedness scores were correlated with the cosine similarity determined by the embeddings models. As word2vec is not capable of representing OOV words, in literature pair terms which are not in vocabulary are commonly not considered for evaluation. To allow for comparison between the word2vec and fastText models, we represent OOV words as null vectors – as originally performed by Bojanowski et al. (2017). However, to determine the difference in performance of in-vocabulary word embeddings and OOV word em-

beddings, we measure correlation with only in-vocabulary terms, and with OOV terms pairs considered and null-imputed for word2vec.

2.4 Extrinsic evaluation

Intrinsic evaluation by itself may provide limited insights and may not represent the true downstream performance (Faruqui et al. 2016; Chiu et al., 2016b). Therefore, we perform extrinsic evaluation using 3 named entity recognition corpora: (i) the BioCreative II Gene Mention task corpus (BC2GM) (Smith et al., 2008) for genes; (ii) the JNLPBA corpus (Kim et al., 2004) annotating proteins, cell lines, cell types, DNA, and RNA; and (iii) the CHEMDNER corpus (Krallinger et al., 2015) which annotates drugs and chemicals, as made available from Luo et al. (2017). Each of these corpora are originally split into a train, development, and test sets – the same splits and sentence ordering were retained here.

The state-of-the-art BiLSTM-CRF neural network architecture (Lample et al., 2016), as implemented in the anago package, was used to train NER models and predict the development set of each corpus for each parameter. Accuracy was determined by the F-score. Each model was run for up to 10 epochs and the best accuracy on the development set was recorded.

2.5 Optimized Embeddings

Hyper-parameters achieving the highest performance for each extrinsic corpus and intrinsic standard were determined for word2vec and fastText. Corpus-specific word2vec and fastText models were trained with the set of optimal hyper-parameters for each corpus, as each corpus annotates different entity classes. For a generalized optimal model, we also trained embeddings on optimal hyper-parameters determined across all cor-

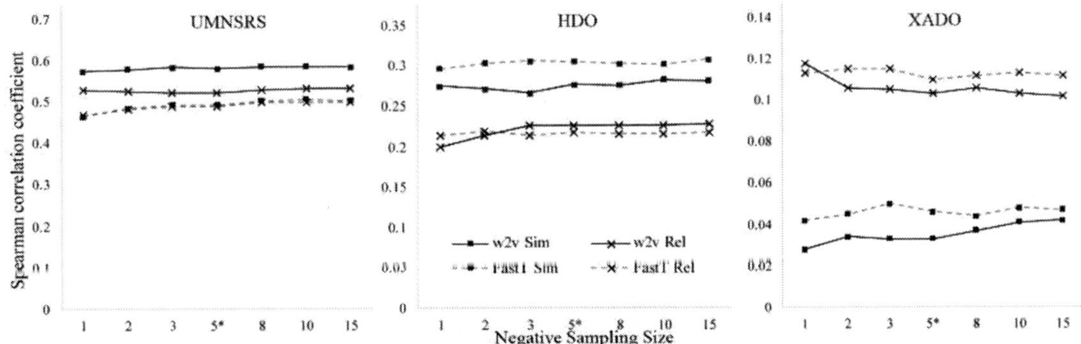

Figure 5: Intrinsic evaluation of negative sampling size in word2vec (w2v) and fastText (FastT) models on UMNSRS, HDO, and XADO datasets (Supp. Table 8).

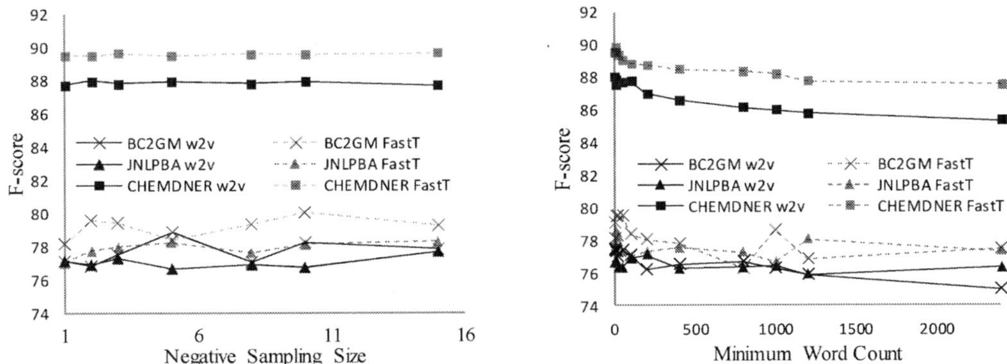

Figure 6: Extrinsic evaluation of negative sampling size in word2vec (w2v) and fastText (FastT) models on BC2GM, JNLPBA and CHEMDNER datasets (Supp. Table 9).

Figure 7: Extrinsic evaluation of minimum word count in word2vec (w2v) and fastText (FastT) models on BC2GM, JNLPBA and CHEMDNER datasets (Supp. Table 11).

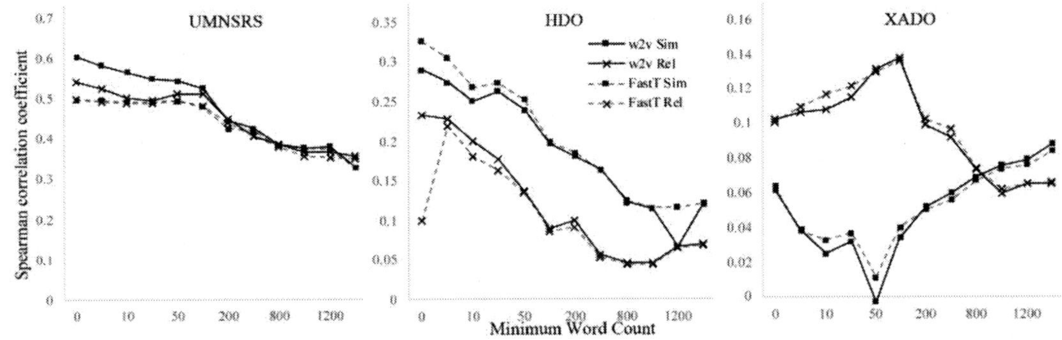

Figure 8: Intrinsic evaluation of the minimum word count in word2vec (w2v) and fastText (FastT) models on UMNSRS, HDO, and XADO datasets (Supp. Table 10).

pora and standards, as well as across intrinsic and extrinsic datasets separately. For the final extrinsic optimized evaluation, the test split was predicted.

3   Results and Discussion

3.1   General trends: word2vec hyperparameter selection

Overall, intrinsic and extrinsic performance of word2vec models (Figure 1-12) obtained similar trends to Chiu et al. (2016a) for the same corpora/standards (i.e. UMNSRS, BC2GM, and JNLPBA), therefore we refer to Chiu et al. (2016a) for further discussion of these trends. Minor differences were recorded for minimum word count (Figure 7-8) and window size (Figure 1-2), where both UMNSRS similarity and relatedness

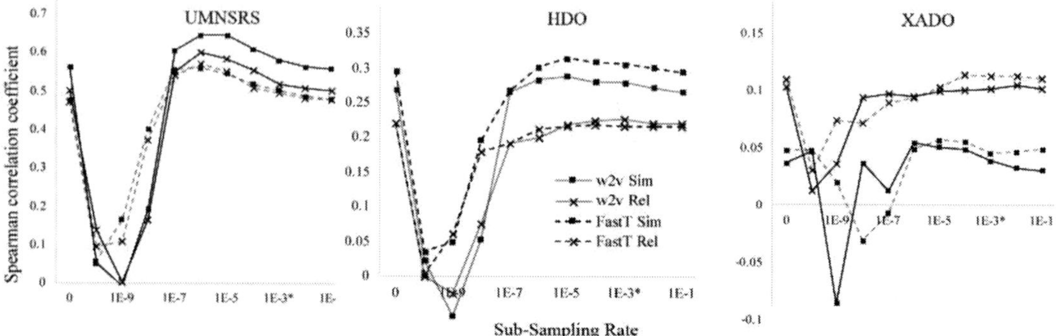

Figure 9: Intrinsic evaluation of sub-sampling rate in word2vec (w2v) and fastText (FastT) models on UMNSRS, HDO, and XADO datasets (Supp. Table 12).

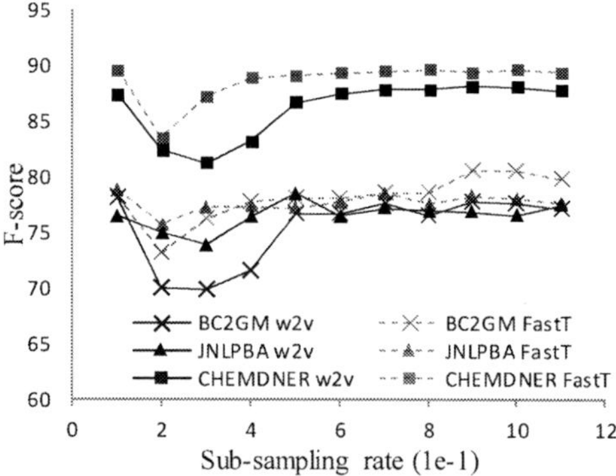

Figure 10: Extrinsic evaluation of sub-sampling rate in word2vec (w2v) and fastText (FastT) models on BC2GM, JNLPBA and CHEMDNER datasets (Supp. Table 13).

decreased with increasing minimum word count, whereas in Chiu et al. (2016a) this was only the case for relatedness.

In intrinsic evaluation of window size, particularly UMNSRS (Figure 1), performance consistently increased with increasing window size. This trend was also reported by Chiu et al. (2016a), where the maximum window size of 30 obtained the highest similarity and relatedness. We reasoned that abstracts generally concern a single topic, therefore predicted that increasing the window size to the average abstract length would capture more relevant information. This was indeed the case, obtaining 0.675 and 0.639 for UMNSRS similarity and relatedness respectively, compared to 0.627 and 0.584 similarity and relatedness respectively reported by Chiu et al. (2016a) for PubMed. As higher intrinsic performance was obtained in our results for similar window sizes, the difference in performance is also contributed to by an increase in the training data and different preprocessing.

In the case of extrinsic evaluation, the best performance was generally obtained with lower window size – a similar trend to that reported in Chiu et al. (2016a).

3.2 General trends: fastText hyper-parameter selection

Except for the character n-gram hyper-parameter, fastText models share the same hyper-parameters with word2vec models. Overall, similar trends in both intrinsic and extrinsic performance were obtained for word2vec and fastText embeddings (Figure 1-12). However, optimal parameters were not necessarily identical, as discussed below.

3.3 Comparison of representations – Intrinsic evaluation

While the overall performance trends with various hyper-parameters for fastText are similar to those obtained by word2vec, we report a number of notable differences.

When intrinsically evaluated with UMNSRS, word2vec representations consistently achieved higher similarity and relatedness compared to fastText for hyper-parameters such as: window size, dimensions and negative sampling, irrespective of the selected hyper-parameters. However, evaluating with HDO and XADO intrinsic datasets, results were more variable. fastText tended to perform similar to or outperform word2vec across negative sampling size, dimensions and window size hyper-parameter ranges.

Differences in performance between datasets may be a result of differences in: (i) number of OOV terms; (ii) rarity of terms; and (iii) term types. As UMNSRS is a manually curated reference list of term pairs with the vocabulary of multiple corpora, including PubMed Central, only up

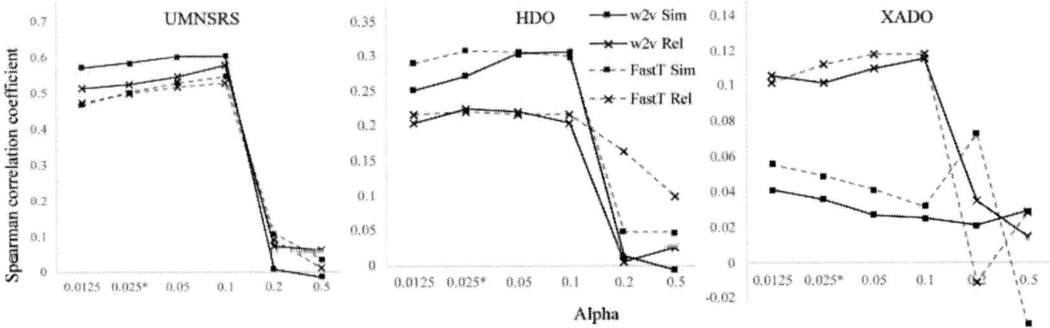

Figure 11: Intrinsic evaluation of the alpha hyper-parameter in word2vec (w2v) and fastText (FastT) models on UMNSRS, HDO, and XADO datasets (Supp. Table 14).

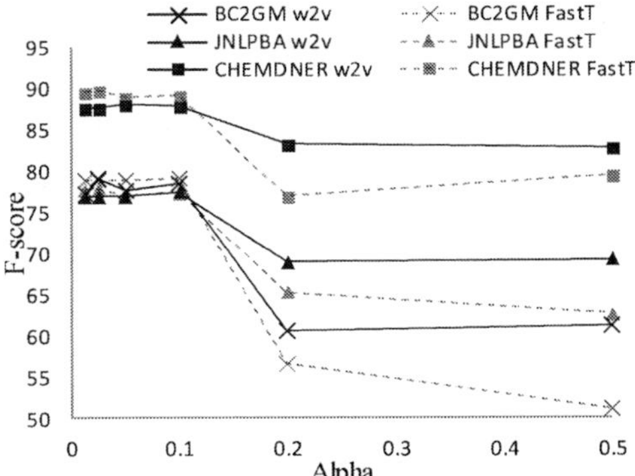

Figure 12: Extrinsic evaluation of alpha in word2vec (w2v) and fastText (FastT) models on BC2GM, JNLPBA and CHEMDNER datasets (Supp. Table 15).

to 9 total tokens were OOV (1.3%; Supp. Table 2). HDO contained up to 5% OOV terms. As OOV terms are represented by null vectors for word2vec models, a decrease in performance with increase in OOV terms is expected.

Skipping OOV term pairs from evaluation (rather than imputing) obtained similar performance trends across datasets, indicating that OOV is not the major contributing factor in such intrinsic performance differences. However, this may also imply that fastText degrades the performance for in-vocabulary terms of the UMNSRS dataset. Similar results were reported by the original authors when assessed on the English WS353 dataset (Bojanowski et al., 2017).

Despite terms being in-vocabulary, the frequency by which these occur in the training dataset may vary. This is indeed the case for UMNSRS and HDO, where UMNSRS has a median rank in-vocabulary frequency 4 times higher than HDO. This may indicate fastText provides better representations for rarer terms. XADO, however, has a

median rank in-vocabulary frequency within 1.3 times of UMNSRS. This implies there are additional contributing factors to such performance differences, including potentially differences in the quality of the ontology graph.

As the intrinsic standards contain various entity classes, differences in representation models' performance (and optimal hyper-parameters) may be dependent on the distribution of entity types. fastText authors reported that fastText outperforms word2vec in languages like German, Arabic, Russian and in rare English words (Bojanowski et al., 2017). This indicates that word2vec and fastText's performance is dependent on the compositionality and word character features, and may therefore be expected to vary between biomedical entity classes.

Biomedical text generally contains terms such as chemicals, genes, proteins and cell-lines which are rich in features such as punctuation, special characters, digits, and mixed-case characters. Such orthographic features have been manually extracted in traditional machine learning methods, or more recently combined with word embeddings, and have been shown to have discriminating power in tasks such as named entity recognition (Galea et al., 2018).

3.4 Comparison of representations – Extrinsic evaluation

When performing named entity recognition as extrinsic evaluation of the word representations models, fastText consistently outperformed word2vec at any hyper-parameter value, and consistently across all 3 corpora (Figures 2,4,6,7,10,12). With 9-13% total OOV tokens, and 14-34% OOV entity tokens (Supp. Table 3, Supp. Fig. 3,4), this indicates the overall likely positive

1,2-dichloromethane	1-(dimethylamino)-2-methyl-3,4-diphenylbutane-1,3-diol	ZNF560
1,2-dichloroethane	8-(N,N-*diethylamino*)oct*yl-3,4*,5-trimethoxybenzoate	*ZNF5*80
*1,2-dichloro*benzene	1,3-*dimethylamy*lamine	*ZNF5*45
*Di*bromo*chloromethane*	8-(*diethylamino*)oct*yl*	*ZNF5*82
*1,2-dichloro*propane	2-cyclohex*yl*-2-hydroxy-2-phen*yl*acetate	*ZNF5*21
water/*1,2-dichloroethane*	*diethylamine*	SOX1

Table 1: Top 5 most similar words to a selection of out-of-vocabulary terms (two chemical systematic names and a protein symbol; top row). Sequences in bold indicate overlap with queried term.

contribution of gained information from computed OOV vectors.

In terms of the specific corpora, the largest performance difference was recorded for genes (BC2GM) and chemical names (CHEMDNER). As these two corpora only tag one entity type, entity variation is lower than JNLPBA which tags 5 entity classes and therefore this may contribute to the dissimilarities in performance difference between the corpora.

In addition to the rich and unique features, outperformance of fastText in extrinsic evaluation may also be attributed to the standardized nomenclature used in biomedical entities which provides additional within-token structure. For example, systematic chemical names follow the IUPAC nomenclature. Prefixes such as *mono*, *di*, and *tri* indicate number of identical substituents in a compound. Similarly, residual groups are represented by prefixes such as *methyl-* and *bromo-*. Additionally, the backbone structure of the molecule is assigned a suffix that indicates structure features (e.g. simple hydrocarbon molecules utilize suffixes to indicate number of single, double or more bonds, where *-ane* indicates single bonds, *-ene* double bonds, *-ynes* triple bonds etc).

With such structure, as fastText is a character-level model, for chemicals such as *1,2-dichloromethane*, most similar words include chemicals which share the substituents and their specific position, defined by the *1,2-dichloro-* prefix (Table 1). Therefore, fastText provides more structurally-similar chemicals, whereas word2vec would treat *1,2-dichloromethane* and *2-dichloromethane* as two completely different/unrelated terms (when excluding context or setting a small window size).

As chemicals can be synthesized and named, it is likely for very specific and big molecules such as *1-(dimethylamino)-2-methyl-3,4-diphenylbutane-1,3-diol* to be OOV. This is a great advantage of character-level embeddings which still enable computing a representation.

Given the highly standardized and structured nomenclature of chemicals, we briefly observed

that fastText models are also able to recall structural analogs when performing analogy tasks. For example, methanol → methanal is an oxidation reaction where an alcohol is converted to an aldehyde, specifically the *-OH* group is converted to a *=O* group. Given ethanol and performing analogy task vector arithmetic, the aldehyde ethanal is returned. Similar results were observed for sulfuric_acid − sulfur + phosphorous, giving phosphoric_acid. Formal evaluation on analogy tasks is required to assess how character-based embeddings perform compared to word2vec.

Genes and proteins have full names as well as short symbolic identifiers which are usually acronymic abbreviations. These are less structured than chemical names, however, as the root portion of the symbols represents a gene family, this accounts for the similarity performance of character-based embeddings. *ZNF560* is an example of OOV protein that was assigned a vector close to ZNF* genes (Table 1) as well as *SOX1*. While *SOX1* does not share character n-grams with *ZNF560*, similarity was determined based on co-occurrence of *ZNF* genes and *SOX1* − genes which are associated with adenocarcinomas (Chang et al., 2015).

While the advantages of character-based similarity for OOV terms are clear, from intrinsic evaluation it appears that for some entities word2vec provides better embeddings. An example of this is when querying *phosphatidylinositol-4,5-bisphosphate* (Supp. Table 16). Whereas the top 5 most similar terms returned by fastText are orthographically, morphologically, and structurally similar, word2vec recalled *PIP2* and *PI(4,5)P2*. These are synonyms of the queried term hence more similar than *phosphatidylinositol-4-phosphate*, for example. A similar result was also observed for genetic variants (SNPs). While fastText returned *rs-* prefixed terms as most similar terms to the reference SNP identifier *rs2243250* (which refers to the SNP Interleukin 4 − 590C/T polymorphism), word2vec recalled terms *590C>T* and *590C/T*; the nucleotide polymorphism specified by the identifier itself (Supp

	UMNSRS						HDO						XADO					
	3	4	5	6	7	8	3	4	5	6	7	8	3	4	5	6	7	8
2	0.443	0.463	0.503	0.532	0.544	0.554	0.219	0.282	0.295	0.296	0.302	0.302	0.031	0.047	0.039	0.032	0.030	0.024
	0.427	0.458	0.497	0.507	0.512	0.516	0.193	0.218	0.219	0.222	0.223	0.224	0.105	0.106	0.106	0.113	0.114	0.114
3		0.487	0.517	0.548	0.560	0.561		0.298	0.307	0.307	0.312	0.312		**0.054**	0.048	0.038	0.032	0.030
		0.478	0.506	0.524	0.530	0.522		0.213	0.217	0.215	0.226	0.225		0.111	0.112	**0.117**	**0.117**	**0.117**
4			0.534	0.562	0.570	0.582			0.313	*0.318*	0.316	0.315			0.040	0.036	0.035	0.030
			0.523	0.539	0.533	0.540			0.218	0.227	0.228	0.227			0.110	0.111	0.113	0.109
5				0.584	0.603	0.596				**0.320**	*0.319*	**0.320**				0.034	0.031	0.029
				0.554	**0.565**	0.552				0.230	0.226	0.228				0.112	0.108	0.109
6					**0.612**	0.607					0.317	*0.319*					0.037	0.035
					0.556	0.549					0.228	**0.234**					0.110	0.108
7						0.601						0.314						0.033
						0.542						0.231						0.102

	BC2GM						JNLPBA						CHEMDNER					
	3	4	5	6	7	8	3	4	5	6	7	8	3	4	5	6	7	8
2	78.96	79.72	79.78	79.91	79.71	**80.26**	*78.20*	77.76	77.99	77.89	77.96	77.96	89.14	89.48	*89.66*	*89.72*	89.58	89.46
3		*79.69*	78.88	78.77	*79.90*	*79.91*		77.83	77.86	77.67	**78.46**	*78.30*		89.48	*89.67*	*89.67*	**89.75**	*89.62*
4			78.94	78.42	78.91	79.05			77.58	76.91	78.00	77.58			89.28	89.37	89.22	89.32
5				77.12	78.67	77.45				76.72	78.04	76.92				89.22	89.13	89.10
6					77.77	77.97					77.82	78.06					89.03	88.81
7						77.73						77.83						88.60

Table 2: Intrinsic (UMNSRS, HDO, XADO; upper row = similarity, lower row = relatedness) and extrinsic (BC2GM, JNLPBA, CHEMDNER) evaluation of the effect of character n-gram ranges on performance. Highest absolute accuracy is indicated in bold and accuracies within the standard error of the highest accuracy is italicized.

		UMNSRS		HDO		XADO		BC2GM	JNLPBA	CHEMDNER
		Sim	Rel	Sim	Rel	Sim	Rel			
int	w2v	**0.726**	**0.690**	0.314	0.237	**0.095**	0.077	76.43	71.84	87.83
	FastT	0.694	0.659	**0.330**	**0.243**	0.074	0.093	76.48	72.47	88.89
ex	w2v	0.506	0.469	0.252	0.184	0.024	**0.120**	77.13	73.61	88.93
	FastT	0.479	0.446	0.283	0.221	0.054	0.116	**79.63**	**74.29**	**90.14**

Table 3. Intrinsic and extrinsic performance for word2vec and fastText models optimized on optimum hyper-parameters from intrinsic (int) and extrinsic (ex) datasets (Supp. Table 27).

Table 19).

Additional examples comparing word2vec and fastText's most similar terms for chemicals, genes and diseases are provided in Supp. Tables 18-22.

From the quantitative results and the above qualitative examples, we observe a trade-off between character sequence similarity and context. The importance of which depends on the entity types – just as different languages benefit differently from word2vec and fastText models (Bojanowski et al., 2017).

3.5 Effect of n-grams size

Intrinsic evaluation shows high variability in the range of n-grams between the different standards (Table 2 & Supp. Table 25). UMNSRS achieves the highest performance (in terms of similarity) with 6-7 n-grams, whereas XADO achieves best results with 3-4 n-grams, and HDO achieves equal performance with ranges: 5-{6,7,8}, 4-6 and 6-8. This indicates the heterogeneity of the terms, both within the reference standards for HDO and XADO, and between standards. This further backs up the difference between the representation models due to entity type differences.

Contrastingly, extrinsic evaluation showed high consistency in n-gram ranges, with all corpora recording highest performance for the ranges 3-7 and 3-8. Within standard error (Supp. Table 23, 24), high performance was also obtained for ranges with lower limit of 2 and 3. Such ranges indicate that both short and long n-grams provide relevant information, complying with the previous discussion and examples for gene nomenclature and chemical naming conventions.

3.6 Optimized Models

Word embeddings trained on individual reference standards' optimal hyper-parameters (Supp. Table 25) achieved 0.733/0.686 similarity/relatedness with word2vec for UMNSRS (Supp. Table 26). This exceeds 0.652/0.601 reported by Chiu et al. (2016a), and the more recent 0.681/0.635 by Yu et al. (2017) achieved by retrofitting representations with knowledgebases, but not 0.75/0.73 by MeSH2Vec using prior knowledge (Jha et al., 2017). We expect further improvement to our models by retrofitting and augmenting prior knowledge.

Corpus-optimized fastText embeddings outperformed word2vec across all extrinsic corpora, recording: 79.33%, 73.30% and 90.54% for BC2GM, JNLPBA, and CHEMDNER (Supp. Table 26). This outperforms Chiu et al. (2016a), Pyysalo et al. (2013) and Kosmopoulos et al. (2015), although differences are also due to different NER architectures used. However, our 90.54% CHEMDNER performance outperforms 89.28% using similar architectures and is close to the 90.84% achieved for attention-based architectures (Luo et al., 2017) - the best performance reported in literature to date.

Optimizing word2vec and fastText representations across all corpora and standards (Supp. Table 28) decreased the performance difference in NER between word2vec and fastText. This is due to the differences in the optimal hyper-parameters between intrinsic and extrinsic data (Supp. Table 29). Based on these differences, and as it had been shown that intrinsic results are not reflective of extrinsic performance (Chiu et al. 2016b), we generated separate word2vec and fastText models optimized on intrinsic and extrinsic datasets separately (Table 3). Again, fastText outperforms word2vec in all NER tasks but only outperforms word2vec for the HDO intrinsic dataset, possibly due to similarity implied from disease suffixes captured by n-grams.

4 Conclusion and future directions

We show that fastText consistently outperforms word2vec in named entity recognition of entities such as chemicals and genes. This is likely to be contributed to by the ability of character-based representations to compute vectors for OOV, and due to the highly structured, standardized and feature-rich nature of such entities.

Intrinsic evaluation indicated that the optimal hyper-parameter set, and hence optimal performance, is highly dataset-dependent. While number of OOV terms and rarity of in-vocabulary terms may contribute to such differences, further investigation is required to determine how the different entity types within the corpora are affected. Similarly, for named entity recognition, investigating the performance differences for each entity class would provide a more fine-grained insight into which classes benefit mostly from fastText, and why.

Empirically, we observed a trade-off between character sequence similarity and context in word2vec and fastText models. It would be interesting to assess how embedding models such as MIMICK, where the word2vec space can be preserved while still being able to generate character-based vectors for OOV terms, compare.

Acknowledgements

We acknowledge financial support by BBSRC (BB/L020858/1), Imperial College Stratified Medicine Graduate Training Programme in Systems Medicine and Spectroscopic Profiling (STRATiGRAD), and Waters corporation.

References

Titipat Achakulvisut, and Daniel E. Acuna. 2016. Pubmed Parser. https://doi.org/10.5281/zenodo.159504

Piotr Bojanowski, Edouard Grave, Armand Joulin and Tomas Mikolov. 2017. Enriching word vectors with subword information. *Transactions of the Association for Computational Linguistics*, 5, pages 135-146. http://aclweb.org/anthology/Q17-1010

Cheng-Chang Chang, Yu-Che Ou, Kung-Liahng Wang, Ting-Chang Chang, Ya-Min Cheng, Chi-Hau Chen, et al. 2015. Triage of Atypical Glandular Cell by SOX1 and POU4F3 methylation: A Taiwanese gynecologic oncology group (TGOG) study. *PLoS ONE* 10(6): e0128705. https://doi.org/10.1371/journal.pone.0128705

Billy Chiu, Gamal Crichton, Anna Korhonen and Sampo Pyysalo. How to train good word embeddings for biomedical NLP. 2016a. In *Proceedings of the 15th Workshop on Biomedical Natural Language Processing*, pages 166-174. https://doi.org/10.18653/v1/W16-2922

Billy Chiu, Anna Korhonen, and Sampo Pyysalo. 2016b. Intrinsic evaluation of word vectors fails to predict extrinsic performance. In *Proceedings of the 1st Workshop on Evaluation Vector Space Representations for NLP*, pages 1-6. https://doi.org/10.18653/v1/W16-2501

Manaal Faruqui, Yulia Tsvetkov, Pushpendre Rastogi, and Chris Dyer. 2016. Problems with evaluation of word embeddings using word similarity tasks. In *Proceedings of the 1st Workshop on Evaluating Vector Space Representations for NLP*, pages 30-35. http://www.aclweb.org/anthology/W16-2506

Dieter Galea, Ivan Laponogov and Kirill Veselkov. 2018. Exploiting and assessing multi-source data for supervised biomedical named entity recognition. *Bioinformatics*, bty152. https://doi.org/10.1093/bioinformatics/bty152

Mourad Gridach. 2017. Character-level neural network for biomedical named entity recognition. *Journal of Biomedical Informatics.* 70, pages 85-91. https://doi.org/10.1016/j.jbi.2017.05.002

Jiangen He and Chaomei Chen. 2018. Predictive effects of novelty measured by temporal embeddings on the growth of scientific literature. *Frontiers in Research Metrics and Analytics*, 3, 9. https://doi.org/10.3389/frma.2018.00009

Kishlay Jha, Guangxu Xun, Vishrawas Gopalakrishnan, and Aidong Zhang. 2017. Augmenting word embeddings through external knowledge-base for biomedical application. *IEEE International Conference on Big Data*, pages 1965-1974. https://doi.org/10.1109/BigData.2017.8258142

Yoon Kim. 2014. Convolutional neural networks for sentence classification. In *Proceedings of the 2014 Conference on Empirical Methods in Natural Language Processing (EMNLP)*, pages 1746-1751. https://doi.org/10.3115/v1/D14-1181

Jin-Dong Kim, Tomoko Ohta, Yoshimasa Tsuruoka, Yuka Tateisi, and Nigel Collier. 2004. Introduction to the bio-entity recognition task at JNLPBA. In *Proceedings of JNLPBA*, pages 70–75. http://www.aclweb.org/anthology/W04-1213

Aris Kosmopoulos, Ion Androutsopoulos, and Georgios Paliouras. 2015. Biomedical semantic indexing using dense word vectors in BioASQ. *Journal of Biomedical Semantics.*

Martin Krallinger, Florian Leitner, Obdulia Rabal, Miguel Vazquez, Julen Oyarzabal and Alfonso Valencia. 2015. CHEMDNER: The drugs and chemical names extraction challenge. *Journal of Cheminformatics*, 7(Suppl 1):S1. https://doi.org/10.1186/1758-2946-7-S1-S1

Guillaume Lample, Miguel Ballesteros, Sandeep Subramanian, Kazuya Kawakami, and Chris Dyer. 2016. Neural architectures for named entity recognition. In *Proceedings of NAACL-HLT,* pages 260-270. https://doi.org/10.18653/v1/N16-1030

Michael Lesk. 1986. Automatic sense disambiguation using machine readable dictionaries: how to tell a pine cone from an ice cream cone. *In Proceedings of the 5th Annual International Conference on Systems Documentation*, pages 24–26. https://doi.org/10.1145/318723.318728

Ling Luo, Zhihao Yang, Pei Yang, Yin Zhang, Lei Wang, Hongfei Lin, and Jian Wang. 2017. An attention-based BiLSTM-CRF approach to document-level chemical named entity recognition. *Bioinformatics*, btx761. https://doi.org/10.1093/bioinformatics/btx761

Tomas Mikolov, Kai Chen, Greg Corrado, and Jeffrey Dean. 2013. Distributed representations of words and phrases and their compositionality. In *Proceedings of the 26th International Conference on Neural Information Processing Systems*, pages 3111-3119

Tomas Mikolov, Edouard Grave, Piotr Bojanowski, Christian Puhrsch, and Armand Joulin. Advances in pre-training distributed word representations. 2018. In *Proceedings of the International Conference on Language Resources and Evaluation.*

Serguei V.S. Pakhomov, Greg Finley, Reed McEwan, Yan Wang, and Genevieve B. Melton. 2016. Corpus domain effects on distributional semantic modeling of medical terms. *Bioinformatics*, 32:23, pages 3635-3644. https://doi.org/10.1093/bioinformatics/btw529

Jeffrey Pennington, Richard Socher, and Christopher D. Manning. 2014. GloVe: Global Vectors for Word Representations. In *Proceedings of the 2014 Conference on Empirical Methods in Natural Language Processing (EMNLP)*, pages 1532-1543. https://doi.org/10.3115/v1/D14-1162

Yuval Pinter, Robert Guthrie, and Jacob Eisenstein. Mimicking word embeddings using subword RNNs. In *Proceedings of the 2017 Conference on Empirical Methods in Natural Language Processing*, pages 102-112. https://doi.org/10.18653/v1/D17-1010

Sampo Pyysalo, Filip Ginter, Hans Moen, Tapio Salakoski, and Sophia Ananiadou. 2013. Distributional semantics resources for biomedical text processing. In *Proceedings of LBM.*

Radim Řehůřek and Petr Sojka. Software Framework for Topic Modelling with Large Corpora. 2010. In *Proceedings of LREC 2010 workshop New Challenges for NLP Frameworks*, pages 46-50. https://doi.org/10.13140/2.1.2393.1847

Lynn Marie Schriml, Cesar Arze, Suvarna Nadendla, Yu-Wei Wayne Chang, Mark Mazaitis, Victor Felix, Gang Feng and Warren Alden Kibbe. 2012. Disease ontology: a backbone for disease semantic integration. *Nucleic Acids Research*, 40: D940-D945. https://doi.org/10.1093/nar/gkr972

Erik Segerdell, Jeff B. Bowes, Nicolas Pollet and Peter D. Vize. 2008. An ontology for *Xenopus* anatomy and development. *BMC Developmental Biology.* 8:92. https://doi.org/10.1186/1471-213X-8-92

Larry Smith, Lorraine K Tanabe, Rie Johnson nee Ando, Cheng-Ju Kuo, I-Fang Chung, Chun-Nan Hsu, Yu-Shi Lin, Roman Klinger, Christoph M Friedrich, Kuzman Ganchev, et al. 2008. Overview of biocreative ii gene mention recognition. *Genome biology*, 9(Suppl 2):1–19. https://doi.org/10.1186/gb-2008-9-s2-s2

Zhibiao Wu and Martha Palmer. 1994. Verb semantics and lexical selection. In *Proceedings of the 32nd Meeting of Association of Computational Linguistics*, pages 33–138. https://doi.org/10.3115/981732.981751

Zhiguo Yu, Byron C. Wallace, Todd Johnson, and Trevor Cohen. 2017. Retrofitting concept vector representations of medical concepts to improve estimates of semantic similarity and relatedness. In *Proceedings of the 16th World Congress on Medical and Health Informatics*

PICO Element Detection in Medical Text via Long Short-Term Memory Neural Networks

Di Jin
MIT
jindi15@mit.edu

Peter Szolovits
MIT
psz@mit.edu

Abstract

Successful evidence-based medicine (EBM) applications rely on answering clinical questions by analyzing large medical literature databases. In order to formulate a well-defined, focused clinical question, a framework called PICO is widely used, which identifies the sentences in a given medical text that belong to the four components: Participants/Problem (P), Intervention (I), Comparison (C) and Outcome (O). In this work, we present a Long Short-Term Memory (LSTM) neural network based model to automatically detect PICO elements. By jointly classifying subsequent sentences in the given text, we achieve state-of-the-art results on PICO element classification compared to several strong baseline models. We also make our curated data public as a benchmarking dataset so that the community can benefit from it.

1 Introduction

The paradigm of evidence-based medicine (EBM) involves the incorporation of current best evidence, such as the reports of randomized controlled trials (RCTs), into decision making for patient care (Sackett, 1997). Such evidence, integrated with the physician's own expertise and patient-specific factors, can lead to better patient outcomes and higher quality health care (Sackett et al., 1996). In practice, successful EBM applications rely on answering clinical questions via analysis of large medical literature databases such as PubMed. And most often, a PICO framework is used to formulate a well-defined, focused clinical question, which decomposes the question into four parts: Participants/Problem (P), Intervention (I), Comparison (C) and Outcome (O) (Richardson et al., 1995).

Typically the analyses that underlie EBM begin by selecting a set of potentially relevant papers, which are then further refined by human judgment to form the evidence base on which the answer to a specific question depends. To facilitate this selection process, it would be advantageous that all papers (or at least their abstracts) can be organized according to the PICO foci. Unfortunately, a significant portion of the medical literature contains either unstructured or sub-optimally structured abstracts, without specifically identified PICO elements. Therefore, we would like to introduce a method to automate the identification of PICO elements in medical abstracts in order to make possible the automated selection of possibly relevant articles for a proposed study.

In this paper, we present a system based on artificial neural networks (ANN) to tackle the issue of extracting PICO elements in medical abstracts as a classification task at the sentence level. Our key contributions are as follows:

1. Previous methods for PICO elements extraction focused on shallow models such as Naive Bayes (NB), Support Vector Machines (SVM) and Conditional Random Fields (CRF), which are limited in modeling capacity. To significantly boost the performance, we propose a Long Short-Term Memory (LSTM) based ANN model to solve this task.

2. Most previous systems detected the PICO elements one by one; thus several classifiers needed to be built and trained separately, which is sub-optimal in efficiency. That approach also cannot take advantage of shared structure among the individual classifiers. In

Proceedings of the BioNLP 2018 workshop, pages 67–75
Melbourne, Australia, July 19, 2018. ©2018 Association for Computational Linguistics

this work we extract PICO components simultaneously from any given medical abstract.

3. In all previous works, the only dataset used for training and test and made public is from (Kim et al., 2011). However, this dataset contains only 1000 abstracts, which is not enough for a ANN based deep learning model to obtain good generalization results. Therefore, we curate a dataset comprising of over tens of thousands of abstracts and make it public as a benchmark dataset so that everyone else can use it.

4. Instead of normally treating PICO detection as a single sentence classification problem, we view it as a sequential sentence classification task, where the sequence of sentences in an abstract is jointly predicted. In this way, the information from the context sentences can be used to help predict the current sentence, which does improve the classification accuracy considerably. Leveraging this strategy, we obtain state-of-the-art PICO elements extraction accuracy, significantly outperforming all previous methods.

2   Related Work

In many previous user studies, the generalized use of the PICO framework or similar schema by clinicians has been validated for its performance improvement on searching literature for clinical questions (Schardt et al., 2007; Boudin et al., 2010c; Znaidi et al., 2015). This has greatly fueled academic interest in the development of systems for automatic PICO element detection. Over the last decade, the research progress for this task can be summarized according to three aspects: models for classification, dataset generation, and task formulation.

Many well-known machine learning techniques have been proposed to build stronger models for this task, including Nave Bayes (NB) (Huang et al., 2013; Boudin et al., 2010a; Demner-Fushman and Lin, 2007), Random Forest (RF) (Boudin et al., 2010a), Support Vector Machine (SVM) (Boudin et al., 2010a; Hansen et al., 2008), Conditional Random Field (CRF) (Kim et al., 2011; Chung, 2009; Chung and Coiera, 2007) and Multi-Layer Perceptron (MLP) (Boudin et al., 2010a; Huang et al., 2011). Also Boudin et al. in

(Boudin et al., 2010b) proposed a location-based weighting strategy as an extension to the language modeling approach inspired by the special distribution pattern of PICO elements in medical abstracts. All these models heavily rely on careful selections of hand-engineered features including lexical features such as bag of words (BOW), stemmed words and cue-words/verbs, and semantic features such as synonyms and hypernyms provided by some ontologies (*e.g.,* WordNet). As an important complement to this task, most recent work from Dernoncourt et al. (Dernoncourt et al., 2016) proposed the model based on currently emerging deep ANN architectures such as LSTM for further performance boosting, as well as to remove the need for hand-crafted features. However, this work has not targeted to address the issue of PICO element detection.

To generate the datasets for both training and test, earlier works mainly relied on manual annotation, which resulted in small corpora on the order of hundreds of abstracts (Demner-Fushman and Lin, 2007; Dawes et al., 2007; Chung, 2009; Kim et al., 2011). Afterwards, later works made use of the structural information embedded in some abstracts for which the authors have clearly stated distinctive sentence headings (Boudin et al., 2010a; Huang et al., 2011, 2013). Specifically, some abstracts contain explicit headings such as "PATIENTS", "SAMPLE" or "OUTCOMES", which can be used to locate sentences corresponding to PICO elements. In this way, tens of thousands of abstracts that contain PICO elements from PubMed can be automatically compiled as a well-annotated dataset, which can increase the size of dataset by two orders of magnitude.

In terms of task formulation, most previous works focused on categorizing one PICO class at a time using an individual classifier (Boudin et al., 2010a; Huang et al., 2013). Therefore, in order to detect all four PICO components, one would need to build and train four individual models, which is inefficient. Furthermore, it is hard to disambiguate the classification label conflicts between different model predictions on the same sentence. These limitations were resolved by working directly on the labels of interest for EBM, allowing multilabel classification instead of binary and allowing sentences that are unrelated to labels of interest to be labeled as an "Other" category (Kim et al., 2011; Demner-Fushman and Lin, 2007). This is a more realistic setting and ought to provide better insight into the performance we should expect for this kind of task.

3   The Proposed Model

First we introduce our notation. We denote scalars in italic lowercase (e.g., k), vectors in bold lowercase (e.g., $\mathbf{s}$) and matrices in italic uppercase (e.g., W). Colon notations $r_{i:j}$ and $\mathbf{s}_{i:j}$ are used to denote the sequence of scalars $(x_i, x_{i+1}, ..., x_j)$ and vectors $(\mathbf{s}_i, \mathbf{s}_{i+1}, ..., \mathbf{s}_j)$.

Our model is composed of three components: the token embedding layer, the sentence-level label inference layer, and the label sequence optimization layer (Figure 1). In the following sections they will be discussed in detail.

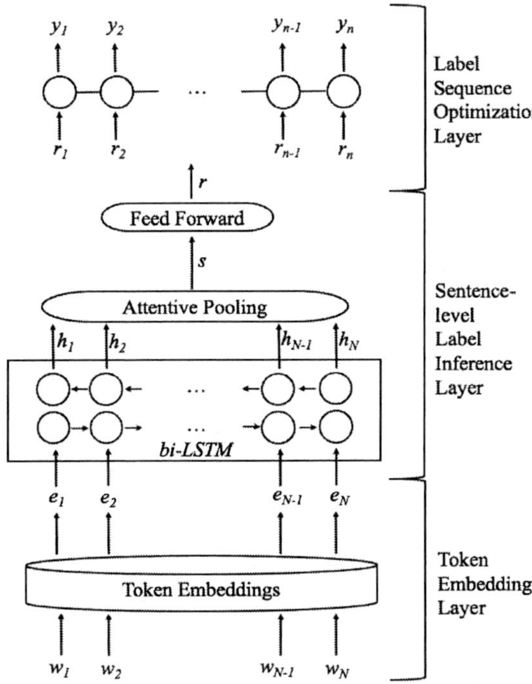

Figure 1: Model architecture. **w**: original token; **e**: token embedding; **h**: bi-LSTM hidden state; **s**: sentence representation vector; **r**: sentence label probability vector; y: predicted sentence label. Replacing bi-LSTM with convolutional neural network (CNN) did not improve the results: we therefore used bi-LSTM.

3.1   Token Embedding Layer

This layer takes as input a given sentence **w** comprising N words $\mathbf{w} = [w_1, w_2, ...w_N]$ and outputs its corresponding vector representation. Token representations are encoded by the column vector in the embedding matrix $W^{word} \in \mathbb{R}^{d^w \times |V|}$, where d^w is the dimension of the word vector and V is the vocabulary of the dataset. Each column $W_i^{word} \in \mathbb{R}^{d^w}$ is the word embedding vector for the ith word in the vocabulary. To transform a certain word w into its corresponding embedding vector e^w, we use the following equation:

$$\mathbf{e}^w = W^{word}\mathbf{v}^w, \tag{1}$$

where $\mathbf{v}^w$ is the one hot vector of word w with dimension of $|V|$ that has 1 at the corresponding index and zero in all other positions. The word embeddings W^{word} can be pre-trained on large unlabeled datasets using unsupervised algorithms such as word2vec (Mikolov et al., 2013), GloVe (Pennington et al., 2014) and fasttext (Bojanowski et al., 2016).

3.2   Sentence-level Label Inference Layer

This layer takes as input the embedding vector **e** of each token in a sentence from the token embedding layer and produces a vector $\mathbf{r} \in \mathbb{R}^l$ to represent the probability that this sentence belongs to each label, where l is the number of labels. To this aim, the sequence of embedding vectors **e** is first input into a bi-directional LSTM (bi-LSTM), which outputs a sequence of hidden states $\mathbf{h}_{1:N}$ ($\mathbf{h} \in \mathbb{R}^{d^h}$) for a sentence of N words with each hidden state corresponding to a token. To form the final representation vector **s** of this sentence, attentive pooling is used, which can be described using the following equations (Yang et al., 2016):

$$\mathbf{u}_i = \tanh(W_s\mathbf{h}_i + \mathbf{b}_s), \tag{2}$$

$$\alpha_i = \frac{\exp(\mathbf{u}_i^\top \mathbf{u}_s)}{\sum_j \exp(\mathbf{u}_j^\top \mathbf{u}_s)}, \tag{3}$$

$$\mathbf{s} = \sum_i \alpha_i \mathbf{h}_i, \tag{4}$$

where $\mathbf{u}_s \in \mathbb{R}^{d^s}$ is the token level context vector used to measure the relevance or importance of each token with respect to the whole sentence, and $W_s \in \mathbb{R}^{d^s \times d^h}$ is the transformation matrix for soft alignment.

The obtained vector **s** is subsequently input to a feed-forward neural network with only one hidden layer, which outputs the corresponding probability vector **r**.

3.3 Label Sequence Optimization Layer

Each medical abstract consists of several sentences with the sentence category following some patterns, such as that the category "Results" is always followed by "Conclusion". Such patterns can yield better classification performance via the conditional random field (CRF) algorithm. Given the sequence of probability vectors $\mathbf{r}_{1:n}$ from the last label inference layer for an abstract of n sentences, this layer outputs a sequence of labels $y_{1:n}$, where y_i represents the predicted label assigned to the ith sentence.

In order to model dependencies between subsequent labels, we incorporate a matrix T that contains the transition probabilities between two subsequent labels; we define $T[i,j]$ as the probability that a token with label i is followed by a token with the label j. The score of a label sequence $y_{1:n}$ is defined as the sum of the probabilities of individual labels and the transition probabilities:

$$s(y_{1:n}) = \sum_{i=1}^{n} \mathbf{r}_i(y_i) + \sum_{i=2}^{n} T[y_{i-1}, y_i]. \quad (5)$$

The score in the above equation can be transformed into the probability of a certain label sequence by taking a softmax operation over all possible label sequences:

$$p(y_{1:n}) = \frac{e^{s(y_{1:n})}}{\sum_{\hat{y}_{1:n} \in Y} e^{s(\hat{y}_{1:n})}}, \quad (6)$$

where Y denotes the set of all possible label sequences. During the training phase, the objective is to maximize the probability of the gold label sequence. While in the testing phase, given an input sequence, the corresponding sequence of predicted labels is chosen as the one that maximizes the score using the Viterbi algorithm (Forney, 1973).

4 Experiments

4.1 Dataset Preparation

The dataset used in this study[1] is curated from MEDLINE, which is a free access database on medical articles. Specifically, we extracted 489,026 abstracts from PubMed by stating the following search limits: 1. Text Availability: Abstract; 2. Languages: English; 3. Publication

---

[1] https://github.com/jind11/PubMed-PICO-Detection

Types: Randomized Controlled Trial (Search conducted on 2017/08/28). Among them, abstracts with structured section headings were selected for automatic annotation of sentence category. Although P, I and O headings were our detection targets, we also annotated the other types of sentences into one of the AIM (A), METHOD (M), RESULTS (R) and CONCLUSION (C) labels to facilitate the use of our CRF label sequence optimization method. Note that, although we have 7 labels in total, we only care about the detection accuracy of the P, I and O labels and thus mainly discuss their performance in the following sections.

In this study, the C component was incorporated into the I category since the "COMPARISON" section also refers to a kind of intervention in an RCT. And in fact, there are very few abstracts with comparison labels found in PubMed.

We annotated a certain section heading into one of the 7 labels based on whether it contains the key words that belong to the assigned label as shown in Table 1 (section headings are only used to generate gold labels and not used for model training and inference). In very rare cases, the section heading of a certain sentence may contain the key words of more than one category, in which case that sentence will be assigned into multi-labels according to Table 1. Table 2 presents a typical abstract example with section headings annotated into the 7 labels. A total of 24,668 abstracts contain at least one of the P/I/O labels. There are 21,198 abstracts with P-labels, 13,712 with I-labels and 20,473 with O-labels (Table 3). Note that, the abstracts in PubMed follow a diversity of rhetorical structure and only a small fraction of them contain PICO elements based on their section headings.

4.2 Training Settings

Ten-fold cross-validation was employed to assess the results statistically, where abstracts were randomly split into 10 equal partitions. Nine of them were used for training and the remaining one for testing. This step repeats for ten rounds. For each round of training, 10% of the training set was randomly extracted as the development set for early stopping, that is, the test set was evaluated at the highest development set performance, which is measured by the average F1 score of all three P/I/O labels.

The token embeddings were pre-trained on a large corpus combining Wikipedia, PubMed and

Category	Heading Name Key Words
Aim (A)	Objective, Background, Purpose, Importance, Introduction, Aim, Rationale, Goal, Context, Hypothesis
Participants (P)	Population, Participant, Sample, Subject, Patient
Intervention (I)	Intervention
Outcome (O)	Outcome, Measure, Variable, Assessment
Method (M)	Method, Setting, Design, Material, Procedure, Process, Methodology
Results (R)	Result, Finding
Conclusion (C)	Conclusion, Implication, Discussion, Interpretation

Table 1: Key words of section headings in structured abstracts for automatic annotation.

Heading Name	Cate.	Sentences
AIMS	A	[...] The aims of the trial were to test for differences between standard 1-and 0.5-mg doses (both twice daily during 8weeks) in (1) abstinence, (2) adherence and (3) side effects.
DESIGN	M	Open-label randomized parallel-group controlled trial with 1-year follow-up. [...]
SETTING	M	Stop-Smoking Clinic of the Virgen Macarena University Hospital in Seville, Spain.
PARTICIPANTS	P	The study comprised smokers (n=484), 59.5% of whom were men with a mean age of 50.67years and a smoking history of 37.5 pack-years.
INTERVENTION	I	Participants were randomized to 1mg (n=245) versus 0.5mg (n=239) and received behavioural support, which consisted of a baseline visit and six follow-ups during 1year.
MEASUREMENTS	O	The primary outcome was continuous self-reported abstinence during 1year, with biochemical verification. [...] Also measured were baseline demographics, medical history and smoking characteristics.
FINDINGS	R	Abstinence rates at 1year were 46.5% with 1mg versus 46.4% with 0.5mg [odds ratio (OR)=0.997; 95% confidence interval (CI) = 0.7-1.43; P=1.0]; [...]
CONCLUSIONS	C	There appears to be no difference in smoking cessation effectiveness between 1mg and 0.5mg varenicline, [...].

Table 2: A typical abstract example with section headings and their corresponding annotated labels. The PMID of this abstract is 28449281.

Category	Abstracts	Sentences
P	21,198	27,695
I	13,712	24,602
O	20,473	32,525

Table 3: Number of times each of the categories P, I and O appear in abstracts and in sentences in the data.

PMC texts (Moen and Ananiadou, 2013) using the word2vec tool[2]. They are fixed during the training phase to avoid over-fitting[3].

The model is trained using the Adam optimization method (Kingma and Ba, 2014). For regularization, dropout is applied to each layer and l_2 regularization is also used. Hyperparameters were optimized via grid search and the best configuration is shown in Table 4. Code for this work is available online[4].

---

[2]https://code.google.com/archive/p/word2vec/

[3]http://bio.nlplab.org/
[4]https://github.com/jind11/LSTM-PICO-Detection

Para.	Para. Name	Value
d^w	Token Embed. Size	200
d^h	LSTM Hidden Size	150
d^s	Attention Vector Size	300
bz	Batch Size	40
lr	Learning Rate	0.001
β	l_2 Regularization Ratio	0.0001

Table 4: Hyperparameters. Batch size refers to the number of abstracts in one batch.

5 Results and Discussion

Table 5 and 6 detail the results of classification for each label in terms of performance scores (precision, recall and F1) and confusion matrix, respectively (for one fold). It can be seen that the classifier is very good at predicting the labels of AIM, RESULTS and CONCLUSION but has difficulty in distinguishing among the labels of PARTICIPANTS, INTERVENTION, OUTCOME and METHOD. Indeed, the PARTICIPANTS, INTERVENTION and OUTCOME sections can be deemed as more specific aspects of the METHOD descriptions, therefore, it is naturally more difficult to tell the P/I/O elements apart from the METHOD section. Since our main goal is to accurately extract the P/I/O components from a given abstract, we will only discuss their performance in the following.

Cate.	p (%)	r (%)	F1 (%)	Support
A	97.7	98.0	97.8	3811
P	88.5	82.8	85.6	2722
I	74.9	81.5	78.1	2331
O	84.5	83.2	83.8	3219
M	87.0	84.2	85.6	5623
R	93.3	96.4	94.8	9236
C	93.8	91.1	92.5	4312
Total	90.1	90.0	90.0	31254

Table 5: Results in terms of precision (p), recall (r) and F-measure (F1) on the test set for each class obtained by our model for one of the ten folds.

Table 7 compares our model against several previously widely-used baseline models. Since there is no benchmarking dataset, we cannot compare with published best models (this is one of the reasons why we want to publish this dataset).

The first baseline is the logistic regression (LR) model that uses the n-gram features extracted from the current sentence for classification. In this

	P	M	C	A	R	O	I
P	2213	197	5	29	84	49	145
M	181	4804	9	40	30	242	317
C	0	6	3904	8	393	1	0
A	4	43	3	3743	6	11	1
R	9	21	175	0	8952	65	14
O	15	277	11	20	136	2688	72
I	40	278	0	0	28	142	1843

Table 6: Confusion matrix obtained by our model for one of the ten folds. Rows correspond to predicted labels, and columns correspond to true labels.

scenario, each sentence is predicted individually without context information from the surrounding sentences considered. Likewise, the second baseline MLP first computes the vector representation for each sentence by taking the max pooling operation of the embeddings of all tokens in the sentence, then classifies the current sentence via a neural network with three hidden layers (hidden layer dimensions are 400, 400 and 200, respectively). On the other hand, the third baseline is a CRF model that also uses n-grams as features (only the first 100 tokens were used for each sentence since most sentences are shorter than 100 tokens) and outputs the most probable label sequence for the whole abstract. Therefore, the CRF baseline takes into account both preceding and succeeding sentences when classifying the current sentence.

As presented by Table 7, the LR baseline performs worst, which is quite reasonable considering that it is still a very shallow model and only uses the local sentence information. As a comparison, the MLP model also only considers the features from the current sentence but performs better than LR because its modeling capacity is much larger. By incorporating the surrounding sentences, the CRF baseline performs even better than MLP system, which verifies that context information is quite useful in sequential classification problems.

Lastly but most importantly, our proposed model performs much better than all the baselines for all three P/I/O labels. The advantages of our model and the reasons for its improved performance are summarized below:

No human-engineered features Our model does not rely on any hand-engineered features that require much domain experience and are quite dif-

Models	P-element (%)			I-element (%)			O-element (%)		
	p	**r**	**F1**	**p**	**r**	**F1**	**p**	**r**	**F1**
LR	66.9	68.5	67.7	55.6	55.0	55.3	65.4	67.0	66.2
MLP	77.8	74.1	75.8	64.3	65.9	64.9	73.8	77.9	75.8
CRF	82.2	77.5	79.8	67.8	70.3	69.0	76.0	76.3	76.2
Our Model	**87.8**	**83.4**	**85.5**	**72.7**	**81.3**	**76.7**	**81.1**	**85.3**	**83.1**

Table 7: Performance in terms of precision (p), recall (r) and F-measure (F1) on the test set with several baselines and our proposed model (average value based on 10 fold cross validation). Since the dataset used here was introduced in this work, there is no previously published method for reference.

ficult to craft.

No n-gram features Unlike many other systems that rely heavily on n-grams, our model simply uses the token embedding vector to represent each token and feeds it into the recurrent neural network (RNN) model for inference. In this way, the pretrained embeddings on large corpora can encode the syntactic and semantic information of words for better language understanding. This can also help combat word scarcity problem. For example, the alternatively spelled tokens "tendonitis" and "tendinitis" are two different unigrams, however, their semantic meanings are the same, and this similarity can be revealed by their corresponding closely parallel embedding vectors.

Joint prediction Instead of predicting each sentence one by one, our model classifies all sentences in one abstract jointly, which improves the overall classification performance by implying the constraints of coherency between subsequent predicted labels. This improvement is clearly evidenced by Table 8.

Sequence modeling An RNN model is good at modeling sequences such as sentences by considering the dependency between tokens, which cannot be accounted for by context-free models such as those using bag of words features. And the long-term memory characteristic of LSTM model further grants the RNN model the ability to cope with long sentences.

Figure 2 presents an example of the transition matrix after the model has been trained, which encodes the transition probability between two subsequent labels. It effectively reflects what label is the most likely one that should follow the current one. For example, a sentence pertaining to the RESULTS is typically followed by a sentence pertaining to the CONCLUSION (1.16), which makes sense. From this transition matrix, we can figure

Model	F1 (%)		
	P	**I**	**O**
Full Model	**85.5**	**76.7**	**83.1**
-sequence optimization	78.2	68.2	78.3

Table 8: Ablation analysis. 10 fold cross validation F1-scores are reported. "-sequence optimization" is our model without the label sequence optimization layer.

out the most probable label sequence: $A \rightarrow M \rightarrow P \rightarrow I \rightarrow O \rightarrow R \rightarrow C$, which is also consistent with our observations.

Table 9 presents a few examples of prediction errors that are related to P/I/O labels. This error analysis suggests that part of the model error comes from the ambiguity between some label pairs, such as O and M, O and R, and I and M. For example, the sentence "Plasma volume and total body haemoglobin were determined at rest." can be deemed as a METHOD description in a general sense, however, it can also be further specified as an OUTCOME. On the other hand, a fair number of sentence labels are indeed debatable. For instance, the sentence "Iron supplementation was given to one group as a substitution remedy, another group was given iron and folic acid and the third group was without supplementation during the collection period." belongs to the PARTICIPANT label according to the gold standard, but it makes more sense that it should be classified as an INTERVENTION.

6 Conclusion

In this work we have presented an LSTM based ANN architecture to detect the PICO elements in medical RCT abstracts. We demonstrated that the use of a more advanced LSTM model and jointly predicting the classes of all sentences in a given text can improve the overall classification perfor-

Sentence	Predicted	Gold
The study included 16 patients who were randomized into one of three 6-month treatment protocols.	P	M
Referral service doing n-of-1 trials at the requests of community and academic physicians.	I	M
Iron supplementation was given to one group as a substitution remedy, another group was given iron and folic acid and the third group was without supplementation during the collection period.	I	P
Plasma urea and creatinine concentrations and angiotensin converting enzyme activity were measured at the start of the study and the end of each treatment period.	O	R
Heart rate was recorded continuously throughout the maneuvre, while blood was sampled for catecholamine determinations prior to the start of straining and again approximately 10 s following the end of straining.	O	I
Plasma volume and total body haemoglobin were determined at rest.	O	M

Table 9: Examples of prediction errors of our model that are related to P/I/O labels. The "Predicted" column indicates the label predicted by our model for a given sentence. The "Gold" column indicates the gold label of the sentence.

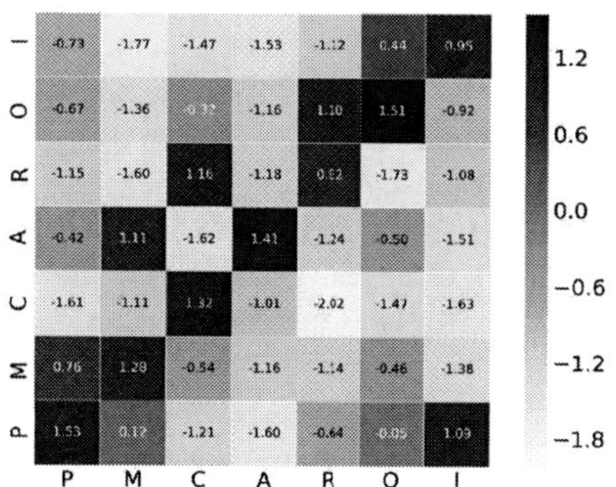

Figure 2: Transition matrix of label sequence. The rows represent the label of the previous sentence, while the columns represent the label of the current sentence.

mance of PICO components. And by publishing our curated dataset for benchmarking, we hope to encourage competition by other approaches than ours and that more effective and efficient methods can be developed in the future.

Acknowledgments

This work was supported by funding grant U54-HG007963 from National Human Genome Research Institute (NHGRI). Thank Matthew McDermott for helping revise the manuscript.

References

Piotr Bojanowski, Edouard Grave, Armand Joulin, and Tomas Mikolov. 2016. Enriching word vectors with subword information. *arXiv preprint arXiv:1607.04606*.

Florian Boudin, Jian-Yun Nie, Joan C Bartlett, Roland Grad, Pierre Pluye, and Martin Dawes. 2010a. Combining classifiers for robust pico element detection. *BMC medical informatics and decision making*, 10(1):29.

Florian Boudin, Jian-Yun Nie, and Martin Dawes. 2010b. Clinical information retrieval using document and pico structure. In *Human Language Technologies: The 2010 Annual Conference of the North American Chapter of the Association for Computational Linguistics*, pages 822–830. Association for Computational Linguistics.

Florian Boudin, Lixin Shi, and Jian-Yun Nie. 2010c. Improving medical information retrieval with pico element detection. In *European Conference on Information Retrieval*, pages 50–61. Springer.

Grace Y Chung. 2009. Sentence retrieval for abstracts of randomized controlled trials. *BMC medical informatics and decision making*, 9(1):10.

Grace Y Chung and Enrico Coiera. 2007. A study of structured clinical abstracts and the semantic classification of sentences. In *Proceedings of the Workshop on BioNLP 2007: Biological, Translational, and Clinical Language Processing*, pages 121–128. Association for Computational Linguistics.

Martin Dawes, Pierre Pluye, Laura Shea, Roland Grad, Arlene Greenberg, and Jian-Yun Nie. 2007. The identification of clinically important elements within medical journal abstracts:

Patient_population_problem, exposure_intervention, comparison, outcome, duration and results (pecodr). *Journal of Innovation in Health Informatics*, 15(1):9–16.

Dina Demner-Fushman and Jimmy Lin. 2007. Answering clinical questions with knowledge-based and statistical techniques. *Computational Linguistics*, 33(1):63–103.

Franck Dernoncourt, Ji Young Lee, and Peter Szolovits. 2016. Neural networks for joint sentence classification in medical paper abstracts. *arXiv preprint arXiv:1612.05251*.

G David Forney. 1973. The viterbi algorithm. *Proceedings of the IEEE*, 61(3):268–278.

Marie J Hansen, Nana Ø Rasmussen, and Grace Chung. 2008. A method of extracting the number of trial participants from abstracts describing randomized controlled trials. *Journal of Telemedicine and Telecare*, 14(7):354–358.

Ke-Chun Huang, I-Jen Chiang, Furen Xiao, Chun-Chih Liao, Charles Chih-Ho Liu, and Jau-Min Wong. 2013. Pico element detection in medical text without metadata: Are first sentences enough? *Journal of biomedical informatics*, 46(5):940–946.

Ke-Chun Huang, Charles Chih-Ho Liu, Shung-Shiang Yang, Furen Xiao, Jau-Min Wong, Chun-Chih Liao, and I-Jen Chiang. 2011. Classification of pico elements by text features systematically extracted from pubmed abstracts. In *Granular Computing (GrC), 2011 IEEE International Conference on*, pages 279–283. IEEE.

Su Nam Kim, David Martinez, Lawrence Cavedon, and Lars Yencken. 2011. Automatic classification of sentences to support evidence based medicine. In *BMC bioinformatics*, volume 12, page S5. BioMed Central.

Diederik P Kingma and Jimmy Ba. 2014. Adam: A method for stochastic optimization. *arXiv preprint arXiv:1412.6980*.

Tomas Mikolov, Ilya Sutskever, Kai Chen, Greg S Corrado, and Jeff Dean. 2013. Distributed representations of words and phrases and their compositionality. In *Advances in neural information processing systems*, pages 3111–3119.

SPFGH Moen and Tapio Salakoski2 Sophia Ananiadou. 2013. Distributional semantics resources for biomedical text processing. In *Proceedings of the 5th International Symposium on Languages in Biology and Medicine, Tokyo, Japan*, pages 39–43.

Jeffrey Pennington, Richard Socher, and Christopher Manning. 2014. Glove: Global vectors for word representation. In *Proceedings of the 2014 conference on empirical methods in natural language processing (EMNLP)*, pages 1532–1543.

W Scott Richardson, Mark C Wilson, Jim Nishikawa, and Robert SA Hayward. 1995. The well-built clinical question: a key to evidence-based decisions. *ACP journal club*, 123(3):A12–A12.

David L Sackett. 1997. *Evidence-based Medicine How to practice and teach EBM*. WB Saunders Company.

David L Sackett, William MC Rosenberg, JA Muir Gray, R Brian Haynes, and W Scott Richardson. 1996. Evidence based medicine: what it is and what it isn't.

Connie Schardt, Martha B Adams, Thomas Owens, Sheri Keitz, and Paul Fontelo. 2007. Utilization of the pico framework to improve searching pubmed for clinical questions. *BMC medical informatics and decision making*, 7(1):16.

Zichao Yang, Diyi Yang, Chris Dyer, Xiaodong He, Alex Smola, and Eduard Hovy. 2016. Hierarchical attention networks for document classification. In *Proceedings of the 2016 Conference of the North American Chapter of the Association for Computational Linguistics: Human Language Technologies*, pages 1480–1489.

Eya Znaidi, Lynda Tamine, and Chiraz Latiri. 2015. Answering pico clinical questions: a semantic graph-based approach. In *Conference on Artificial Intelligence in Medicine in Europe*, pages 232–237. Springer.

Coding Structures and Actions with the COSTA Scheme in Medical Conversations

Nan Wang[†‡♯], **Yan Song**[♠], **Fei Xia**[‡]
[†]University of California-Los Angeles, CA, USA
[‡]University of Washington, WA, USA
[♯]Hunan University, Hunan, China
[♠]Tencent AI Lab
nwang3@ucla.edu

Abstract

This paper describes the COSTA scheme for coding structures and actions in conversation. Informed by Conversation Analysis, the scheme introduces an innovative method for marking multi-layer structural organization of conversation and a structure-informed taxonomy of actions. In addition, we create a corpus of naturally occurring medical conversations, containing 318 video-recorded and manually transcribed pediatric consultations. Based on the annotated corpus, we investigate 1) treatment decision-making process in medical conversations, and 2) effects of physician-caregiver communication behaviors on antibiotic over-prescribing. Although the COSTA annotation scheme is developed based on data from the task-specific domain of pediatric consultations, it can be easily extended to apply to more general domains and other languages.

1 Introduction

Conversational understanding has been investigated for long by various fields of study such as philosophy of language (Austin, 1962; Searle, 1969, 1985; Wittgensein, 1953), sociology (Schütz, 1967; Sacks, 1992; Garfinkel, 1967; Goffman, 1983), and artificial intelligence (Grosz and Sidner, 1986; Core and Allen, 1997; Perrault. and Allen, 1980; Pollack, 1986) .

Conversational structures are at the heart of the inquiry. Drawing from the philosophical and sociological views of conversational understanding (Schütz, 1967; Wittgensein, 1953; Weber, 1991), Conversation Analysis (CA) was developed to study the systematic organization of conversation and answer the question: *'How is conversation made possible?'*(Heritage, 1984; Schegloff, 2007; Sacks et al., 1974). In artificial intelligence, researchers also explored various theories and practices in analyzing conversation structures, based on which intelligent dialog systems can be developed to assist human with various types of tasks (Core and Allen, 1997; Carletta et al., 1997; Grosz and Sidner, 1986; Jurafsky et al., 1997; Stolcke et al., 2000; Mayfield et al., 2014). In medicine, research shows that a thorough understanding of physician-patient communication structure is important for delivering quality health care and achieving optimal health outcomes (Heritage and Maynard, 2006; Zolnierek and Dimatteo, 2009; Stivers, 2007).

Despite the enormous contribution that existing research has made to advance our knowledge in conversational structures and understanding, limitations exist and opportunities stand for future research. For CA, although the theory and practices of analyzing conversational structures and actions exist, there has not been any synthesized scheme to analyze the hierarchical structure of complete conversations; nor is there any corpus in which such information is annotated. In artificial intelligence, although existing studies recognized the role of structures and actions in conversation understanding and developed annotation schemes to code such information, most of them has only implemented structural annotations at a shallow layer. Moreover, due to a lack of appropriate language resources and tools, research on medical communication in clinical setting remains limited.

Motivated by these challenges, we propose COSTA (COnversational STructures and Actions) – a scheme for coding hierarchical structures and actions in conversations, and a corpus of medical conversation with such annotations.

76

Proceedings of the BioNLP 2018 workshop, pages 76–86
Melbourne, Australia, July 19, 2018. ©2018 Association for Computational Linguistics

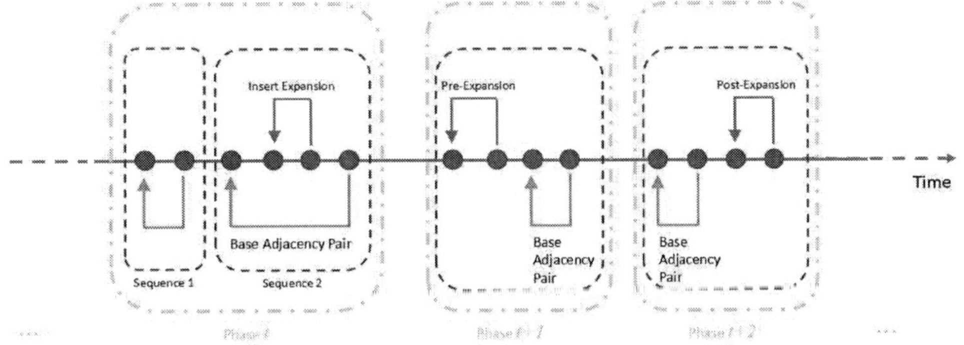

Figure 1: A schematic representation of hierarchical structure of conversation. Blue nodes are turns following a chronological order (the horizontal axis). The arrows link two turns in an adjacency pair. Base adjacency pairs are marked by green arrows; adjacency pairs in sequence expansions are marked by gray arrows. Sequences are marked by blue boxes, and phases are marked by yellow boxes.

2 Conversation Analysis

The COSTA scheme is informed by the sociological theory of conversation analysis (CA). Although CA resembles discourse structure theories such as Rhetorical Structure Theory (RST) (Mann and Thompson, 1988) and that of Penn Discourse Tree Bank (PDTB) (Prasad et al., 2008) in a sense that utterances are considered as structurally organized, what distinguishes CA is that its theory is based on dialogic text rather than monological text (e.g., news articles, academic articles, etc.). Conversation is viewed as organized with 'interaction' orders (Weber, 1991); by contrast, monological text does not take into account recipients' reactions in its immediate context. This means that these two types of discourse are distinctively different and might need to be analyzed with different structural frameworks.

Using naturally occurring conversational data, CA aims to investigate the methods and resources that participants systematically use and rely on to produce intelligible actions and make sense of each other (Heritage, 1984).

Two of the major dimensions of CA involve *sequence organization* and *action formation*. *Sequence organization* addresses questions such as how successive turns are formed up to be 'coherent' with the prior turn, and relatedly, how the overall composition of a conversation gets structured, what those structures are, and how the placement in the overall structure informs the construction and understanding of the talk (Schegloff, 2007). *Action formation* refers to the problem as to how the resources of language, the body, the environment of the interaction, and position in the interaction are fashioned into conformations designed to be and recognizable by the recipient as particular action (Schegloff, 2007).

2.1 Conversational Structures

In CA, structural organization of conversation can be conventionally analyzed at three layers (Schegloff, 2007).

(1) *Turn*: Turns are segmented at each change of speakership. A turn is analyzed in terms of how it is designed to implement some social actions (e.g., a question, a proposal, etc. (Drew, 2014)).

(2) *Sequence organization*: Sequence organization examines how successive turns are formed up to be 'coherent' with the prior turn to accomplish some courses of social actions (e.g., question-answer, proposal-acceptance, greeting-greeting, etc. (Schegloff, 2007)). Relatedly, *adjacency pairs* are the most basic unit of sequence organization (Schegloff, 2007). The idea is that social actions are produced to either initiate a possible sequence of action or to respond to an already initiated action (Stivers, 2014). By initiating a sequence of actions, social actors impose a normative obligation on co-interactants to provide a type-fitted response at the first possible opportunity (Stivers, 2014; Sacks, 1992). Yet, an adjacency pair may, but need not, be expanded, with one or multiple forms of sequence *expansions* (i.e., pre-, insert, and post-expansion) (Schegloff, 2007; Stivers, 2014). Therefore, a cluster of turns in conversation can be analyzed as to whether they form up a coherent sequence with one base adjacency pair and its expansions.

(3) *Overall organization*: A single conversation is viewed as conducted to accomplish some social

TID	APP	SL	PS	PR	Speech Text	Action	Outcome
58	1-B	0	P4	D	感冒 了 . (He's) got a cold.		
59	2-B	58		M	嗯 . Ok.		
60	1-B	0	P5	D	吃 点 药 吧 ? Take some oral medicine, ok?	B2	
61	1	60		M	啊 ? Huh?	C0	
62	2	61		D	先 吃 点 药，好 吧 ? Take some oral medicine first, ok?		
63	2-B	60		M	哦，好 . Oh, ok.		
64	3-B	60		D	嗯 . Ok.		
65	1-B	0	P5	M	开 点 青霉素 给 我们 好吧? Could you prescribe us some Penicillin?	A1	
66	2-B	65		D	嗯，行 . Ok. Alright.		D1

Table 1: An example of annotated excerpt. TID: Turn ID; PR: Participant Role (M: Mother, D: Doctor); APP: Adjacency Pair Part; SL: Sequence Link; PS: Phase; Action: Conversational Action; Outcome: Prescribing Outcome. Note: Prescribing outcome is annotated at the last turn of a conversation. D1 is included in this table for illustration purpose.

activities, and the social activities can be viewed as involving multiple, normatively ordered sequences of actions (Sacks, 1992; Robinson, 2014; Sacks and Schegloff, 1973). For example, the social activity of telling a trouble to a friend usually involves approaching, arriving at, delivering, working up, and exiting from the trouble (Jefferson, 1988); the activity of dealing with acute medical concerns involves presenting, gathering information about, diagnosing and treating the concern in the American primary care settings (Robinson, 2003, 2014).

Based on the CA theory, we synthesize and formalize the CA analytical practices by developing an annotation scheme of conversational structures which has the following four layers:

(1) Turn: A conversation is segmented at each change of speakership. A turn consists of all its construction units before the speakership changes.

(2) Adjacency Pair: Turns are analyzed as to whether they form up adjacency pairs. An adjacency pair has two parts - a first pair part (FPP) initiates an action, and a second pair part (SPP) responds to a FPP action. For instance, FPP could be a request and SPP is a response to the request.

(3) Sequence of Actions: Turns are also analyzed as to whether they form up a sequence of actions. A sequence is composed of a base adjacency pair and zero or more expansions.

(4) Phase: At a highest level, a conversation may consist of several ordered phases. For instance, a medical conversation may include phases for history taking, diagnosis, treatment, etc. A phase consists of one or more sequences of actions.

This hierarchical structural organization of conversation is illustrated in Figure 1. In this figure, blue nodes are turns in a conversation in a chronological oder. Yellow boxes represent phases in a conversation; blue boxes represent sequences. Within a sequence, an arrow links two turns in an adjacency pairs - green arrows represent base adjacency pairs, whereas gray arrows represent that the adjacency pairs are expansions of a base adjacency pair.

These concepts can be further illustrated with the examples in Table 1. Table 1 is a short excerpt of a medical conversation in which the physician and the mother are engaged in an activity of dealing with the patient's acute respiratory tract infection symptoms.

Phase: The excerpt contains two phases in a medical conversation: Turns 58-59 belong to a diagnosis phase, in which a diagnosis of the patient's condition is provided and received; Turns 60-66 are part of a treatment phase, in which a treatment recommendation is offered and accepted. Note that a phase can contain multiple sequences. For example, there are two sequences in Turns 60-66, in which two treatment recommendations are offered and received (Turns 60-63 and 64-66, respectively).

Sequence: The example contains three sequence. Two of them, Turns 58-59 and 65-66 each contain only one adjacency pair. The third one, Turns

60-64, contains two adjacency pairs: Turns 60-63-64 is the base adjacency pair;[1] Turns 61-62 is an insert expansion of the bases pair, as the mother and the physicians deal with repairing a hearing problem with the physician's turn (Schegloff et al., 1977).

Adjacency pair and Turn: Each Chinese line in Table 1 is a turn, and they form multiple adjacency pairs. For instance, Turn 65-66 forms an adjacency pair, where the mother initiates a request for a Penicillin prescription in Turn 65 and the physician grants it in Turn 66, thereby fulfilling the expectation set up by the request.

2.2 Conversational Actions

Definition of action has long been of considerable interest to many fields. In CA, the central sense of action is the ascription and assignment of 'a main job' that the turn is doing (*i.e., what the response must deal with in order to count as an adequate next turn; whether the turn fits to the overall contextual environment or not*) (Levinson, 2014).

The structural placement of a turn thus is essential for action recognition and ascription in conversations. First, action ascription is informed by the sequential position of a turn in a local adjacency pair (e.g., question-answer, offer-acceptance). A first pair part (FPP), by projecting a matched second pair part (SPP), maps an action onto the second. Thus, the same utterance might be understood as different actions by virtue of its location. For example, Turn 58 *'He's got a cold.'* in Table 1 is understood as delivering a diagnosis, rather than providing an account, because of its sequential context as being an initiating action in the diagnosis phase, rather than an answer responding to a question (e.g., *'Why is him not here today?'*). In sum, CA views the positioning of an utterance in the ongoing conversation as fundamental to the understanding of its meaning as performing some actions. Social actors rely on their shared knowledge or commensense about the sequential context to make sense of each other. This structure-informed theory about conversational actions thus distinguishes CA from other approaches such as Speech Act Theory, which exclusively focuses on the surface composition of an utterance.

In this study, we use this structure-informed taxonomy of action to identify the conversational actions that are hypothesized to affect the prescribing decision outcome of the medical visits. This will be explained in Section 3.3.

Item	Number
Total Number of Visits	318
Total Number of Hospitals	5
Total Number of Physicians	9
Total Number of Patients	318
Average length of a visit	4.9 minutes
Total length of the recordings	26 hours

Table 2: Statistics of the raw data. Total number of patients are calculated by those accompanied by caregivers.

3 Corpus Construction and Annotation Scheme

How do we annotate structures and actions in conversation? In this section, we describe the corpus that we constructed for the study and the annotation procedure of the COSTA scheme.

3.1 Video-recording and Transcription

We created a corpus containing 318 medical conversations between pediatricians and patients/caregivers, collected from five hospitals in China in 2013.

Raw Data: The raw data are video-recordings of the medical conversations. Due to its pediatric setting, the conversations were mostly between physicians and patients' caregivers. We call each conversation (i.e., a recording of a complete medical visit) a *visit*. Table 2 shows raw data statistics.

Transcribing: The video-recordings were transcribed to capture both what was said and how it was said in the conversation. The conversation was segmented into turns at each speakership change in two passes. The first pass transcribed the verbatim words of a turn; the second pass transcribed speech production features (e.g., intonations, overlapping, etc.), as well as non-verbal activities (e.g., nodding, coughing, etc.). Example of the transcript is in the *Speech Text* column in Table 1. Details of the transcribing symbols are described in (Jefferson, 2004).

Five undergraduate students and one graduate student transcribed the data. Each conversation was transcribed by two annotators and verified by a third. The inter-annotator agreement was 91%.[2]

---

[1]While an adjacency pair typically contains two turns, there are exceptions such as the one here, where Turn 64 is a sequence closing third turn (see Section 3.2).

[2]The character error rate was 8.9% when one transcript was treated as the reference and the other as system output.

Ethical Consideration: Research procedures were reviewed and approved by the UCLA IRB (Ref: IRB#13-000748). All identifiable information were removed.

3.2 Structure Annotations

To annotate structures in conversation, we create five attributes: Turn ID (TID), Participant Role (PR), Adjacency Pair Part (APP), Sequential Link (SL), and Phase (PS). The first four are at the turn level, and the last one is at the sequence level.

TID (Turn ID) is a sequential number automatically assigned to a turn, indicating the temporal position of the turn in a conversation.

PR (Participant Role) marks the speakership of a turn, using labels from a pre-defined label set (which is task-specific). For example, in Table 1, Label D stands for Doctor, and M stands for Mother. The PR label is particularly informative when there are more than two participants.

APP (Adjacency pair part) marks the position of a turn in an adjacency pair and it normally has one of the two values:

- *1* marks an initiating action (FPP).
- *2* marks a responding action (SPP).

This can be illustrated in Table 1, lines 58-59. In addition to 1 and 2, APP can have other values:

- *0* marks a turn occupied by a noticeable silence or some non-verbal activities.
- *3* marks a turn as 'sequence closing third (SCT)'. SCT is in fact a minimal form of post-expansion of an adjacency pair, indicating that no further talk is projected beyond this turn. However, it is ritually used and viewed as part of the base adjacency pair, making it a three-part exception of the adjacency pair (Schegloff, 2007). For example, in Table 1 (lines 60–64), a treatment recommendation is delivered at line 60 and accepted at line 63. This sequence can be considered as completed with the second pair part turn fulfills the expectation of the first pair part. Following this, the physician produces an acknowledgment token *'ok'* at line 64, indicating no further talk projected related to the sequence. This turn is thus marked as 3 in the APP attribute.

Although a sequence is ideally composed of a two-part adjacency pair (the minimal form), it can

be and is usually expanded, and thereby consist of one base adjacency pair and one or more expansions.

To distinguish a base adjacency pair from its expansions, we attach label B to the APP value of the turns in the base adjacency pair, such as the pair formed by Turns 58-59.

Given that an adjacency pair can be expanded with other turns (e.g., by an insert expansion) and some adjacency pairs can be incomplete (e.g., a question is not answered), APP labels alone will not be sufficient to indicate which turns form an adjacency pair and which adjacency pairs form a sequence. The *SL* attribute is created to solve this problem.

SL (Sequential link) is a pointer to another turn in the same sequence, indicating the dependency-like relation between two turns. The SL values are set according to the following rules:

- *Rule 1*: In an adjacency pair, the non-FPP (e.g., SPP and SCT) always points to its corresponding FPP. That is, the SL value of an non-FPP turn of an adjacency pair is the TID of its corresponding FPP.
- *Rule 2(a)*: The base adjacency pair in a sequence is like the root of a dependency structure; therefore, the SL of the FPP of the base adjacency pair is set to 0.
- *Rule 2(b)*: If a sequence includes any forms of expansion, the expansion pair 'depends' on the base pair; therefore, the SL value of the FPP of an expansion pair is the TID of the FPP of the base adjacency pair.

To illustrate an example of a sequence with an insert sequence, we can look at Turns 60–64 in Table 1. At Turn 60, the physician initiates a recommendation, which sets up an expectation for the mother's acceptance. However, the mother displays a hearing problem before she finally accepts it at Turn 63. In this sequence, Turns 60 and 63 are FPP and SPP of the base adjacency pair, respectively; Turns 61 and 62 are FPP and SPP of an insert expansion of the base adjacency pair. The SL values of Turns 62-64, 60, and 61 are set according to Rule 1, 2(a), and 2(b), respectively.

Note that although not shown in Table 1, expansion adjacency pair can possibly be further expanded with its own expansions. In such cases, the rules above still apply. As a result, the conversational structure of a sequence is a tree, and it is very similar to the dependency structure for a sentence: the SL attribute is just like the dependency arc, indicating the dependency of the non-FPP turns on FPP turns and that of the FPP of an expansion pair on the FPP of the base adjacency pair. While we are not using dependency type on the arc, the type can be easily inferred from the APP attribute including label -B.

PS (Phase) indicates the nature of sequence (i.e., what phase a sequence belongs to) in a conversation, and it is marked at the first turn in a sequence.

The labels for PS are task-specific and the ones that we used for this corpus are: *P0:* Consultation opening, *P1:* Problem presentation, *P2:* History taking, *P3:* Physical examination, *P4:* Diagnosis, *P5:* Treatment, *P6:* Addressing additional concerns, *P7:* Consultation closing.

In Table 1, Turn 58 is the start of a sequence of actions for delivering diagnosis, thus its PS label is P4. Similarly, Turn 60 and Turn 65 are the start of two action sequences of physician's treatment recommendations, thus their PS labels are both P5. Note that phases can go back and forth. Therefore, a P7 label can precede a P6 Label.

In sum, PS marks the natures of and boundaries of sequences; SL marks the relations of turns within a sequence (similar to a dependency tree); and APP indicates the role of a turn within an adjacency pair.

Based on the CA theory, this multi-layer structure annotation scheme is not only salient in indicating a turn's position in a conversation, but also important for determining the type of action that a turn undertakes (Stivers, 2014; Schegloff, 2007; Sacks, 1992). The hierarchical structural information thus forms a fine-grained contextual constraint for the way of a turn in conversation can be understood. Therefore, by incorporating our shared knowledge or commonsense about the context of a turn in conversation, the COSTA annotation scheme is capable of dealing with problems such as comprehending indirect speech actions, as it no long relies on the surface composition of a turn to classify its action type.

As this is preliminary work, we used code-recode procedure to test the agreement of the structural and action annotations. The overall agreement achieved 94.43% among the APP, SL, and PS attributes[3].

3.3 Task-specific Annotations

Besides examining conversational structures, we also examine the decision-making process of antibiotic treatment in the specific clinical context of pediatric consultations. This task is motivated by the fact that antibiotic over-prescribing and bacterial resistance is a big global public health crises today, and the problem is particularly severe in China in the pediatric settings (Li et al., 2012; Laxminarayan et al., 2013).

Several kinds of physician-patient/caregiver conversational actions are annotated, as well as prescribing outcome of the visits. For example, the task-specific annotations are marked on the last two columns of Table 1, and explained below:

Caregivers' advocacy for antibiotics (A) is marked in the turn where a caregiver advocates for antibiotic treatment in the medical visits. This attribute has four possible values, indicating a varying degree of overtness of the advocating actions:

- *A1:* Explicit request for antibiotics (e.g., *Can you prescribe me some antibiotics?*)
- *A2:* Statement of desire for antibiotics (e.g., *Her mother wants to put her on antibiotics.*)
- *A3:* Inquiry about antibiotics (e.g., *Does he need antibiotics?*)
- *A4:* Evaluation of treatment effectiveness (e.g., *Antibiotics always work well for her.*)

Physicians' treatment recommendation (B) is used for a turn where physician makes a treatment recommendation. This attribute has three possible values, indicating a varying degree of physician authoritarian style in delivering the treatment recommendation.

- *B1:* Pronouncement (e.g., *She has to take some antibiotics now.*)
- *B2:* Proposal (e.g., *How about we put her on antibiotics?*)
- *B3:* Offer (e.g., *If you'd like, I can prescribe you some antibiotics.*)

Response to treatment recommendation or **Response to antibiotic advocacy (C)** is used for a turn if it contains a response to either an antibiotic treatment advocacy (A) or a treatment recommendation (B). Such a turn normally appears immediately after a turn with an A or B action. Two possible values are: *C1:* Acceptance and *C0:* Non-acceptance [4].

---

[3]Since PR is assigned during the transcribing process and

TID is assigned automatically after the transcribing process, they were excluded from the test.

[4]Partial or full rejection are annotated as non-acceptance.

Item	Total	Avg.
Characters	468,162	1472.2
Words	270,042	849.2
Turns	39,216	123.3
Non-verbal turns	5,815	18.3
Adjacency pairs	20,123	63.3
Sequences	9114	28.7

Table 3: Statistics of the annotated corpus. Total number of visit is 318. Avg. refers to that average number is calculated per visit.

Prescribing Outcome (D) marks whether antibiotics are prescribed in a visit. This label is annotated at the end of the conversation as a derived result. It has two possible values: *D1:* Antibiotic treatment and *D0:* Non-antibiotic treatment .

The overall code-recode agreement of the task-specific annotations achieved 97% among the four types of behavior[5].

4   Results

In this section, we first present basic statistics of our corpus; next, we report our findings on 1) the process of treatment decision-making in medical consultation, and 2) the association between physician-patient/caregiver conversational behaviors and antibiotic prescribing outcome in medical consultations.

4.1   Corpus Statistics

In total, our corpus contains 318 manually transcribed conversations, among which 187 are acute visits and 131 are follow-up visits. Table 3 summarizes the statistics of the corpus. The corpus contains nearly 40K turns with 470K Chinese characters, which is considerably large in terms of manually annotated natural human conversations. The Chinese sentences are then automatically word segmented with an in-house CRF model. On average, each visit has three participants (physician might talk to more than one caregiver), and the turns in a visit form 63 adjacency pairs, which in turn form 29 action sequences.

4.2   Treatment Decision-making Process

To investigate the process of treatment decision-making in medical consultation, we focus on the interactive process in which a physician's treatment recommendation is accepted. We found that a physician's treatment recommendation is

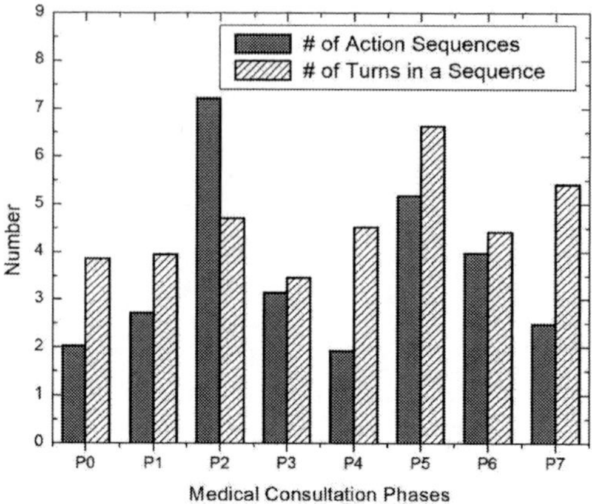

Figure 2: Average number of action sequences and average number of turns in a sequence in medical consultation phases. *P0*: Opening *P1*: Problem presentation *P2*: History taking *P3*: Physical exams *P4*: Diagnosis *P5*: Treatment *P6*: Addressing additional concerns *P7*: Closing.

not always immediately accepted in the next turn; rather, it can be resisted or rejected by a patient or caregiver. In doing so, the patient or caregiver has the opportunity to negotiate for a treatment that is in line with their own wants. As a result, this could lead to rather expanded treatment recommendation action sequence shapes.

After examining our corpus, we found that physicians' treatment recommendations are resisted by caregivers 41% of the time. On average, a treatment recommendation action sequence takes 6.63 turns for its completion. In comparison, other actions in a medical consultation are usually less expanded. A history-taking action sequence takes 4.70 turns, and a problem presentation action sequence takes 3.95 turns to complete on average.

In our corpus, the average number of turns of an action sequence is the greatest in treatment phase (P5) throughout all phases in medical consultation. Figure 2 shows the distribution of the average number of turns for an action sequence and average number of actions in each phase of medical consultations. The long sequence suggests that the treatment phase is where communication problems (understanding or accepting physician's recommendations) are most likely to occur.

4.3   Association between Conversational Behaviors and Antibiotic Prescribing

From the annotated corpus, we can collect various statistics to study the association between physician/caregiver behavior and antibiotics prescribing outcome. Table 4 shows the distribution of advo-

---

[5]See Chilisa and Preece (2005) for details of the code-recode strategy. The overall code-recode agreement was calculated based on the average of the four task-specific labels

Advocating Action Type	# of Visits	% of Visits
A1 Requests	10	5.3
A2 Statements	14	7.5
A3 Inquiries	50	26.7
A4 Evaluations	26	13.9
Total	100	54.0

Table 4: Frequency and distribution of caregiver advocacy for antibiotics. # of Visit refers to number of visits in which advocacy is observed. % of Visit refers to proportion of visits in which advocacy is observed out of 187 visits.

Prescriptions	V.w.A.	V.w/o.A.	Total	%
Antibiotics	72	39	111	59.4
Non-antibiotics	28	48	76	40.6
Total # of Visits	100	87	187	100.0
%	53.5	46.5	100.0	

Table 5: Frequency and distribution of prescribing outcomes by occurrence of advocacy in number of visits. V.w.A.:visits with advocacy; V.w/o.A.:visits without advocacy. The rightmost column shows the percentage of antibiotic prescriptions out of a total of 187 visits. The bottom row shows the percentage of visits in which caregiver advocacy is observed out of the 187 visits.

cating actions that Chinese caregivers use to advocate for antibiotics. Table 5 shows the distribution of antibiotics prescribing outcomes by occurrence of caregivers' advocacy for antibiotics. The result reveals that caregiver advocacy for antibiotic treatment is significantly associated with antibiotic prescribing outcome. What is more troubling about this finding is that while caregiver advocacy for antibiotic treatment occurred in 54% of the acute visits in our corpus, similar kind of caregiver advocacy for antibiotics were observed only 9% of the time in the similar setting of American pediatric consultations (Stivers, 2002).

In addition, we found that physicians tend to use less authoritarian styles of treatment recommendations (i.e., B2 and B3 combined) than more authoritarian ones (i.e., B1). Table 6 shows distribution of the three types of treatment recommendation actions in the Chinese pediatric context. Moreover, in response to caregivers' advocacy for antibiotic treatment, physicians more frequently resist it than grant it, as shown in Table 7. These findings indicate that physicians play a less dominant role in antibiotic over-prescribing in the medical visits; in contrast, caregivers have a significant influence on the prescribing outcomes.

Multivariate logistic regression results reveal that caregiver advocacy for antibiotic treatment significantly increases the likelihoods of antibiotic prescribing in a visit – caregivers' advocacy was associated with 9.23 times increased likelihoods of antibiotic prescription (Odds Ratio (OR) = 9.23, 95% Confidence Interval(CI): 3.30-33.08); whereas physician's response to caregivers' advocacy has a significant effect on the prescribing outcome – physicians' resistance to caregivers' advocacy reduced the likelihoods of antibiotic prescriptions by 77% (OR=0.23, 95%CI: 0.06-0.68), controlling for the socio-demographic variables in our model.

5   Discussion

Conversational structures have been recognized as critical for conversational understanding in both sociology and artificial intelligence. Although past research has made enormous contributions to this important inquiry, no annotation scheme exists, with which the hierarchical structural organizations of conversation can be captured. Motivated by this gap, we developed the COSTA and created a corpus annotated with this scheme.

5.1   Related Theories and Schemes

Informed by Conversation Analysis (CA), the theoretical framework of the COSTA annotation scheme is largely in line with the existing discourse structure theories and annotation schemes. Although the existing theories have recognized that utterances in conversation have higher-level forms of hierarchical structures (Grosz and Sidner, 1986; Carletta et al., 1997), most have only implemented annotations of conversational structures at turn level and between a pair of turns (e.g., by distinguishing Forward Communicative Function and Backward Communicative Function (Core and Allen, 1997; Jurafsky et al., 1997)).

In addition, the COSTA annotation scheme also presents an innovative method for annotating actions in conversation. Most of the existing annotation schemes of dialog acts for conversations (Core and Allen, 1997; Jurafsky et al., 1997; Stolcke et al., 2000) and particularly, for medical dialogue (Hoxha et al., 2016; Mayfield et al., 2014) were based on Speech Act Theory (SAT); however, the SAT has long been criticized for being difficult in dealing with indirect dialog acts. Different from the SAT, the CA theory considers the sequential position of a turn as critical for action recognition and ascription. The COSTA annotation scheme thus 1) allows multi-layer annotation

Action Type	# of Visits	%
B1 (Pronouncements)	61	41.5
B2 (Proposals)	65	44.2
B3 (Offers)	21	14.3
Total	147	100

Table 6: Frequency and distribution of physicians' treatment recommendations, by three recommendation actions. Percentage of action is out of a total of 147 visits, in which physicians rather than caregivers initiated treatment discussions.

Response to Advocacy	Visit	%
C0 (Non-acceptance)	65	65.0
C1 (Acceptance)	35	35.0
Total	100	100

Table 7: Frequency and distribution of physicians' response to caregiver advocacy for antibiotics. Percentage is out of a total of 100 visits where caregiver advocacy is observed.

at a turn, and 2) depends on the multi-layer structural annotations of a turn for action taxonomy. It thus offers great flexibilities in annotating indirect actions.

5.2 Applications to Different Domains

The COSTA annotation scheme can be used for both general domains and for task-specific domains. While the values for TID, APP, and SL are likely to remain the same for different domains, the values for PR and PS and additional attributes such as A-D labels as described in Section 3.3 are task-specific. In addition,, because the CA theory about conversational structures and actions applies to both ordinary conversation and task-specific conversation, we believe that the same scheme with slight customization (e.g., using a different label set for PS) can accommodate analysis of conversational structures and actions in other task-specific service settings such as airliner hotlines, 911 call centers, etc. Furthermore, since social norms underlying conversations do not tend to vary significantly across cultures, the COSTA scheme can be applied to languages other than Chinese.

5.3 Applications of the Corpus

Although research in medicine has long been concerned with effective communication between physicians and patients, related language resources are still lacking. Our corpus is one of the first to have multi-layer structure annotations of complete natural conversations, in the task-specific setting of physician-patients/caregivers medical consultations.

The findings regarding structural shape of a typ-

ical medical consultation and the process through which a treatment decision is made can be applied to research and practices in medicine and beyond. For example, communication effectiveness can be improved by focusing on phases that are identified as critical in medical consultations (e.g., treatment phase in which sequences are most expanded). In addition, intervention programs can be developed to reduce antibiotic over-prescribing by training physicians to resist caregivers' pressure more effectively. Moreover, the rich information of the corpus can be valuable for building intelligent dialogue system for applications in clinical setting (Campillos et al., 2016).

6 Conclusion

In this paper, we propose a general scheme for annotating multi-layer conversational structures and actions and use that scheme to build a corpus of medical conversations in Chinese pediatric settings. First, our work extends the theory and practice of the sociological field of conversation analysis (CA) by creating an annotation scheme for coding conversational structures and actions. Second, we create a corpus of naturally occurring conversations between physicians and caregivers. The corpus can be used not only for research of general purposes such as conversational understanding, modeling human social behavior of cooperation and coordination, but also for more specific purposes such as identifying risk factors for antibiotic prescribing. Third, we demonstrate that conversational behavior indeed affects medical decisions. We hope our findings can be used to train physicians for effective communication.

For future work, we want to test the usefulness of the scheme in other domains. In addition, we plan to extend COSTA to mark turn construction unit (TCU) [6] . We plan to release the dataset once it is completed.

References

John L. Austin. 1962. *How to Do Things with Words*. William James Lectures. Oxford University Press.

Leonardo Campillos, Dhouha Bouamor, Eric Bilinski, Anne-Laure Ligozat, Pierre Zweigenbaum, and So-

---

[6]In this version, we treated turn as the smallest unit of conversation, but as discussed in (Sacks et al., 1974), a turn may consist of several TCUs (e.g., one TCU answers the question in the previous turn, and another TCU initiates a question).

phie Rosset. 2016. Integrating a dialogue system into a virtual patient consultation.

Jean Carletta, Amy Isard, Stepen Isard, C. Jacquelin Kowtko, Gwyneth Doherty-Sneddon, and Anne H. Anderson. 1997. The Reliability of a Dialogue Structure Coding Scheme. *Computational Linguistics* 23(1):13–31.

Bagele Chilisa and Julia Preece. 2005. *Research methods for adult educators in Africa.* UNESCO Institute for Education; Pearson Education, Hamburg, Germany; Cape Town, South Africa.

Mark Core and James Allen. 1997. Coding Dialogs with the DAMSL Annotation Scheme. In *Proceedings of AAAI Fall Symposium on Communicative Action in Humans and Machines.*

Paul Drew. 2014. Turn Design. In J. Sidnell and Tanya Stivers, editors, *The Handbook of Conversation Analysis*, John Wiley & Sons, West Sussex, UK, pages 131–149.

Harold Garfinkel. 1967. *Studies in Ethnomethodology.* Prentice-Hall Inc.

Erving Goffman. 1983. The interaction order: American sociological association, 1982 presidential address. *American Sociological Review* 48(1):1–17.

Barbara Grosz and Candace Sidner. 1986. Attention, Intentions, and the Structure of discourse. *Computational Linguistics* 12(3):175–204.

John Heritage. 1984. A Change-of-State Token and Aspects of its Sequential Placement. In J. Maxwell Atkinson, , and John Heritage, editors, *Structures of social action: Studies in Conversation Analysis*, Cambridge University Press, Cambridge, U.K., chapter 13, pages 299–345.

John Heritage and Douglas W. Maynard. 2006. *Communication in Medical Care: Interaction between Primary Care Physicians and Patients.* Cambridge University Press, Cambridge.

Julia Hoxha, Praveen Chandar, Zhe He, James Cimino, David Hanauer, and Chunhua Weng. 2016. DREAM: Classification Scheme for Dialog Acts in Clinical Research Query Mediation. *Journal of Biomedical Informatics* 59:89–101.

Gail Jefferson. 1988. On the sequential organization of trouble-talk in ordinary conversation. *Social Problems* 35(4):418–441.

Gail Jefferson. 2004. Glossary of transcript symbols with an introduction. In Gene H. Lerner, editor, *Conversation Analysis: Studies from the First Generation*, John Benjamins, Amsterdam / Philadelphia, chapter 2, pages 13–31.

Daniel Jurafsky, Elizabeth Shriberg, and Debra Biasca. 1997. Switchboard SWBD-DAMSL Shallow-Discourse-Function Annotation Coders Manual, Draft 13. Technical report, University of Colorado, Boulder.

R. Laxminarayan, A. Duse, C. Wattal, AK. Zaidi, HF. Wertheim, V. Sumpradit, E. Vlieghe, GL. Hara, IM. Gould, H. Goossens, C. Greko, AD. So, M.Bigdeli, G. Tomson, W. Woodhouse, E. Ombaka, AQ. Peralta, FN. Qamar, F. Mir, S. Kariuki, ZA. Bhutta, A. Coates, R. Bergstrom, GD. Wright, ED. Brown, and O. Cars. 2013. Global antibiotic consumption 2000 to 2010: an analysis of national pharmaceutical sales data. *Lancet Infectious Diseases* 13(12):1057–1098.

Stephen Levinson. 2014. Action Formation and Action Ascription. In J. Sidnell and Tanya Stivers, editors, *The Handbook of Conversation Analysis*, John Wiley & Sons, West Sussex, UK, pages 103–130.

Yongbin Li, Jing Xu, Fang Wang, Bin Wang, Liqun Liu, Wanli Hou, Hong Fan, Yeqing Tong, Juan Zhang, and Zuxun Lu. 2012. Overprescribing in china, driven by financial incentives, results in very high use of antibiotics, injections, and corticosteroids. *Health Affairs (Project Hope)* 31(5):1075–1082.

William C. Mann and Sandra A. Thompson. 1988. Rhetorical Structure Theory: toward a Functional Theory of Text Organization. *Text: Interdisciplinary Journal for the Study of Discourse* 8(3):243–281.

Elijah Mayfield, M Barton Laws, Ira B. Wilson, and Carolyn Penstein Ros'e. 2014. Automating annotation of information-giving for analysis of clinical conversation. *Journal of American Medical Informatics Association* 21(e1):e122–e128.

Raymond C. Perrault. and James Allen. 1980. A Plan-based Analysis of Indirect Speech Acts. *Computational Linguistics* 6(3):167–182.

Martha Elizabeth Pollack. 1986. *Inferring Domain Plans in Question–answering.* Ph.D. thesis, Philadelphia, PA, USA. UMI order no. GAX86–14850.

Rashmi Prasad, Nikhil Dinesh, Alan Lee, Livio Robaldo Aravind Joshi Eleni Miltsakaki, and Bonnie Webber. 2008. The Penn Discourse Treebank 2.0. In *Proceedings of the 6th International Conference on Language Resources and Evaluation (LREC).*

Jeffery D. Robinson. 2003. An interactional structure of medical activities during acute visits and its implications for patient's participation. *Health Communication* 15(1):27–57.

Jeffery D. Robinson. 2014. Overall Structural Organization. In J. Sidnell and Tanya Stivers, editors, *The Handbook of Conversation Analysis*, John Wiley & Sons, West Sussex, UK, pages 257–280.

Harvey Sacks. 1992. *Lectures on Conversation,* volume 1 & 2. Basil Blackwell.

Harvey Sacks and Emanuel A. Schegloff. 1973. Opening up closings. *Semiotica* 8(4):289–327.

Harvey Sacks, Emanuel A. Schegloff, and Gail Jefferson. 1974. A Simplest Systematics for the Organization of Turn-taking for Conversation. *Language* 50(4, Part 1):696–735.

Emanuel A. Schegloff. 2007. *Sequence Organization in Interaction: Volume 1: A Primer in Conversation Analysis*. Cambridge University Press.

Emanuel A. Schegloff, Gail Jefferson, and Harvey Sacks. 1977. The Preference for Self-Correction in the Organization of Repair in Conversation. *Language* 53(2):361–382.

Alfred Schütz. 1967. *Phenomenonology of the Social World*. Northwestern University Press.

John R. Searle. 1969. *Speech Acts: An Essay in the Philosophy of Language*. Cambridge University Press.

John R. Searle. 1985. *Expression and Meaning: Studies in the Theory of Speech Acts*. Cambridge University Press.

Tanya Stivers. 2002. Participating in decisions about treatment: Overt parent pressure for antibiotic medication in pediatric encounters. *Social Science & Medicine* 54(7):1111–1130.

Tanya Stivers. 2007. *Prescribing under pressure: Parent-physician conversations and antibiotics*. Oxford University Press, London.

Tanya Stivers. 2014. Sequence Organization. In J. Sidnell and Tanya Stivers, editors, *The Handbook of Conversation Analysis*, John Wiley & Sons, West Sussex, UK, pages 191–209.

Andreas Stolcke, Klaus Ries, Noah Coccaro, Elizabeth Shriberg, Rebecca Bates, Daniel Jurafsky, Paul Taylor, Rachel Martin, Carol Van Ess-Dykema, and Marie Meteer. 2000. Dialogue Act Modeling for Automatic Tagging and Recognition of Conversational Speech. *Computational Linguistics* 26(3):339–373.

Max Weber. 1991. The Nature of Social Action. In W. G. Runiciman Weber, editor, *Selections in Translation*, Cambridge.

Ludwig Wittgensein. 1953. *Philosophical Investigations*. Oxford University Press.

Kelly B. Haskard Zolnierek and M. Ribin Dimatteo. 2009. Physician communication and patient adherence to treatment: A meta-analysis. *Medical care* 47(8):826–834.

A Neural Autoencoder Approach for Document Ranking and Query Refinement in Pharmacogenomic Information Retrieval

Jonas Pfeiffer[1,2]   Samuel Broscheit[1]   Rainer Gemulla[1]   Mathias Göschl[2]

[1]University of Mannheim, Mannheim, Germany
[2]Molecular Health GmbH, Heidelberg, Germany
{lastname}@informatik.uni-mannheim.de
Mathias.Goeschl@molecularhealth.com
Jonas.Pfeiffer.90@Gmail.com

Abstract

In this study, we investigate learning-to-rank and query refinement approaches for information retrieval in the pharmacogenomic domain. The goal is to improve the information retrieval process of biomedical curators, who manually build knowledge bases for personalized medicine. We study how to exploit the relationships between genes, variants, drugs, diseases and outcomes as features for document ranking and query refinement. For a supervised approach, we are faced with a small amount of annotated data and a large amount of unannotated data. Therefore, we explore ways to use a neural document auto-encoder in a semi-supervised approach. We show that a combination of established algorithms, feature-engineering and a neural auto-encoder model yield promising results in this setting.

1 Introduction

Personalized medicine strives to relate genomic detail to patient phenotypic conditions (such as disease, adverse reactions to treatment) and to assess the effectiveness of available treatment options (Brunicardi et al., 2011). For computer-assisted decision making, knowledge bases need to be compiled from published scientific evidence. They describe *biomarker* relationships between key entity types: *Disease, Protein/Gene, Variant/Mutation, Drug,* and *Patient Outcome (Outcome)* (Manolio, 2010). While automated information extraction has been applied to simple relationships — such as *Drug-Drug* (Asada et al., 2017) or *Protein-Protein* (Peng and Lu, 2017); (Peng et al., 2015); (Li et al., 2017) interaction — with adequate precision and recall, clinically actionable biomarkers need to satisfy rigorous quality criteria set by physicians and therefore call upon manual data curation by domain experts.

To ascertain the timeliness of information, curators are faced with the labor-intensive task to identify relevant articles in a steadily growing flow of publications (Lee et al., 2018). In our scenario, curators iteratively refine search queries in an electronic library, such as PubMed.[1] The information the curators search for, are biomarker-facts in the form of {*Gene(s) - Variant(s) - Drug(s) - Disease(s) - Outcome*}. For example, a curator starts with a query consisting of a single gene, e.g. $q_1 = \{PIK3CA\}$, and receives a set of documents D_1. After examining D_1, the curator identifies the variants *H1047R* and *E545K*, which yields queries $q_2 = \{PIK3CA, H1047R\}$ and $q_3 = \{PIK3CA, E545K\}$ that lead to D_2 and D_3. As soon as studies are found that contain the entities in a biomarker relationship, the entities and the studies are entered into the knowledge base. This process is then repeated until, theoretically, all published literature regarding the gene *PIK3CA* has been screened.

Our goal is to reduce the amount of documents which domain experts need to examine. To achieve this, an information retrieval system should rank documents high that are relevant to the query and should facilitate the identification of relevant entities to refine the query.

Classic approaches for document ranking, like tf-idf (Luhn, 1957); (Spärck Jones, 1972), or bm25 (Robertson and Zaragoza, 2009), and, for example, the Relevance Model (Lavrenko and Croft, 2001) for query refinement are established techniques in this setting. They are known to be robust and do not require data for training. However, as they are based on a bag-of-words model (BOW),

---

[1]https://www.ncbi.nlm.nih.gov/pubmed

Proceedings of the BioNLP 2018 workshop, pages 87–97
Melbourne, Australia, July 19, 2018. ©2018 Association for Computational Linguistics

they cannot represent a semantic relationship of entities in a document. This, for example, yields search results with highly ranked review articles that only *list* query terms, without the desired relationship between them. Therefore, we investigate approaches that model the semantic relationships between biomarker entities. This can either be addressed by combining BOW with rule-based filtering, or by supervised learning, i.e. learning-to-rank (LTR).

Our goal is, to tailor document ranking and query refinement to the task of the curator. This means that a document ranking model should assign a high rank to a document that contains the query entities in a biomarker relationship. A query refinement model should suggest additional query terms, i.e. biomarker entities, to the curator that are relevant to the current query. Given the complexity of entity relationships and the high variety of textual realizations this requires either effective feature engineering, or large amounts of training data for a supervised approach. The in-house data set of Molecular Health consists of 5833 labeled biomarker-facts, and 24 million unlabeled text documents from PubMed. Therefore, a good solution is to exploit the large amount of unlabeled data in a semi-supervised approach. Li et al. (2015) have shown that a neural auto-encoder with LSTMs (Hochreiter and Schmidhuber, 1997) can encode the syntactics and semantics of a text in a dense vector representation. We show that this representation can be effectively used as a feature for semi-supervised learning-to-rank and query refinement.

In this paper, we describe a feature engineering approach and a semi-supervised approach. In our experiments we show that the two approaches are, in comparison, almost on par in terms of performance and even improve in a joint model. In Section 2 we describe the neural auto-encoder, and then proceed in Section 3 to describe our models for document ranking and in Section 4 the models for query refinement.

2   Neural Auto-Encoder

In this study, we use an unsupervised method to encode text into a dense vector representation. Our goal is to investigate if we can use this representation as an encoding of the relations between biomarker entities.

Following Sutskever et al. (2014) Cho et al.

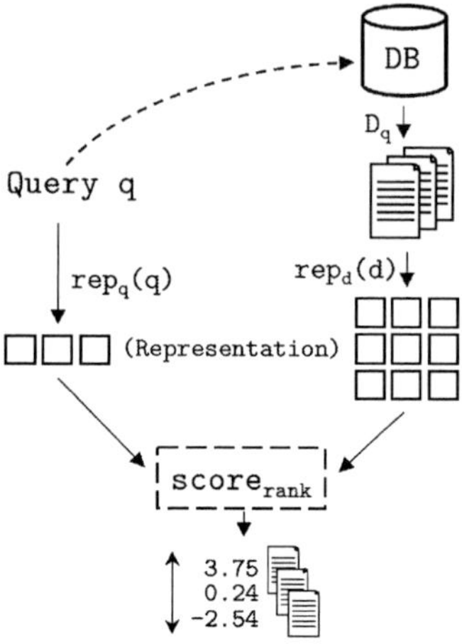

Figure 1: Document Ranking

(2014), Dai and Le (2015), and Li et al. (2015) we implemented a text auto-encoder with a Sequence-to-Sequence approach. In this model an encoder Enc produces a vector representation $\mathbf{v} = Enc(d)$ of an input document $d = [\mathbf{w}_1, \mathbf{w}_2, \ldots, \mathbf{w}_n]$, with $\mathbf{w}_i$ being word embedding representations (Mikolov et al., 2013). This dense representation $\mathbf{v}$ is then fed to a decoder Dec, that tries to reconstruct the original input, i.e. $\hat{d} = Dec(\mathbf{v})$. During training we minimize $error(\hat{d}, d)$. After training we only use the $Enc(d)$ to encode the text. We want to explore if we can use Enc to encode the documents *and* the query. We will use the output of the document encoder Enc as features for a document ranking model and for a query refinement model.

3   Document Ranking

Information retrieval systems rank documents in the order that is estimated to be most useful to a user query by assigning a numeric score to each document. Our pipeline for document ranking is depicted in Figure 1: Given a query q, we first retrieve a set of documents D_q that contain all of the query terms. Then, we compute a representation $rep_q(q)$ for the query q, and a representation $rep_d(d)$ for each document $d \in D_q$. Finally, we compute the score with a ranker model $score_{rank}$.

For rep_d we need to find a representation for an arbitrary number of entity-type combinations, because a fact can consist of e.g. 3 *Genes*, 4 *Variants*, 1 *Drug*, 0 *Diseases* and 0 *Outcomes*. In the following, we describe several of the settings for $rep_q(q)$, $rep_d(d)$ and the ranker model.

3.1 Bag-of-Words Models

We have implemented two commonly used BOW models tf-idf and bm25. For these models the text representations $rep_q(q)$ and $rep_d(d)$ is the vector space model.

3.2 Learning-to-Rank Models

For the learning-to-rank models, we chose a multilayer perceptron (MLP) as the scoring function $score_{rank}$. In the following we explain how $rep_q(q)$ and $rep_d(d)$ are computed.

Feature Engineering Model We created a set of basic features: encoding the frequency of entity types, distance features between entity types, and context words of entities. In this model, features are query dependent and are computed on-demand by a feature function $f(q, d)$.

The algorithm to compute the distance feature is as follows: Given query q with entities $e \in q$ and document $d = [w_1, w_2, \ldots, w_n]$, with w being words in the document. Let $type(e)$ be the function that yields the entity type, f.ex. $type(e) = Gene$. Then, if $e_i, e_j \in q$ and there exists a $w_k = e_i, w_l = e_j$ then we add $|l - k|$ to the bucket of $\{type(e_i), type(e_j)\}$. To summarize the collected distances we compute $min()$, $max()$, $mean()$ and $std()$ over all collected distances for each bucket separately.

For the context words feature, we collected in a prior step a list of the top 20 words for each $\{type(e_i), type(e_j)\}$ bucket, i.e. we collect words that are between $w_k = e_i$ and $w_l = e_j$ if $|k - l| <$ 10. We remove stop words, special characters and numbers from this list and also manually remove words using domain knowledge. The top 20 of remaining words for each $\{type(e_i), type(e_j)\}$ bucket are used as boolean indicator features.

Auto-Encoder Model In this model we use the auto-encoder from Section 2 to encode the query and the document. The input to the score function is the element-wise product, denoted by $\odot$, of the query encoding $rep_q(q) = Enc(q) = \mathbf{q}$ and the document encoding $rep_d(d) = Enc(d) = \mathbf{d}$:

$$score_{rank}(\mathbf{d}, \mathbf{q}) = MLP(\mathbf{d} \odot \mathbf{q}) \quad (1)$$

To encode the queries we compose a pseudo text using the query terms. The input to the auto-encoder Enc are the word embeddings of the pseudo text for $rep_q(q)$ and the word embeddings of the document terms for $rep_d(d)$.

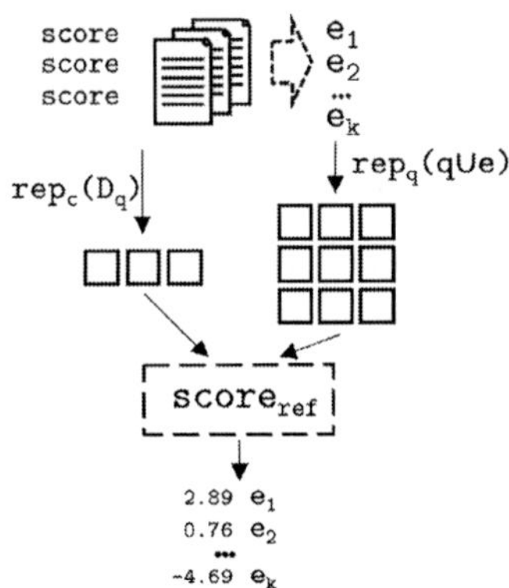

Figure 2: Query Refinement

4 Query Refinement

Query refinement finds additional terms for the initial query q that better describe the information need of the user (Nallapati and Shah, 2006). In our approach we follow Cao et al. (2008), in which the ranked documents D_q are used as pseudo relevance feedback. Our goal is to suggest relevant entities e that are contained in D_q and that are in a biomarker relationship to q. Therefore, we define a scoring function $score_{ref}$ for ranking of candidate entities e to the query q with respect to the retrieved document set D_q. See Figure 2 for a sketch of the query refinement pipeline. In the following, we describe several of the scoring functions.

4.1 Bag-of-Words Models

We implemented the two classic query refinement models the Rocchio algorithm (Rocchio and Salton, 1965) and Relevance Model.

4.2 Auto-Encoder Model

In this model, we also try to exploit the auto-encoding of the query. The idea is as follows: (i) Given a list of documents and their scores $D_q = [(\mathbf{d}_1, s_1), (\mathbf{d}_2, s_2), \ldots, (\mathbf{d}_n, s_n)]$ for a query q from the previous step, we use the ranking score s_i as pseudo-relevance feedback to create a pseudo document $rep_c(\mathbf{D}, \hat{\mathbf{s}}) = \sum_i^n \mathbf{d}_i \hat{s}_i = \hat{\mathbf{d}}$. The scores s are normalized so that they are non-negative and $\sum_i \hat{s}_i = 1$, see Appendix A.1. (ii) From all entities $e_i \in D_q \backslash q$ we generate new query encoding $rep_q(q \cup e_i) = Enc(q \cup e_i) = \hat{\mathbf{q}}_{e_i}$ (iii) We rank the entities based on the pseudo document using the scoring function

$$score_{ref}(\hat{\mathbf{q}}_{e_i}, \hat{\mathbf{d}}) = MLP(\hat{\mathbf{d}} \odot \hat{\mathbf{q}}_{e_i}) \quad (2)$$

i.e. we propose those entities as a query refinement that agree with the most relevant documents.

5 Experiments

In this section, we first explain our evaluation strategy to assess the performance of the respective models for document ranking and query refinement. Subsequently, we describe the settings for the data and the results of the experiments that we have conducted.

5.1 Evaluation Protocol

Document ranking models are evaluated by their ability to rank relevant documents higher than irrelevant documents. Query refinement models are evaluated both, by their ability to rank relevant query terms high, and by the recall of retrieved relevant documents when the query automatically is refined by the 1st, 2nd and 3rd proposed query term. We evaluate our models using mean average precision (MAP) (Manning et al., 2008, Chapter 11) and normalized discounted cumulative gain (nDCG) (Järvelin and Kekäläinen, 2000).

For both the document ranking and query refinement approach we interpret a biomarker-fact as a perfect query and the corresponding papers as the true-positive (or relevant) papers associated with this query. In this way, we use the curated facts as document level annotation for our approach. Because we want to assist the iterative approach of curators in which they refine an initially broad query to ever narrower searches, we need to create valid partial queries and associated relevant documents to mimic this procedure. Therefore, we generate sub-queries, which are partials of the facts. We generated two data sets: one for document ranking and one for query refinement. For document ranking, we generated all distinct subsets of the facts. For query refinement, we defined the eliminated entities (of the sub-query generation process) as true-positive refinement terms. For both data sets, we use all associated documents, of the original biomarker-fact, as true-positive relevant documents.

5.2 Data

Unlabeled Data   As unlabeled data we use ~24 Million abstracts of PubMed. To automatically annotate PubMed abstracts with disambiguated biomarker entities, we use a tool set that has been developed together with biomedical curators. It employs "ProMiner"[2] (Hanisch et al., 2005) for *Protein/Gene* and *Disease* entity recognition and regular expression based text scanning using synonyms from "DrugBank"[3] and "PubChem"[4] for the identification of *Drugs* and manually edited regular expressions, relating to "HGVS"[5] standards, to retrieve *Variants*. We restricted the PubMed documents to include at least one entity of type *Gene*, *Drug* and *Disease* leaving us with 2.7 Million documents. Additionally we replaced the text of every disambiguated entity with its id.

Labeled Data   As labeled data we use a knowledge base that contains a set of 5833 hand curated {*Gene(s) - Variant(s) - Drug(s) - Disease(s) - Outcome*} biomarker-facts and their associated papers that domain experts extracted from ~1200 fulltext documents. We only keep facts in which the disambiguated entities are fully represented in the available abstracts. This restricted our data set to a set of 1181 distinct facts. The 4 top curated genes are *EGFR* (29%), *BRAF* (13%), *KRAS* (8%), and *PIK3CA* (5%).

Cross Validation   To exploit all of our labeled data for training and testing we do 4-fold cross-validation. Because in our scenario a curator starts with an initial entity of type *Gene* we have generated our validation and test sets based on *Gene*s, instead of randomly sampling facts. This also guarantees us to never have the same sub-query included in the training, validation and test set. In total we have built 12 different splits of our data set basing the validation and test set each on a different gene. The respective training sets are built with all remaining facts that do not include the validation and test gene. Statistics can be found in Table 1.

5.3 Training of Embeddings and Auto-Encoder

The training data for the embeddings and the auto-encoder are the PubMed abstracts described in the previous Section 5.2. We trained the embeddings with Skip-Gram. For the vocabulary, we keep the top 100k most frequent words, while making sure all known entities are included. We use a window size of 10 and train the embeddings for 17

---

[2]https://www.scai.fraunhofer.de/en/business-research-areas/bioinformatics/products/prominer.html
[3]https://www.drugbank.ca
[4]https://pubchem.ncbi.nlm.nih.gov
[5]http://www.hgvs.org

Validation		Testing		Train
Gene	#Data	Gene	#Data	#Data
BRAF	1135	EGFR	1822	4386
BRAF	1075	KRAS	968	6310
BRAF	1088	PIK3CA	549	6944
EGFR	1700	BRAF	1241	4386
EGFR	1475	KRAS	968	5536
EGFR	1605	PIK3CA	549	5957
KRAS	573	EGFR	1822	5536
KRAS	804	BRAF	1241	6310
KRAS	778	PIK3CA	549	7754
PIK3CA	298	EGFR	1822	5957
PIK3CA	382	BRAF	1241	6944
PIK3CA	367	KRAS	968	7754

Table 1: Statistics about the train/validation/test splits

epochs. We normalize digits to "0" and lowercase all words. Tokenization is done by splitting on white space and before and after special characters.

For both, the encoder and the decoder, we use two LSTM cells per block with hidden size 400 each. We skip unknown tokens and feed the words in reverse order for the decoder following Sutskever et al. (2014). The auto-encoder was trained for 15 epochs using early stopping.

5.4 Document Ranking

In this section, we describe the document ranking models, their training and then discuss the results of their evaluation.

Models We evaluate the BOW models (*tf-idf*, *bm25*) (Section 3.1) and the two LTR models using feature engineering (*feat*) and the auto-encoded features (*auto-rank*) (Section 3.2). We also evaluate an additional set of models to investigate if maybe simpler solutions can be competitive. See Table 2 for an overview over all ranking models.

(a.) A simpler solution than learning a MLP for a score function is to compute the similarity between $\mathbf{q} = Enc(q)$ and the document encoding $\mathbf{d} = Enc(d)$. Therefore, we use the cosine similarity as scoring function between the vector representations $\mathbf{q}$ and $\mathbf{d}$ (*auto/cosine*).

(b.) Instead of encoding the documents and queries with the auto-encoder, we encode the documents and queries based on their tf-idf weighted embeddings, i.e. $\mathbf{q} = \sum tfidf(q_i) * \mathbf{w}_{q_i}$. Similarly

to the auto-rank model, the input to the classifier MLP is the element-wise product of the query encoding and the document encoding (*emb*).

(c.) Due to promising results of the *auto-rank* model, *bm25*, and the *feat* model, we also tested combinations of them. We tested the concatenation of the the *bm25* score with the *auto-rank* features (*auto-rank + bm25*) as well as the concatenation of *feat* with the *auto-rank* features (*auto-rank + feat*).

Training We train our models with Adam (Kingma and Ba, 2014) and tune the initial learning rate, the other parameters are kept default of TensorFlow[6]. We use a pairwise hinge loss (Chen et al., 2009) and compare relevant documents with irrelevant documents.

The ranking score function is parameterized by a MLP for which the number of layers is a hyperparameter which is tuned using grid-search. The input layer size is based on the number of input features. To limit the total number of parameters, we decrease the layer size while going deeper, i.e. layer i has size $l_i = \frac{b-i+1}{b}|u|$, with b being the depth of the network, $|u|$ the number of input features. For activation functions between layers we use ReLU (Glorot et al., 2011).

We employ grid search over the hyperparameters: dropout: $[0.3, 0.2, 0.1, 0.0]$, number of layers: $[1, 2, 3, 4]$, learning rates for Adam: $[0.0005, 0.001]$, batch size: $[40, 60]$, max 400 epochs. We conducted hyper-parameter tuning for each model and validation/test split separately. The best parameters for the models using the auto-encoded features were: 1 layer, dropout $p \in [0.3, 0.2]$ with a batch size of 60 and learning rate 0.0005. The feat models were best with 1-2 layers, dropout 0.0 and a learning rate at 0.001.

The BOW models as well as *auto/cosine* were only computed for the respective validation and test sets.

Results In Table 3 we have listed the average MAP and nDCG scores of the test sets. The *tf-idf* model is outperformed by most of the other models. However, *bm25*, which additionally takes the length of a document into account, performs very well. *tf-idf* and *bm25* have the major benefit of fast computation.

The *feat* model slightly outperforms the *auto-*

---

[6]Tensorflow V 1.3 https://www.tensorflow.org/api_docs/python/tf/train/AdamOptimizer

Model	$rep_q(q)$	$rep_d(d)$	score
tf-idf	BOW	BOW	dot product
bm25	BOW	BOW + doc length	bm25
emb	$\mathbf{q} = $ tf-idf BOW $\cdot\mathbf{w}$	$\mathbf{q} = $ tf-idf BOW $\cdot\mathbf{w}$	$MLP(\mathbf{q} \odot \mathbf{d})$
feat		$f(q,d)$	$MLP(f(q,d))$
auto/cosine	$\mathbf{q} = Enc(q)$	$\mathbf{d} = Enc(d)$	$cos(Enc(q), Enc(d))$
auto	$\mathbf{q} = Enc(q)$	$\mathbf{d} = Enc(d)$	$MLP(\mathbf{q} \odot \mathbf{d})$
auto + bm25	$\mathbf{q} = Enc(q)$, BOW	$\mathbf{d} = Enc(d)$, BOW + doc length	$MLP(\mathbf{q} \odot \mathbf{d},$ bm25$)$
auto + feat	$\mathbf{q} = Enc(q)$	$f(q,d)$ $\quad$ $\mathbf{d} = Enc(d)$	$MLP(\mathbf{q} \odot \mathbf{d}, f(q,d))$

Table 2: Query and document representation for ranking models

Test	Metric	tf-idf	bm25	emb	feat	auto/ cosine	auto-rank	auto-rank + bm25	auto-rank + feat
EGFR	MAP	0.289	0.632	0.310	0.575	0.054	0.545	0.588	**0.699**
	nDCG	0.424	0.728	0.460	0.695	0.129	0.653	0.716	**0.810**
KRAS	MAP	0.327	0.610	0.466	0.609	0.058	0.575	0.774	**0.820**
	nDCG	0.456	0.723	0.592	0.712	0.145	0.688	0.867	**0.914**
BRAF	MAP	0.342	0.656	0.427	0.704	0.063	0.563	0.702	**0.812**
	nDCG	0.480	0.751	0.572	0.802	0.163.	0.671	0.820	**0.901**
PIK3CA	MAP	0.341	0.633	0.486	0.625	0.079	0.541	0.779	**0.810**
	nDCG	0.473	0.729	0.617	0.718	0.171	0.656	0.859	**0.895**

Table 3: Test Scores Document Ranking

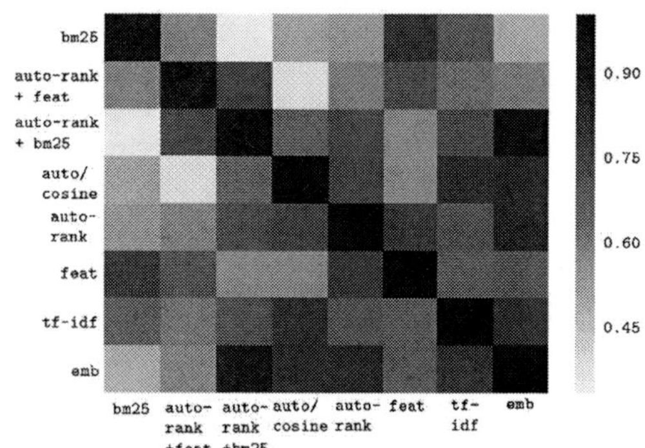

Figure 3: Correlation of Document Ranking Models

rank model. The distance features are a strong indicator for the semantic dependency between entities. These relationships need to be learned in the *auto-rank* model.

The cosine similarity of a query and a document (*auto/cos*) does not yield a good result. This shows that the auto-encoder has learned many features, most of which do not correlate with our task. We also find that *emb* does not yield an equal performance to *auto-rank*. The combination of the *auto-rank + feat* model is slightly better than the *auto-rank + bm25* model, both of which have the overall best performance. This shows, that the auto-encoder learns something orthogonal to term frequency and document length. The best model with respect to document ranking is the *auto-rank + feat* model.

In Figure 3 we show the correlation between the different models. Interestingly, the *bm25* and the *feat* strongly correlate. However, the scores of *bm25* do not correlate with the scores of the combination of *auto-rank* and *bm25*. This indicates, that the model does not primarily learn to use the *bm25* score but also focuses on the the the auto-encoded representation. This underlines the hypothesis that the auto-encoder is able to represent latent features of the relationship of the query terms in the document.

Influence of the Data It is interesting to see that the learned models do not perform well for the *EGFR* set. The reason for this might be that testing on it reduces the amount of training data substantially, as *EGFR* is the best curated gene and thus the largest split of the data set.

In a manual error analysis we compared the

rankings of four of the best models (*bm25*, *feat*, *auto-rank*, and *auto-rank + bm25*). We observe cases where the *auto-rank* model is unable to detect, when similar entities are used, i.e. entities like *Neoplasm* and *Colorectal Neoplasm*. In these cases the *bm25* helps, as it treats different words as different features. However, both *bm25* and the *feat* models rank reviews high, that only *list* query terms. For example, when executing the query {*BRAF, PIK3CA, H1047R*}, these models rank survey articles high (i.e. (Li et al., 2016); (Janku et al., 2012); (Chen et al., 2014)).

The *auto-rank* model on the other hand ranks those document high, for which each entities are listed in a semantic relationship (i.e. (Falchook et al., 2013)).

5.5 Query Refinement

In this section we describe our training approaches for query refinement and discuss our findings. The pseudo relevance feedback for the query refinement is based on the ranked documents from the previous query. For our experiments we chose the second best document ranker (*auto-rank + bm25*) from the previous experiments, because our prototype implementation for *auto-rank + feat* was computationally too expensive.

Models We combined both the auto-encoder features with the candidate terms of the respective BOW models (*auto-ref + rocch + relev*). In order to identify if the good results of this combination are due to the BOW models, or if the auto-encoded features have an effect, we trained a MLP with the same amount of parameters, but only use the features of the two BOW models as input (*rocchio + relev*).

Training For training, we use the same settings for query refinement as for document ranking and again use a pairwise hinge loss. Here we compare entities that occur in the facts with randomly sampled entities which occur in the retrieved documents.

Due to limitations in time we were only able to test our query refinement models on one validation/test split. We chose to use the split data set of genes *KRAS* and *PIK3CA* for validation and testing respectively. We have restricted our models to only regard the top 50 ranked documents for refinement.

Results To evaluate the ranking of entity terms we have computed nDCG@10, nDCG@100 and MAP, see Table 4 for the results. We also compute Recall@k of relevant documents for automatically refined queries using the 1st, 2nd and 3rd ranked entities. The scores can be found in Table 5.

Tables 4 and 5 show that the Relevance Model outperforms the Rocchio algorithm in every aspect. Both models outperform the auto encoder approach (*auto-ref*). We suspect that summing over the encodings distorts the individual features too much for a correct extraction of relevant entities to be possible.

The combination of all three models (*auto-ref + rocchio + relevance*) outperforms the other models in most cases. Especially the performance for ranking of entity terms is increased using the auto-encoded features. However, it is interesting to see that the *rocchio + relevance* model outperforms the recall for second and third best terms. This indicates that for user-evaluated term suggestions, the inclusion of the auto-encoded features is advisable. For automatic query refinement however, in average, this is not the case.

Variants	**Diseases**	**Drugs**
H1047R	Color. Neop.	Lapatinib
~~V600E~~	Liposarcoma	Mitomycin
~~T790M~~	Adenocarcin.	Linsitinib
E545K	Glioblastoma	Dactolisib
E542K	Stomach Neop.	Pictrelisib

Table 6: Refinement Terms for Query {PIK3CA}

Query Refinement Example In Table 6 we show the top ranked entities of type *Variants*, *Diseases* and *Drugs* for the query {*PIK3CA*}. While the diseases and the drugs are all relevant, *V600E* and *T790M* are in fact not variants of the gene *PIK3CA*.

However, when refining the query {*PIK3CA, V600E, BRAF, H1047R, Dabrafenib*}, the top ranked diseases are [*Melanoma, Neoplasms, Carcinoma Non Small Cell Lung (CNSCL), Thyroid Neoplasms, Colorectal Neoplasms*]. Using *Melanoma* for refinement, retrieves the top ranked paper (Falchook et al., 2013) which perfectly includes all these entities in a biomarker relationship.

Metrics	rocchio	relevance model	auto-ref	rocchio + relevance	auto-ref + rocchio + relevance
nDCG@10	0.232	0.274	0.195	0.341	**0.464**
nDCG@100	0.360	0.397	0.329	0.439	**0.536**
MAP	0.182	0.223	0.156	0.270	**0.386**

Table 4: Ranked Entity Scores for KRAS Validation and PIK3CA Testing

Metrics	Top n-th Entity	rocchio	relevance model	auto-ref	rocchio + relevance	auto-ref + rocchio + relevance
Recall@10	1	0.594	0.603	0.272	0.574	**0.696**
	2	0.535	0.561	0.339	**0.580**	0.522
	3	0.533	0.544	0.366	**0.555**	0.458
Recall@40	1	0.683	0.691	0.307	0.680	**0.779**
	2	0.603	0.633	0.374	**0.649**	0.586
	3	0.610	0.623	0.402	**0.626**	0.522

Table 5: Refinement Recall Scores for KRAS Validation and PIK3CA Testing

6 Related Work

The focus of research in this domain has primarily targeted the extraction of entity relations. Peng and Lu (2017), Peng et al. (2015), and Li et al. (2017) try to extract *Protein-Protein* relationships. Asada et al. (2017) try to extract *Drug-Drug* interaction and Lee et al. (2018) target the extraction of *Mutation-Gene* and *Mutation-Drug* relations. Jameson (2017) have derived a document ranking approach for PubMed documents using word embeddings trained on all PubMed documents. Xu et al. (2017) propose using auto-encoders on the vector-space model in a supervised setting for information retrieval and show that it improves performance. The quality of biomedical word embeddings was investigated by Th et al. (2015) and Chiu et al. (2016). Dogan et al. (2017) have developed an open source data set to which we would like to adapt our approach. Sheikhshab et al. (2016) have developed a novel approach for tagging genes which we would like to explore.

Palangi et al. (2016) use LSTMs to encode the query and document and use the cosine similarity together with the click-through data as features for ranking in a supervised approach. Cao et al. (2008) define a distance based feature-engineered supervised learning approach to identify good expansion terms. They try to elaborate if the selected terms for expansion are useful for information retrieval by identifying if the terms are actually related to the initial query. Nogueira and Cho (2017) have introduced a reinforcement learning approach for query refinement using logging data. They learn a representation of the text and the query using RNNs and CNNs and reinforce the end result based on recall of a recurrently expanded query. Sordoni et al. (2015) have developed a query reformulation model based on sequences of user queries. They have used a Sequence-to-Sequence model using RNNs to encode and decode queries of a user.

7 Conclusion

We have considered several approaches for document ranking and query refinement by investigating classic models, feature engineering and, due to the large amount of unlabeled data, a semi-supervised approach using a neural auto-encoder.

Leveraging the large amounts of unlabeled data to learn an auto-encoder on text documents yields semantically descriptive features that make subsequent document ranking and query refinement feasible. The combination with BOW features increases the performance substantially, which for our experiments, outputs the best results, for both document ranking and query refinement.

We were able to achieve promising results, however, there is a wide range of Sequence-to-Sequence architectures and text encoding strategies, therefore, we expect that there is room for improvement.

References

Masaki Asada, Makoto Miwa, and Yutaka Sasaki. 2017. Extracting drug-drug interactions with attention cnns. In *BioNLP 2017, Vancouver, Canada, August 4, 2017*, pages 9–18.

Francis Charles Brunicardi, Richard A. Gibbs, David A. Wheeler, John Nemunaitis, William Fisher, John Goss, and Changyi Johnny Chen. 2011. Overview of the development of personalized genomic medicine and surgery. *World Journal of Surgery*, 35:1693–1699.

Guihong Cao, Jian-Yun Nie, Jianfeng Gao, and Stephen Robertson. 2008. Selecting good expansion terms for pseudo-relevance feedback. In *Proceedings of the 31st Annual International ACM SIGIR Conference on Research and Development in Information Retrieval, SIGIR 2008, Singapore, July 20-24, 2008*, pages 243–250.

Jing Chen, Fang Guo, Xin Shi, Lihua Zhang, Aifeng Zhang, Hui Jin, and Youji He. 2014. BRAF V600E mutation and KRAS codon 13 mutations predict poor survival in Chinese colorectal cancer patients. *BMC Cancer*, 14(1):802.

Wei Chen, Tie-Yan Liu, Yanyan Lan, Zhiming Ma, and Hang Li. 2009. Ranking measures and loss functions in learning to rank. In *Advances in Neural Information Processing Systems 22: 23rd Annual Conference on Neural Information Processing Systems 2009. Proceedings of a meeting held 7-10 December 2009, Vancouver, British Columbia, Canada.*, pages 315–323.

Billy Chiu, Gamal K. O. Crichton, Anna Korhonen, and Sampo Pyysalo. 2016. How to train good word embeddings for biomedical NLP. In *Proceedings of the 15th Workshop on Biomedical Natural Language Processing, BioNLP@ACL 2016, Berlin, Germany, August 12, 2016*, pages 166–174.

Kyunghyun Cho, Bart van Merrienboer, Çaglar Gülçehre, Dzmitry Bahdanau, Fethi Bougares, Holger Schwenk, and Yoshua Bengio. 2014. Learning phrase representations using RNN encoder-decoder for statistical machine translation. In *Proceedings of the 2014 Conference on Empirical Methods in Natural Language Processing, EMNLP 2014, October 25-29, 2014, Doha, Qatar, A meeting of SIGDAT, a Special Interest Group of the ACL*, pages 1724–1734.

Andrew M. Dai and Quoc V. Le. 2015. Semi-supervised sequence learning. In *Advances in Neural Information Processing Systems 28: Annual Conference on Neural Information Processing Systems 2015, December 7-12, 2015, Montreal, Quebec, Canada*, pages 3079–3087.

Rezarta Islamaj Dogan, Andrew Chatr-aryamontri, Sun Kim, Chih-Hsuan Wei, Yifan Peng, Donald C. Comeau, and Zhiyong Lu. 2017. Biocreative VI precision medicine track: creating a training corpus for mining protein-protein interactions affected by mutations. In *BioNLP 2017, Vancouver, Canada, August 4, 2017*, pages 171–175.

Gerald S. Falchook, Jonathan C. Trent, Michael C. Heinrich, Carol Beadling, Janice Patterson, Christel C. Bastida, Samuel C. Blackman, and Razelle Kurzrock. 2013. BRAF Mutant Gastrointestinal Stromal Tumor: First report of regression with BRAF inhibitor dabrafenib (GSK2118436) and whole exomic sequencing for analysis of acquired resistance. *Oncotarget*, 4(2).

Xavier Glorot, Antoine Bordes, and Yoshua Bengio. 2011. Deep sparse rectifier neural networks. In *Proceedings of the Fourteenth International Conference on Artificial Intelligence and Statistics, AISTATS 2011, Fort Lauderdale, USA, April 11-13, 2011*, pages 315–323.

Daniel Hanisch, Katrin Fundel, Heinz-Theodor Mevissen, Ralf Zimmer, and Juliane Fluck. 2005. Prominer: rule-based protein and gene entity recognition. *BMC Bioinformatics*, 6(S-1).

Sepp Hochreiter and Jürgen Schmidhuber. 1997. Long short-term memory. *Neural Computation*, 9(8):1735–1780.

Anthony Jameson. 2017. A tool that supports the psychologically based design of health-related interventions. In *Proceedings of the 2nd International Workshop on Health Recommender Systems co-located with the 11th International Conference on Recommender Systems (RecSys 2017), Como, Italy, August 31, 2017.*, pages 39–42.

Filip Janku, Jennifer J. Wheler, Aung Naing, Vanda M. T. Stepanek, Gerald S. Falchook, Siqing Fu, Ignacio Garrido-Laguna, Apostolia M. Tsimberidou, Sarina A. Piha-Paul, Stacy L. Moulder, J. Jack Lee, Rajyalakshmi Luthra, David S. Hong, and Razelle Kurzrock. 2012. PIK3CA Mutations in Advanced Cancers: Characteristics and Outcomes. *Oncotarget*, 3(12).

Kalervo Järvelin and Jaana Kekäläinen. 2000. IR evaluation methods for retrieving highly relevant documents. In *SIGIR 2000: Proceedings of the 23rd Annual International ACM SIGIR Conference on Research and Development in Information Retrieval, July 24-28, 2000, Athens, Greece*, pages 41–48.

Diederik P. Kingma and Jimmy Ba. 2014. Adam: A method for stochastic optimization. *CoRR*, abs/1412.6980.

Victor Lavrenko and W. Bruce Croft. 2001. Relevance-based language models. In *SIGIR 2001: Proceedings of the 24th Annual International ACM SIGIR Conference on Research and Development in Information Retrieval, September 9-13, 2001, New Orleans, Louisiana, USA*, pages 120–127.

Kyubum Lee, Byounggun Kim, Yonghwa Choi, Sunkyu Kim, Won-Ho Shin, Sunwon Lee, Sungjoon Park, Seongsoon Kim, Aik Choon Tan, and Jaewoo Kang. 2018. Deep learning of mutation-gene-drug relations from the literature. *BMC Bioinformatics*, 19(1):21:1–21:13.

Gang Li, Cathy H. Wu, and K. Vijay-Shanker. 2017. Noise reduction methods for distantly supervised biomedical relation extraction. In *BioNLP 2017, Vancouver, Canada, August 4, 2017*, pages 184–193.

Jiwei Li, Minh-Thang Luong, and Dan Jurafsky. 2015. A Hierarchical Neural Autoencoder for Paragraphs and Documents.

Wan-Ming Li, Ting-Ting Hu, Lin-Lin Zhou, Yi-Ming Feng, Yun-Yi Wang, and Jin Fang. 2016. Highly sensitive detection of the PIK3CA H1047R mutation in colorectal cancer using a novel PCR-RFLP method. *BMC Cancer*, 16(1):454.

Hans Peter Luhn. 1957. A statistical approach to mechanized encoding and searching of literary information. *IBM Journal of Research and Development*, 1(4):309–317.

Christopher D. Manning, Prabhakar Raghavan, and Hinrich Schütze. 2008. *Introduction to information retrieval*. Cambridge University Press.

Teri A. Manolio. 2010. Genomewide Association Studies and Assessment of the Risk of Disease. *New England Journal of Medicine*, 363(2):166–176.

Tomas Mikolov, Ilya Sutskever, Kai Chen, Gregory S. Corrado, and Jeffrey Dean. 2013. Distributed representations of words and phrases and their compositionality. In *Advances in Neural Information Processing Systems 26: 27th Annual Conference on Neural Information Processing Systems 2013. Proceedings of a meeting held December 5-8, 2013, Lake Tahoe, Nevada, United States.*, pages 3111–3119.

Ramesh Nallapati and Chirag Shah. 2006. Evaluating the quality of query refinement suggestions in information retrieval.

Rodrigo Nogueira and Kyunghyun Cho. 2017. Task-oriented query reformulation with reinforcement learning. In *Proceedings of the 2017 Conference on Empirical Methods in Natural Language Processing, EMNLP 2017, Copenhagen, Denmark, September 9-11, 2017*, pages 574–583.

Hamid Palangi, Li Deng, Yelong Shen, Jianfeng Gao, Xiaodong He, Jianshu Chen, Xinying Song, and Rabab K. Ward. 2016. Deep sentence embedding using long short-term memory networks: Analysis and application to information retrieval. *IEEE/ACM Trans. Audio, Speech & Language Processing*, 24(4):694–707.

Yifan Peng, Samir Gupta, Cathy H. Wu, and K. Vijay-Shanker. 2015. An extended dependency graph for relation extraction in biomedical texts. In *Proceedings of the Workshop on Biomedical Natural Language Processing, BioNLP@IJCNLP 2015, Beijing, China, July 30, 2015*, pages 21–30.

Yifan Peng and Zhiyong Lu. 2017. Deep learning for extracting protein-protein interactions from biomedical literature. In *BioNLP 2017, Vancouver, Canada, August 4, 2017*, pages 29–38.

Stephen E. Robertson and Hugo Zaragoza. 2009. The probabilistic relevance framework: BM25 and beyond. *Foundations and Trends in Information Retrieval*, 3(4):333–389.

Joseph Rocchio and Gerard Salton. 1965. Information search optimization and interactive retrieval techniques. In *AFIPS '65 (Fall, part I)*.

Golnar Sheikhshab, Elizabeth Starks, Aly Karsan, Anoop Sarkar, and Inanç Birol. 2016. Graph-based semi-supervised gene mention tagging. In *Proceedings of the 15th Workshop on Biomedical Natural Language Processing, BioNLP@ACL 2016, Berlin, Germany, August 12, 2016*, pages 27–35.

Alessandro Sordoni, Yoshua Bengio, Hossein Vahabi, Christina Lioma, Jakob Grue Simonsen, and Jian-Yun Nie. 2015. A hierarchical recurrent encoder-decoder for generative context-aware query suggestion. In *Proceedings of the 24th ACM International Conference on Information and Knowledge Management, CIKM 2015, Melbourne, VIC, Australia, October 19 - 23, 2015*, pages 553–562.

Karen Spärck Jones. 1972. A Statistical Interpretation of Term Specificity and its Retrieval. *Journal of Documentation*, 28(1):11–21.

Ilya Sutskever, Oriol Vinyals, and Quoc V. Le. 2014. Sequence to sequence learning with neural networks. In *Advances in Neural Information Processing Systems 27: Annual Conference on Neural Information Processing Systems 2014, December 8-13 2014, Montreal, Quebec, Canada*, pages 3104–3112.

Muneeb Th, Sunil Kumar Sahu, and Ashish Anand. 2015. Evaluating distributed word representations for capturing semantics of biomedical concepts. In *Proceedings of the Workshop on Biomedical Natural Language Processing, BioNLP@IJCNLP 2015, Beijing, China, July 30, 2015*, pages 158–163.

Bo Xu, Hongfei Lin, Yuan Lin, and Kan Xu. 2017. Learning to rank with query-level semi-supervised autoencoders. In *Proceedings of the 2017 ACM on Conference on Information and Knowledge Management, CIKM 2017, Singapore, November 06 - 10, 2017*, pages 2395–2398.

A Supplemental Material

A.1 Normalizing Document Scores for Query Refinement

Because we used a hinge loss instead of cross entropy loss in the ranking model, we cannot interpret the scores $\mathbf{s}$ of the ranker as logits. While we do not know the magnitude of the ranker score, we do, however, expect the scores to be positive for relevant documents. If however many documents have scores below zero, this should also be regarded. Based on this, we have defined a normalization setting of the document scores:

$$s_{min} = \min(\min(\mathbf{s}), 0.0) \tag{3}$$

$$\hat{\mathbf{s}} = \frac{\mathbf{s} - s_{min}}{\sum_i^{|\mathbf{s}|}(s_i - s_{min})} \tag{4}$$

Biomedical Event Extraction Using Convolutional Neural Networks and Dependency Parsing

Jari Björne and Tapio Salakoski

Department of Future Technologies, University of Turku

Turku Centre for Computer Science (TUCS)

Faculty of Science and Engineering, FI-20014, Turku, Finland

`firstname.lastname@utu.fi`

Abstract

Event and relation extraction are central tasks in biomedical text mining. Where relation extraction concerns the detection of semantic connections between pairs of entities, event extraction expands this concept with the addition of trigger words, multiple arguments and nested events, in order to more accurately model the diversity of natural language.

In this work we develop a convolutional neural network that can be used for both event and relation extraction. We use a linear representation of the input text, where information is encoded with various vector space embeddings. Most notably, we encode the parse graph into this linear space using dependency path embeddings.

We integrate our neural network into the open source Turku Event Extraction System (TEES) framework. Using this system, our machine learning model can be easily applied to a large set of corpora from e.g. the BioNLP, DDI Extraction and BioCreative shared tasks. We evaluate our system on 12 different event, relation and NER corpora, showing good generalizability to many tasks and achieving improved performance on several corpora.

1 Introduction

Detection of semantic relations is a central task in biomedical text mining where information is retrieved from massive document sets, such as scientific literature or patient records. This information often consists of statements of interactions between named entities, such as signaling pathways between proteins in cells, or the combinatorial effects of drugs administered to a patient. *Relation* and *event* extraction are the primary methods for retrieving such information.

Relations are usually described as typed, sometimes directed, pairwise links between defined named entities. Automated relation extraction aims to develop computational methods for their detection.

Event extraction is a proposed alternative for relation extraction. Events differ from relations in that they can connect together more than two entities, that they have an annotated trigger word (usually a verb) and that events can act as arguments of other events. In the GENIA corpus, a sentence stating "The binding of proteins A and B is regulated by protein C" would be annotated with two nested events *REGULATION(C, BIND(A, B))*. While events can capture the semantics of text more accurately, their added complexity makes their extraction a more complicated task.

Many methods have been developed for relation extraction, with various kernel methods such as the graph kernel being widely used (Mooney and Bunescu, 2006; Giuliano et al., 2006; Airola et al., 2008). For the more complex task of event extraction approaches such as pipeline systems (Björne, 2014; Miwa et al., 2010), semantic parsing (McClosky et al., 2011) and joint inference (Riedel and McCallum, 2011) have been explored.

In recent years, the advent of deep learning has resulted in advances in many fields, and relation and event extraction are no exception. Considerable performance increases have been gained with methods such as convolutional (Zeng et al., 2014) and recurrent neural networks (Miwa and Bansal, 2016). Some proposed systems have relied entirely on word embeddings (Quan et al., 2016), while others have developed various network architectures for utilizing parse graphs as an additional source of information (Collobert et al., 2011; Liu et al., 2015; Xu et al., 2015; Ma et al.,

98

Proceedings of the BioNLP 2018 workshop, pages 98–108

Melbourne, Australia, July 19, 2018. ©2018 Association for Computational Linguistics

2015; Peng et al., 2017a).

In this work we present a new convolutional neural network method for extraction of both events and relations. We integrate our network as a classification module into the Turku Event Extraction System (Björne, 2014)[1], allowing it to be easily applied to corpora or texts stored in the TEES XML format. Our neural network model is characterized by a unified representation of input examples that can be applied to detection of both keywords as well as their relations.

2 Materials and Methods

2.1 Corpora

We develop and evaluate our method on a large number of event and relation corpora (See Table 1). These corpora originate from three BioNLP Shared Tasks (Kim et al., 2009, 2011; Nédellec et al., 2013), the two Drug–Drug Interaction (DDI) Extraction tasks (Segura-Bedmar et al., 2011, 2013) and the recent BioCreative VI Chemical–Protein relation extraction task (Krallinger et al., 2017). The BioNLP corpora cover various domains of molecular biology and provide the most complex event annotations. The DDI and BioCreative corpora use pairwise relation annotations, and one of the DDI corpora defines also a drug named entity recognition (NER) task.

All of these corpora are used in the TEES XML format and are installed or generated with the TEES system. The corpora are parsed with the TEES preprocessing pipeline, which utilizes the BLLIP parser (Charniak and Johnson, 2005) with the McClosky biomodel (McClosky, 2010), followed by conversion of the constituency parses into dependency parses with the Stanford Tools (de Marneffe et al., 2006). These tools generate the deep parse graph which is used as the source for our dependency path features.

2.2 TEES Overview

The TEES system is based around a graph representation of events and relations. Named entities and event triggers are nodes, and relations and event arguments are the edges that connect them. An event is modelled as a trigger node and its set of outgoing edges. For a detailed overview of TEES we refer to Björne (2014).

TEES works as a pipeline method that models relation and event extraction as four consecutive

classification tasks (See Figure 2). The first stage is *entity detection* where each word token in a sentence is classified as an entity or a negative, generating the nodes of the graph. This stage is used in NER tasks as well as for event trigger word detection. The second stage is *edge detection* where relations and event arguments are predicted for all valid pairs of named entity and trigger nodes. For relation extraction tasks where entities are given as known data this is the only stage used.

In the entity detection stage TEES predicts a maximum of one entity per word token. However, since events are *n*-ary relations, event nodes may overlap. The *unmerging* stage duplicates event nodes by classifying each candidate event as a real event or not. Finally, *modifier detection* can be used to detect event modality (such as speculation or negation) on corpora where this is annotated.

2.3 Neural Network Overview

In TEES the four classification stages are implemented as multiclass classification tasks using the $\text{SVM}^{multiclass}$ support vector machine (Tsochantaridis et al., 2005) and a rich set of features derived mostly from the dependency parse graph.

We develop our convolutional neural network method using the Keras (Chollet et al., 2015) package with the TensorFlow backend (Dean et al., 2015). We extend the TEES system by replacing the SVM-based classifier modules with our network, using various vector space embeddings as input features. Our neural network design follows a common approach in NLP where the input sequence is processed by parallel convolutional layers (Kim, 2014; Zeng et al., 2014; Quan et al., 2016).

We use the same basic network structure for all four TEES classification stages (See Figure 1). The input examples are modelled in the context of a sentence window, centered around the candidate entity, relation or event. The sentence is modelled as a linear sequence of word tokens. Each word token is mapped to relevant vector space embeddings. These embeddings are concatenated together, resulting in an n-dimensional vector for each word token.

This merged input sequence is processed by a set of 1D convolutions with window sizes 1, 3, 5 and 7. Global max pooling is applied for each convolutional layer and the resulting features are merged together into the convolution output vec-

---

[1]http://jbjorne.github.io/TEES/

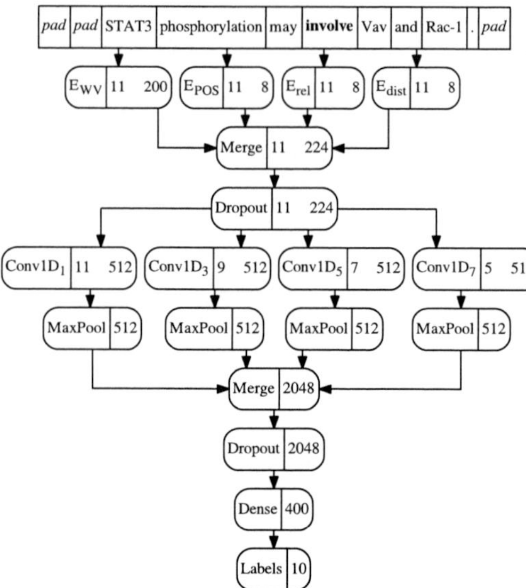

Figure 1: We use the same network architecture for all four classification tasks, with only the input embeddings and the number of predicted labels changing between tasks. This figure demonstrates *entity detection* where the word "involve" is being classified as belonging to one or more of 10 entity types. The input sentence is padded for the fixed example length (11 in this case) and the word tokens are mapped to corresponding embeddings, in this example Word Vectors, POS tags, relative positions and distances. The embedding vectors are merged together before the convolutional layers, and the results of the convolution operations are likewise merged before the dense layer, after which the final layer shows the predicted labels.

tor. This output vector is fed into a dense layer of 200–800 neurons, which is connected to the final classification layer where each label is represented by one neuron. The classification layer uses sigmoid activation, and the other layers use relu activation. Classification is performed as multilabel classification where each example may have 0–n positive labels. We use the *adam* optimizer with *binary crossentropy* and a learning rate of 0.001.

Dropout of 0.1–0.5 is applied at two stages in the system to increase generalization. Weights are learned for all input embeddings except for the word vectors, where we use the original weights as-is to ensure generalization to words outside the task's training corpus.

2.4 Input Embeddings

All of the features used by our system are represented as embeddings, sets of vectors where each unique input item (such as a word string) maps to its own n-dimensional vector. The type and number of embeddings we use varies by classification task and is used to model the unique characteristics of each task (See Figure 2). The pre-made word vectors we use are 200-dimensional and the rest of the embeddings (learnt from the input corpus) are 8-dimensional.

Word Vectors are the most important of these embeddings. We use word2vec (Mikolov et al., 2013) vectors induced on a combination of the English Wikipedia and the millions of biomedical research articles from PubMed and PubMed Central by Pyysalo et al. (2013)[2].

POS (Part-of-speech) tags generated by the BLLIP parser are used to define the syntactic categories of the words.

Entity features are used in cases where such information is already available, as in relation extraction where the pairs of entities are already known, or in event extraction where named entities are predefined.

Distance features follow the approach proposed by Zeng et al. (2014) where the relative distances to tokens of interest are mapped to their own vectors.

Relative Position features are used to mark whether tokens are located (B)efore, (A)fter or in the (M)iddle of the classified structure, or if they form a part of it as entities, event triggers or arguments. These features aim to model the context of the example in a manner somewhat similar to the shallow linguistic kernel of Giuliano et al. (2006).

Path Embeddings describe the shortest undirected path from a token of interest to another token in the sentence. Multiple sets of vectors (0–4), one for directed dependencies at each distance, are used for the dependencies of the path. For example, if paths of depth 4 are used, a shortest path of three directed dependencies connecting two tokens of interest could be modelled with four embedding vectors e.g. $\leftarrow$*dobj, nsubj*$\rightarrow$, *nn*$\rightarrow$, *NULL*. Our path embeddings are inspired by the concept of distance embeddings used by Zeng et al. (2014): Since it is possible to model linear distances between tokens in the input sentence, it

---

[2]`http://evexdb.org/pmresources/`
`vec-space-models/`

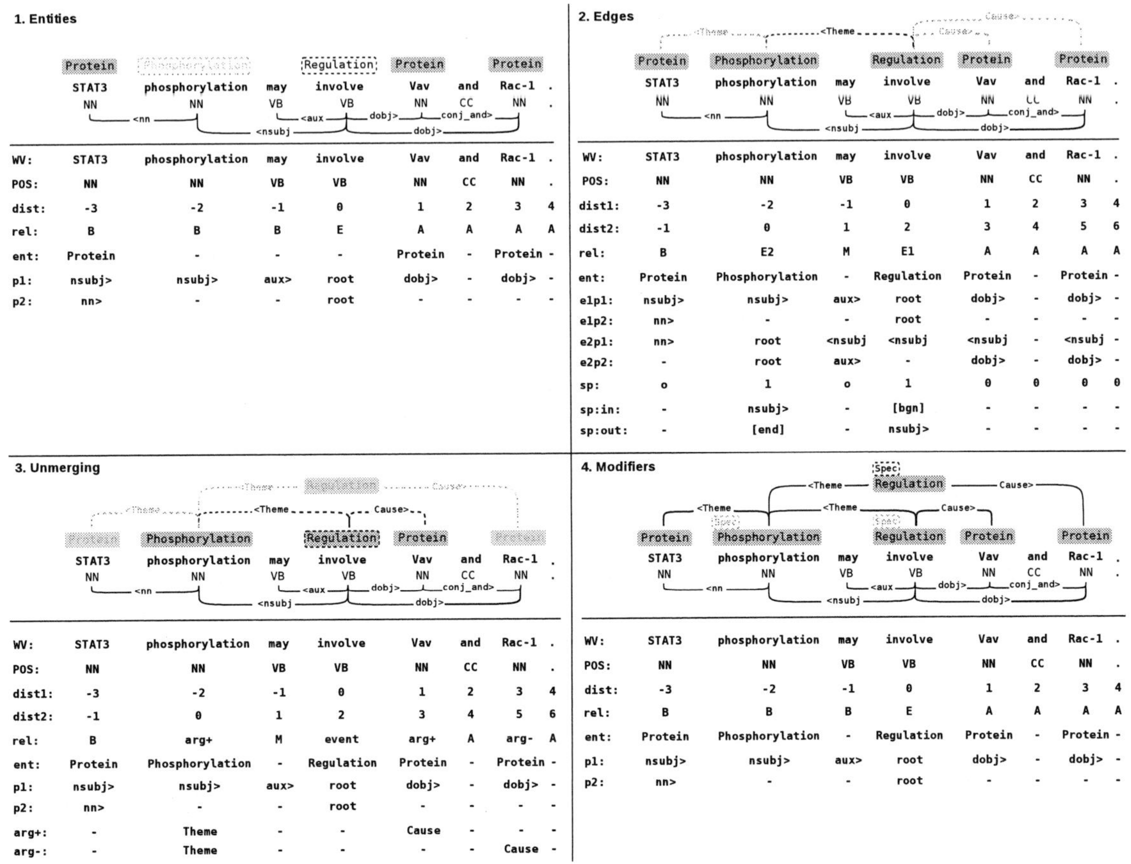

Figure 2: System stages. The TEES pipeline performs event extraction in four consecutive stages, generating first 1. the nodes (entities) and 2. edges (relations) of the event graph, which is then "pulled apart" in 3. unmerging, followed optionally by 4. modifier detection. The example being classified is shown with a dotted line in each image, and other examples in the same sentence with light gray dotted lines. We replace the four SVM classification stages in the TEES pipeline with convolutional neural networks. In place of the rich feature representations we use a sentence model where word token and dependency parse information is represented by embeddings. The Word Vector, POS and entities features are straightforwardly produced from the information of each token. The distance and relative position features model the position of the token in the sentence. The path features mark the dependencies connecting each token to a token of interest (candidate entity or relation endpoint). The shortest path features mark the set of dependencies forming the shortest path for a candidate relation. In the unmerging stage candidate event arguments are also used as features.

is also possible to model any other distance between these tokens, in our case paths in the dependency parse.

Shortest Path Embeddings follow the approach of the n-gram features used in methods such as the TEES SVM system. The shortest path consists of the tokens and dependencies connecting the two entities of a candidate relation. For each token on the path we define two embeddings, one for the incoming and one for the outgoing dependency. For example, if the shortest path would consist of three tokens and the two dependencies connecting them, the shortest path embedding vectors for the three tokens could be e.g. *([begin], ←nsubj), (←nsubj, dobj→), (dobj→, [end])*. Thus, our shortest path embeddings can be seen as a more detailed extrapolation of the "on dep path embeddings" of Fu et al. (2017).

Event Argument Embeddings are used only in the unmerging stage where predicted entities and edges are divided into separate events.

2.5 Parameter Optimization

When developing our system, we use the training set for learning, and the development set for parameter validation. We use the early stopping approach where the network is trained until the validation loss no longer decreases. We train for up to 500 epochs, stopping once validation loss has no longer decreased for 10 consecutive epochs.

Neural network models can be very sensitive to the initial random weights. Despite the relatively large training and validation sets, our model exhibits performance differences of several percentage points with different random seeds. In the current TensorFlow backend it is not possible to efficiently fix the random seed[3], and in any case this would be unhelpful, as the impact of any given seed varies with the training data. Instead, we compensate for the inherent randomness of the network by training multiple models with randomized initializations and use as the final model the one which achieved the best performance on the validation set (measured using the micro-averaged F-score).

In addition to the random seed and optimal epoch, neural networks depend on a large number of hyperparameters. We use the process of training multiple randomized models also for parameter optimization. In addition to varying the random seed, we pick a random combination of hyperparameters from the ranges to be optimized, so that different models are randomized both in terms of initialization and the parameters. We test sizes of 200, 400 and 800 for the final dense layer, filter sizes of 128, 256 and 512 for the convolutional layers and dropout values of 0.1, 0.2 and 0.5. In addition, we experiment with path depths of 0–4 for the path embeddings.

Training a single model can still be prone to overfitting if the validation set is too small. To improve generalizability, we explore the use of model ensembles. Instead of using the best randomized model as the final one, we take n-best models, ranked with micro-averaged F-score on the validation set, and use their averaged predictions. These ensemble predictions are calculated for each label as the average of all the models' predicted confidence scores.

With SVMs or random forests it is possible to "refit" a classifier after parameter selection, by retraining on the combined training and optimization sets, and this approach is also used by the TEES SVM classifiers. With the neural network, we cannot retrain with the validation set, as there would be no remaining data for detecting the optimal epoch. We approach also this issue using model ensembles. As the final source of randomization, we randomly redistribute the training and validation set documents before training each model. In this manner, the n-best models will together cover a larger part of the training data.

By training a large set of randomized models and using the n-best ones, we aim to address the effect of random initialization, parameter optimization and coverage of training data using the same process. However, with the size of the corpora used, training even a single model is relatively time consuming. In practice we are able to train only around 20 models for each of the four stages of the classification pipeline. Thorough parameter optimization comparable to the SVM system is thus not computationally feasible with the neural network, but good performance on varied corpora indicates that the current approach is at least adequate.

3 Results and Discussion

The results of applying our proposed system on the various corpora are shown in Table 2. We com-

---

[3] https://github.com/keras-team/keras/issues/2280

Corpus	Domain	E	I	S
GE09	Molecular Biology	10	6	11380
GE11	Molecular Biology	10	6	14958
EPI11	Epigenetics and PTM:s	16	6	11772
ID11	Infectious Diseases	11	7	5118
REL11	Entity Relations	1	2	11351
DDI11	Drug–Drug Interactions	-	1	5806
DDI13 9.1	Drug NER	4	-	9605
DDI13 9.2	Drug–Drug Interactions	-	4	10239
GE13	Molecular Biology	15	6	8369
CG13	Cancer Genetics	42	9	5938
PC13	Pathway Curation	24	9	5040
CP17	Chemical–Protein Int.	-	5	6249

Table 1: The corpora used in this work are listed with their domain, number of event and entity types (E), number of event argument and relation types (I), and number of sentences (S).

pare our method with previous results from the shared tasks for which these corpora were introduced as well as with later research. In the next sections we analyse our results for the different corpus categories.

3.1 The BioNLP Event Extraction Tasks

The BioNLP Event Extraction tasks provide the most complex corpora with often large sets of event types and at times relatively small corpus sizes. The GENIA corpora from 2009 and 2011 have been the subject of most event extraction research. Our proposed method achieves F-scores of 57.84 and 58.10 on GE09 and GE11, respectively. Compared to the best reported results of 58.27 (Miwa et al., 2012) and 58.07 (Venugopal et al., 2014), our method shows similar performance on these corpora.

Our CNN reimplementation of TEES outperforms the original TEES SVM system on all the BioNLP corpora. In addition, we achieve to the best of our knowledge the highest reported performance on the GE11, EPI11, REL11, CG13 and PC13 BioNLP Shared Task corpora.

The annotations for the test sets of the BioNLP Shared Task corpora are not provided, instead the users upload their predictions to the task organizers' servers for evaluation. While this method provides very good protection against overfitting and data leaks, unfortunately many of these evaluation servers are no longer working. Thus, we were able to evaluate our system on only a subset of all existing BioNLP Shared Task corpora.

3.2 The Drug–Drug Interactions Tasks

There have been two instances of the Drug–Drug Interactions Shared Task. The first one in 2011 concerned the detection of untyped relations for adverse drug effects. Unlike the other corpora, no official evaluator system or program exists for this corpus so we use our own F-score calculation. The lower performance compared to the original TEES system warrants further examination, but in any case the DDI11 corpus has been mostly superseded by the more detailed DDI13 corpora.

On the DDI13 corpora task 9.1, drug named entity recognition, our CNN system performs better than the original TEES entry, but neither of these TEES versions can detect more than single-token entities so they are not well suited for this task. Nevertheless, this result demonstrates the potential applicability of our method also to NER tasks.

Of all the DDI corpora the DDI13 task 9.2 corpus, typed relation extraction, has been the subject of much neural network based research in the past few years. A large variety of methods have been tested, and good results have been achieved by highly varying network models, some of which use no parsing or graph-like features, such as the multichannel convolutional neural network of Quan et al. (2016) which combines multiple sets of word vectors and achieves an F-score of 70.21. The highest result of 73.5 so far has been reported by Lim et al. (2018) who used a binary tree-LSTM model ensemble, with which our system achieves minutely higher, in practice comparable performance. Most recent DDI13 systems use corpus-specific rules for filtering negative candidate relations from the training data, which usually results in performance gains. As we aim to develop a generic method easily applicable to any corpus we did not implement these DDI filtering rules.

3.3 The CHEMPROT Task

Of all the evaluated corpora the CHEMPROT corpus used in the BioCreative VI Chemical–Protein relation extraction task is the most recent. Thus it provides an interesting point of comparison with current methods in relation extraction. All of our models outperform the task winning system combination of Peng et al. (2017b), with our mixed five model ensemble achieving a 5 pp increase over the shared task winning result. The CHEMPROT corpus is relatively large compared to its low number of five relation types, possibly making learning

Corpus	P	R	F	Team / System	Method
GE09	58.48	46.73	51.95 *	TEES (Björne et al., 2009)	SVM
	-	-	55.96	6 x system ensemble (Kim et al., 2009)	Metasystem
	63.19	50.28	56.00	EventMine (Miwa et al., 2010)	SVM
	-	-	57.4	M3+enju (Riedel and McCallum, 2011)	Joint inference
	65.19	52.67	**58.27**	EventMine+PR+FE (Miwa et al., 2012)	SVM
	63.08	53.96	58.16	BioMLN (Venugopal et al., 2014)	MLN and SVM
	64.94	48.08	55.25	*Ours* (single)	CNN
	67.58	48.02	56.15	*Ours* (5 x ensemble)	CNN
	69.87	49.34	57.84	*Ours* (mixed 5 x ensemble)	CNN
GE11	57.65	49.56	53.30 †	TEES (Björne and Salakoski, 2011)	SVM
	64.75	49.41	56.04 *	FAUST (Riedel et al., 2011)	Joint inference / parser
	63.48	53.35	57.98	EventMine-CR (Miwa et al., 2012)	SVM
	63.61	53.42	58.07	BioMLN (Venugopal et al., 2014)	MLN and SVM
	66.46	48.96	56.38	Stacked Generalization (Majumder et al., 2016)	SVM
	64.86	50.53	56.80	*Ours* (single)	CNN
	68.76	49.97	57.87	*Ours* (5 x ensemble)	CNN
	69.45	49.94	**58.10**	*Ours* (mixed 5 x ensemble)	CNN
EPI11	53.98	52.69	53.33 *	TEES (Björne and Salakoski, 2011)	SVM
	54.42	54.28	54.35	EventMine multi-corpus (Miwa et al., 2013)	SVM
	65.97	45.79	54.06	*Ours* (single)	CNN
	65.40	48.84	55.92	*Ours* (5 x ensemble)	CNN
	64.93	50.00	**56.50**	*Ours* (mixed 5 x ensemble)	CNN
ID11	48.62	37.85	42.57 †	TEES (Björne and Salakoski, 2011)	SVM
	65.97	48.03	55.59 *	FAUST (Riedel et al., 2011)	Joint inference / parser
	-	-	55.6	M3+Stanford (Riedel and McCallum, 2011)	Joint inference
	61.33	58.96	**60.12**	EventMine EasyAdapt (Miwa et al., 2013)	SVM
	65.53	48.17	55.52	*Ours* (single)	CNN
	70.51	49.69	58.30	*Ours* (5 x ensemble)	CNN
	66.48	50.66	57.50	*Ours* (mixed 5 x ensemble)	CNN
REL11	37.0	47.5	41.6 †	VIB - UGhent (Van Landeghem et al., 2011)	SVM
	68.0	50.1	57.7 *	TEES (Björne and Salakoski, 2011)	SVM
	70.87	59.56	64.72	*Ours* (single)	CNN
	76.30	48.09	58.99	*Ours* (5 x ensemble)	CNN
	73.65	61.17	**66.83**	*Ours* (mixed 5 x ensemble)	CNN
DDI11	58.04	68.87	62.99 †	TEES (Björne et al., 2011)	SVM
	60.54	71.92	**65.74** *	WBI (Thomas et al., 2011)	kernels + CBR
	69.83	55.49	61.84	*Ours* (single)	CNN
	69.78	57.21	62.88	*Ours* (5 x ensemble)	CNN
	77.57	49.93	60.75	*Ours* (mixed 5 x ensemble)	CNN
DDI13 9.1	73.7	57.9	64.8 †	TEES (Björne et al., 2013)	SVM
	73.4	69.8	**71.5** *	WBI-NER (Rocktäschel et al., 2013)	CRF
	72	63	67	*Ours* (single)	CNN
	71	63	67	*Ours* (5 x ensemble)	CNN
	73	63	68	*Ours* (mixed 5 x ensemble)	CNN
DDI13 9.2	73.2	49.9	59.4 †	TEES (Björne et al., 2013)	SVM
	64.6	65.6	65.1 *	FBK-irst (Chowdhury and Lavelli, 2013)	kernels
	75.99	62.25	70.21	Multichannel CNN (Quan et al., 2016)	CNN
	73.4	69.6	71.48	Joint AB-LSTM Model (Sahu and Anand, 2017)	LSTM
	74.1	71.8	72.9	Hierarchical RNNs (Zhang et al., 2017)	RNN
	77.8	69.6	73.5	One-Stage Model Ensemble (Lim et al., 2018)	RNN
	75.80	70.38	72.99	PM-BLSTM (Zhou et al., 2018)	LSTM
	75.29	66.29	70.51	*Ours* (single)	CNN
	78.60	64.15	70.64	*Ours* (5 x ensemble)	CNN
	80.54	67.62	**73.51**	*Ours* (mixed 5 x ensemble)	CNN
GE13	56.32	46.17	50.74 †	TEES (Björne and Salakoski, 2013)	SVM
	58.03	45.44	50.97 *	EVEX (Hakala et al., 2013)	TEES + rerank
	59.24	48.95	**53.61**	BioMLN (Venugopal et al., 2014)	MLN and SVM
	58.95	40.29	47.87	*Ours* (single)	CNN
	62.18	42.29	50.34	*Ours* (5 x ensemble)	CNN
	65.78	44.38	53.00	*Ours* (mixed 5 x ensemble)	CNN
CG13	64.17	48.76	55.41 *	TEES (Björne and Salakoski, 2013)	SVM
	55.82	48.83	52.09 †	EventMine (Miwa and Ananiadou, 2013)	SVM
	60.45	51.34	55.52	*Ours* (single)	CNN
	63.92	51.00	56.74	*Ours* (5 x ensemble)	CNN
	66.55	50.77	**57.60**	*Ours* (mixed 5 x ensemble)	CNN
PC13	55.78	47.15	51.10 †	TEES (Björne and Salakoski, 2013)	SVM
	53.48	52.23	52.84 *	EventMine (Miwa and Ananiadou, 2013)	SVM
	58.31	47.08	52.10	*Ours* (single)	CNN
	58.66	48.49	53.09	*Ours* (5 x ensemble)	CNN
	62.16	50.34	**55.62**	*Ours* (mixed 5 x ensemble)	CNN
CP17	66.08	56.62	60.99 †	TEES (Mehryary et al., 2017)	SVM
	56.10	67.84	61.41 †	RSC (Corbett and Boyle, 2017)	LSTM
	72.66	57.35	64.10 *	NCBI (Peng et al., 2017b)	SVM, CNN, and RNN
	71.40	61.86	66.28	*Ours* (single)	CNN
	74.38	60.44	66.69	*Ours* (5 x ensemble)	CNN
	75.13	65.07	**69.74**	*Ours* (mixed 5 x ensemble)	CNN

Table 2: Results. Performance is shown in Precision, Recall and F-score, measured on the corpus test set for related work and our TEES CNN method (single best model, 5 model ensemble, or mixed 5 model ensemble with randomized train/validation set split). Shared task winning results are indicated with * and shared task participant results with †. The highest F-score for each corpus is shown in bold. All of our results except for DDI11 are evaluated using the official evaluation program or server of each task.

Corpus	0	1	2	3	4
GE11 (n)	69.85	71.45	72.34	72.87	72.64
GE11 (e)	88.10	88.38	88.92	88.51	88.68
DDI13 (e)	59.46	59.23	58.22	56.50	54.37
CP17 (e)	54.55	55.46	56.51	56.98	56.25

Table 3: The effect of path embeddings. The impact of using increasing depths of paths for embeddings is shown in terms of averaged F-score on the development set for entity (n) and edge (e) detection.

easier for our system.

3.4 Effect of Deep Parsing

Compared to neural networks which use either only word vectors, or model parse structure at the network level (e.g. graph-LSTMs), an interesting aspect of our method is that it can function both with and without parse information. By turning dependency path features on and off we can evaluate the impact of deep parsing on the system (See Table 3).

The path embeddings have the most impact on GE11 entity detection, where these paths link the entity candidate token to each other token. In GE11 event argument extraction the role of the path context embeddings is diminished. Surprisingly, on the DDI13 9.2 relation corpus path embeddings reduce performance, perhaps due to very long sentences and very indirect relations between the entity pairs. However, on another relation corpus, the CHEMPROT corpus, the path embeddings again increase performance, perhaps indicating that the CHEMPROT relation annotations follow more closely sentence syntax.

3.5 Computational Requirements

Our system improves on performance compared to the SVM-based TEES, but at the cost of increased computational requirements. The neural network effectively requires a specialized GPU for training and even then training times can be an issue.

For example, training the original TEES system on a four-core CPU for the GE09 task takes about 3 hours and classification of the test set with this model can be done in four minutes. For comparison, our GE09 neural network with 20 models for all four stages takes around nine hours to train on a Tesla P100 GPU. However, test set classification with a single model takes only about three minutes and using a five model ensemble about ten minutes.

Thus, while training the proposed method is much slower, classification can be performed relatively quickly. While the hardware and time requirements are much higher than with the SVM system, our proposed system can for some corpora achieve performance increases of even 10 pp. In most applications such gains are likely worth the increased computational requirements.

4   Conclusions

We have developed a convolutional neural network system that together with different vector space embeddings can be applied to diverse text classification tasks. We replace the TEES system's event extraction pipeline components with this network and demonstrate considerable performance gains on a set of large event and relation corpora, achieving state-of-the-art performance on many of them and the best reported performance on the GE11, EPI11, REL11, CG13, PC13, DDI13 9.2 and CP17 corpora.

To the best of our knowledge our system represents the first application of neural networks to extraction of complex events from the BioNLP GENIA corpora. Our system uses a unified linear sentence representation where graph analyses such as dependency parses are fully included using our dependency path embeddings, and we demonstrate that these path embeddings can increase the performance of the convolutional model. Unlike systems where separate subnetworks are used to model graph structures, our network receives all of the information through the unified linear representation, allowing the whole model to learn from all the features.

The Turku Event Extraction System provides a unified approach for utilizing a large number of event and relation extraction corpora. As we integrate our proposed convolutional neural network method into the TEES system, it can be used as easily as the original TEES system, with the framework handling tasks such as preprocessing and format conversions. Our Keras-based neural network implementation can also be extended and modified, allowing continued experimentation on the wide set of corpora supported by TEES. We publish our method and our trained neural network models as part of the TEES open source project[4].

---

[4]`https://github.com/jbjorne/TEES/wiki/`
`TEES-CNN-BioNLP18`

Acknowledgments

We thank CSC – IT Center for Science Ltd for providing computational resources.

References

Antti Airola, Sampo Pyysalo, Jari Björne, Tapio Pahikkala, Filip Ginter, and Tapio Salakoski. 2008. All-paths graph kernel for protein-protein interaction extraction with evaluation of cross-corpus learning. *BMC bioinformatics*, 9(11):S2.

Jari Björne. 2014. *Biomedical Event Extraction with Machine Learning*. Ph.D. thesis, University of Turku.

Jari Björne, Antti Airola, Tapio Pahikkala, and Tapio Salakoski. 2011. Drug-drug interaction extraction from biomedical texts with svm and rls classifiers. *Proceedings of DDIExtraction-2011 challenge task*, pages 35–42.

Jari Björne, Juho Heimonen, Filip Ginter, Antti Airola, Tapio Pahikkala, and Tapio Salakoski. 2009. Extracting complex biological events with rich graph-based feature sets. In *Proceedings of the Workshop on Current Trends in Biomedical Natural Language Processing: Shared Task*, pages 10–18. Association for Computational Linguistics.

Jari Björne, Suwisa Kaewphan, and Tapio Salakoski. 2013. Uturku: drug named entity recognition and drug-drug interaction extraction using svm classification and domain knowledge. In *Second Joint Conference on Lexical and Computational Semantics (* SEM), Volume 2: Proceedings of the Seventh International Workshop on Semantic Evaluation (SemEval 2013)*, volume 2, pages 651–659.

Jari Björne and Tapio Salakoski. 2011. Generalizing biomedical event extraction. In *Proceedings of the BioNLP Shared Task 2011 Workshop*, pages 183–191. Association for Computational Linguistics.

Jari Björne and Tapio Salakoski. 2013. Tees 2.1: Automated annotation scheme learning in the bionlp 2013 shared task. In *Proceedings of the BioNLP Shared Task 2013 Workshop*, pages 16–25.

Eugene Charniak and Mark Johnson. 2005. Coarse-to-fine n-best parsing and MaxEnt discriminative reranking. In *Proceedings of the 43rd Annual Meeting of the Association for Computational Linguistics (ACL'05)*, pages 173–180. Association for Computational Linguistics.

François Chollet et al. 2015. Keras. https://github.com/keras-team/keras.

Md Faisal Mahbub Chowdhury and Alberto Lavelli. 2013. Fbk-irst: A multi-phase kernel based approach for drug-drug interaction detection and classification that exploits linguistic information. In *Second Joint Conference on Lexical and Computational Semantics (* SEM), Volume 2: Proceedings of the Seventh International Workshop on Semantic Evaluation (SemEval 2013)*, volume 2, pages 351–355.

Ronan Collobert, Jason Weston, Léon Bottou, Michael Karlen, Koray Kavukcuoglu, and Pavel Kuksa. 2011. Natural language processing (almost) from scratch. *Journal of Machine Learning Research*, 12(Aug):2493–2537.

P. Corbett and J. Boyle. 2017. Improving the learning of chemical-protein interactions from literature using transfer learning and word embeddings. In *Proceedings of the BioCreative VI Workshop*, pages 180–183.

J Dean, R Monga, et al. 2015. Tensorflow: Large-scale machine learning on heterogeneous systems. *TensorFlow. org. Google Research. Retrieved*, 10.

Lisheng Fu, Thien Huu Nguyen, Bonan Min, and Ralph Grishman. 2017. Domain adaptation for relation extraction with domain adversarial neural network. In *Proceedings of the Eighth International Joint Conference on Natural Language Processing (Volume 2: Short Papers)*, volume 2, pages 425–429.

Claudio Giuliano, Alberto Lavelli, and Lorenza Romano. 2006. Exploiting shallow linguistic information for relation extraction from biomedical literature. In *11th Conference of the European Chapter of the Association for Computational Linguistics*.

Kai Hakala, Sofie Van Landeghem, Tapio Salakoski, Yves Van de Peer, and Filip Ginter. 2013. EVEX in ST'13: Application of a large-scale text mining resource to event extraction and network construction. In *Proceedings of the BioNLP Shared Task 2013 Workshop*, pages 26–34, Sofia, Bulgaria. Association for Computational Linguistics.

Jin-Dong Kim, Tomoko Ohta, Sampo Pyysalo, Yoshinobu Kano, and Jun'ichi Tsujii. 2009. Overview of BioNLP'09 Shared Task on Event Extraction. In *Proceedings of the BioNLP 2009 Workshop Companion Volume for Shared Task*, pages 1–9, Boulder, Colorado. Association for Computational Linguistics.

Jin-Dong Kim, Sampo Pyysalo, Tomoko Ohta, Robert Bossy, Ngan Nguyen, and Jun'ichi Tsujii. 2011. Overview of BioNLP Shared Task 2011. In *Proceedings of BioNLP Shared Task 2011 Workshop*, pages 1–6, Portland, Oregon, USA. Association for Computational Linguistics.

Yoon Kim. 2014. Convolutional neural networks for sentence classification. *arXiv preprint arXiv:1408.5882*.

M. Krallinger, O. Rabal, S.A. Akhondi, M.P. Pérez, J. Santamaría, G.P. Rodríguez, G. Tsatsaronis, A. Intxaurrondo, J.A. López, U. Nandal, E. Van Buel, A. Chandrasekhar, M. Rodenburg, A. Laegreid,

M. Doornenbal, J. Oyarzabal, A. Lourenço, and A. Valencia. 2017. Overview of the biocreative vi chemical-protein interaction track. In *Proceedings of the BioCreative VI Workshop*, pages 141–146.

Sangrak Lim, Kyubum Lee, and Jaewoo Kang. 2018. Drug drug interaction extraction from the literature using a recursive neural network. *PloS one*, 13(1):e0190926.

Yang Liu, Furu Wei, Sujian Li, Heng Ji, Ming Zhou, and Houfeng Wang. 2015. A dependency-based neural network for relation classification. *arXiv preprint arXiv:1507.04646*.

Mingbo Ma, Liang Huang, Bing Xiang, and Bowen Zhou. 2015. Dependency-based convolutional neural networks for sentence embedding. *arXiv preprint arXiv:1507.01839*.

Amit Majumder, Asif Ekbal, and Sudip Kumar Naskar. 2016. Biomolecular event extraction using a stacked generalization based classifier. In *Proceedings of the 13th International Conference on Natural Language Processing*, pages 55–64.

David McClosky. 2010. *Any domain parsing: automatic domain adaptation for natural language parsing*. Ph.D. thesis, Department of Computer Science, Brown University.

David McClosky, Mihai Surdeanu, and Christopher D Manning. 2011. Event extraction as dependency parsing. In *Proceedings of the 49th Annual Meeting of the Association for Computational Linguistics: Human Language Technologies-Volume 1*, pages 1626–1635. Association for Computational Linguistics.

Marie-Catherine de Marneffe, Bill MacCartney, and Christopher Manning. 2006. Generating typed dependency parses from phrase structure parses. In *Proceedings of LREC-06*, pages 449–454.

Farrokh Mehryary, Jari Björne, Tapio Salakoski, and Filip Ginter. 2017. Combining support vector machines and LSTM networks for chemical-protein relation extraction. In *Proceedings of the BioCreative VI Workshop*, pages 175–179.

Tomas Mikolov, Ilya Sutskever, Kai Chen, Greg S Corrado, and Jeff Dean. 2013. Distributed representations of words and phrases and their compositionality. In C. J. C. Burges, L. Bottou, M. Welling, Z. Ghahramani, and K. Q. Weinberger, editors, *Advances in Neural Information Processing Systems 26*, pages 3111–3119. Curran Associates, Inc.

Makoto Miwa and Sophia Ananiadou. 2013. NaCTeM EventMine for BioNLP 2013 CG and PC tasks. In *Proceedings of the BioNLP Shared Task 2013 Workshop*, pages 94–98, Sofia, Bulgaria. Association for Computational Linguistics.

Makoto Miwa and Mohit Bansal. 2016. End-to-end relation extraction using lstms on sequences and tree structures. *arXiv preprint arXiv:1601.00770*.

Makoto Miwa, Sampo Pyysalo, Tadayoshi Hara, and Jun'ichi Tsujii. 2010. A comparative study of syntactic parsers for event extraction. In *Proceedings of the 2010 Workshop on Biomedical Natural Language Processing*, BioNLP '10, pages 37–45, Stroudsburg, PA, USA. Association for Computational Linguistics.

Makoto Miwa, Sampo Pyysalo, Tomoko Ohta, and Sophia Ananiadou. 2013. Wide coverage biomedical event extraction using multiple partially overlapping corpora. *BMC bioinformatics*, 14(1):175.

Makoto Miwa, Paul Thompson, and Sophia Ananiadou. 2012. Boosting automatic event extraction from the literature using domain adaptation and coreference resolution. *Bioinformatics*, 28(13):1759–1765.

Raymond J Mooney and Razvan C Bunescu. 2006. Subsequence kernels for relation extraction. In *Advances in neural information processing systems*, pages 171–178.

Claire Nédellec, Robert Bossy, Jin-Dong Kim, Jung-Jae Kim, Tomoko Ohta, Sampo Pyysalo, and Pierre Zweigenbaum. 2013. Overview of BioNLP Shared Task 2013. In *Proceedings of the BioNLP Shared Task 2013 Workshop*, pages 1–7, Sofia, Bulgaria. Association for Computational Linguistics.

Nanyun Peng, Hoifung Poon, Chris Quirk, Kristina Toutanova, and Wen-tau Yih. 2017a. Cross-sentence n-ary relation extraction with graph lstms. *arXiv preprint arXiv:1708.03743*.

Yifan Peng, Anthony Rios, Ramakanth Kavuluru, and Zhiyong Lu. 2017b. Chemical-protein relation extraction with ensembles of svm, cnn, and rnn models. In *Proceedings of the BioCreative VI Workshop*, pages 147–150.

S. Pyysalo, F. Ginter, H. Moen, T. Salakoski, and S. Ananiadou. 2013. Distributional semantics resources for biomedical text processing. In *Proceedings of LBM 2013*, pages 39–44.

Chanqin Quan, Lei Hua, Xiao Sun, and Wenjun Bai. 2016. Multichannel convolutional neural network for biological relation extraction. *BioMed research international*, 2016.

Sebastian Riedel and Andrew McCallum. 2011. Fast and robust joint models for biomedical event extraction. In *Proceedings of the Conference on Empirical Methods in Natural Language Processing*, pages 1–12. Association for Computational Linguistics.

Sebastian Riedel, David McClosky, Mihai Surdeanu, Andrew McCallum, and Christopher D. Manning. 2011. Model Combination for Event Extraction in BioNLP 2011. In *Proceedings of BioNLP Shared Task 2011 Workshop*, pages 51–55, Portland, Oregon, USA. Association for Computational Linguistics.

Tim Rocktäschel, Torsten Huber, Michael Weidlich, and Ulf Leser. 2013. Wbi-ner: The impact of domain-specific features on the performance of identifying and classifying mentions of drugs. In *Second Joint Conference on Lexical and Computational Semantics (* SEM), Volume 2: Proceedings of the Seventh International Workshop on Semantic Evaluation (SemEval 2013)*, volume 2, pages 356–363.

Sunil Kumar Sahu and Ashish Anand. 2017. Drug-drug interaction extraction from biomedical text using long short term memory network. *arXiv preprint arXiv:1701.08303*.

I. Segura-Bedmar, P. Martínez, and D. Sánchez-Cisneros. 2011. The 1st DDIExtraction-2011 challenge task: Extraction of Drug-Drug Interactions from biomedical texts. In *Proceedings of the 1st Challenge Task on Drug-Drug Interaction Extraction 2011: 7 Sep 2011; Huelva, Spain*, pages 1–9.

Isabel Segura-Bedmar, Paloma Martínez, and María Herrero Zazo. 2013. SemEval-2013 Task 9 : Extraction of Drug-Drug Interactions from Biomedical Texts (DDIExtraction 2013). In *Second Joint Conference on Lexical and Computational Semantics (*SEM), Volume 2: Proceedings of the Seventh International Workshop on Semantic Evaluation (SemEval 2013)*, pages 341–350, Atlanta, Georgia, USA. Association for Computational Linguistics.

Philippe Thomas, Mariana Neves, Illés Solt, Domonkos Tikk, and Ulf Leser. 2011. Relation Extraction for Drug-Drug Interactions using Ensemble Learning. In *Proc. of the 1st Challenge task on Drug-Drug Interaction Extraction (DDIExtraction 2011) at SEPLN 2011*, page 11–18, Huelva, Spain.

Ioannis Tsochantaridis, Thorsten Joachims, Thomas Hofmann, and Yasemin Altun. 2005. Large margin methods for structured and interdependent output variables. *Journal of Machine Learning Research (JMLR)*, 6(Sep):1453–1484.

Sofie Van Landeghem, Thomas Abeel, Bernard De Baets, and Yves Van de Peer. 2011. Detecting entity relations as a supporting task for biomolecular event extraction. In *Proceedings of the BioNLP Shared Task 2011 Workshop*, pages 147–148. Association for Computational Linguistics.

Deepak Venugopal, Chen Chen, Vibhav Gogate, and Vincent Ng. 2014. Relieving the computational bottleneck: Joint inference for event extraction with high-dimensional features. In *Proceedings of the 2014 Conference on Empirical Methods in Natural Language Processing (EMNLP)*, pages 831–843.

Yan Xu, Lili Mou, Ge Li, Yunchuan Chen, Hao Peng, and Zhi Jin. 2015. Classifying relations via long short term memory networks along shortest dependency paths. In *Proceedings of the 2015 Conference on Empirical Methods in Natural Language Processing*, pages 1785–1794.

Daojian Zeng, Kang Liu, Siwei Lai, Guangyou Zhou, and Jun Zhao. 2014. Relation classification via convolutional deep neural network. In *Proceedings of COLING 2014, the 25th International Conference on Computational Linguistics: Technical Papers*, pages 2335–2344.

Yijia Zhang, Wei Zheng, Hongfei Lin, Jian Wang, Zhihao Yang, and Michel Dumontier. 2017. Drug–drug interaction extraction via hierarchical rnns on sequence and shortest dependency paths. *Bioinformatics*.

Deyu Zhou, Lei Miao, and Yulan He. 2018. Position-aware deep multi-task learning for drug–drug interaction extraction. *Artificial intelligence in medicine*.

BioAMA: Towards an End to End BioMedical Question Answering System

Vasu Sharma*, Nitish Kulkarni*, Srividya Pranavi Potharaju*,
Gabriel Bayomi*, Eric Nyberg, Teruko Mitamura
Language Technologies Institute
School Of Computer Science
Carnegie Mellon University
`[vasus, nitishkk, spothara, gbk, ehn, teruko] @cs.cmu.edu`

Abstract

In this paper, we present a novel Biomedical Question Answering system, *BioAMA*: "Biomedical Ask Me Anything" on task 5b of the annual BioASQ challenge (Balikas et al., 2015). We focus on a wide variety of question types including factoid, list based, summary and yes/no type questions that generate both exact and well-formed 'ideal' answers. For summary-type questions, we combine effective IR-based techniques for retrieval and diversification of relevant snippets for a question to create an end-to-end system which achieves a ROUGE-2 score of 0.72 and a ROUGE-SU4 score of 0.71 on ideal answer questions (7% improvement over the previous best model). Additionally, we propose a novel Natural Language Inference (NLI) based framework to answer the yes/no questions. To train the NLI model, we also devise a transfer-learning technique by cross-domain projection of word embeddings. Finally, we present a two-stage approach to address the factoid and list type questions by first generating a candidate set using NER taggers and ranking them using both supervised and unsupervised techniques.

1 Introduction

In the era of ever advancing medical sciences and the age of the internet, a remarkable amount of medical literature is constantly being posted online. This has led to a need for an effective retrieval and indexing system which can allow us to extract meaningful information from these vast knowledge sources. One of the most effective and natural ways to leverage this huge amount of data in real life is to build a Question Answering (QA) system which will allow us to directly query this data and extract meaningful and structured information in a human readable form.

Our key novel contributions are as follows:

1. We achieve state of the art results in automatic evaluation measures for the ideal answer questions in Task 5b of the BioASQ dataset, yielding a 7% improvement over the previous state of the art system (Chandu et al., 2017).

2. We introduce a novel NLI-based approach for answering the yes/no style questions in the BioASQ dataset. We model this as a Textual Entailment (TE) problem and use Hierarchical Convolutional Neural Network based Infersent models (Conneau et al., 2017) to answer the question. To address the challenge of inadequate training data, we also introduce a novel embedding projection technique which allows for effective transfer learning from models trained on larger datasets with a different vocabulary to work well on the much smaller BioASQ dataset.

3. We present two-stage approach to answer factoid and list type questions. By using an ensemble of biomedical NER taggers to generate a candidate answer set, we devise unsupervised and supervised ranking algorithms to generate the final predictions.

4. We improve upon the MMR framework for relevant sentence selection from the chosen snippets that was introduced in the work of Chandu et al. (2017). We experiment with a number of more informative similarity metrics to replace and improve upon the baseline Jaccard similarity metric.

109

Proceedings of the BioNLP 2018 workshop, pages 109–117
Melbourne, Australia, July 19, 2018. ©2018 Association for Computational Linguistics

2   Relevant Literature

Biomedical Question answering has always been a hot topic of research among the QA community at large due to the relative significance of the problem and the challenge of dealing with a non standard vocabulary and vast knowledge sources. The BioASQ challenge has seen large scale participation from research groups across the world. One of the most prominent among such works is from Chandu et al. (2017) who experiment with different biomedical ontologies, agglomerative clustering, Maximum Marginal Relevance (MMR) and sentence compression. However, they only address the ideal answer generation with their model. Peng et al. (2015) in their BioASQ submission use a 3 step pipeline for generating the exact answers for the various question types. The first step is question analysis where they subdivide each question type into finer categories and classify each question into these subcategories using a rule based system. They then perform candidate answer generation using POS taggers and use a word frequency-based approach to rank the candidate entities. Wiese et al. (2017) propose a neural QA based approach to answer the factoid and list type questions where they use FastQA: a machine comprehension based model (Weissenborn et al., 2017) and pre-train it on the SquaD dataset (Rajpurkar et al., 2016) and then finetune it on the BioASQ dataset. They report state of the art results on the Factoid and List type questions on the BioASQ dataset. Another prominent work is from Sarrouti and Alaoui (2017) who handle the generation of the exact answer type questions. They use a sentiment analysis based approach to answer the yes/no type questions making use of SentiWord-Net for the same. For the factoid and list type questions they use UMLS metathesaurus and term frequency metric for extracting the exact answers.

3   The BioASQ challenge

BioASQ challenge (Balikas et al., 2015) is a large scale biomedical question answering and semantic indexing challenge, which has been running as an annual competition since 2013. We deal with the Phase B of the challenge which deals with large scale biomedical question answering. The dataset provides a set of questions and snippets from PubMed, which are relevant to the specific question. It also provides users with a question type and urls of the relevant PubMed articles it-self. The 5b version of this dataset consists of 1,799 questions in 3 distinct categories:

1. **Factoid type**: This question type has a single entity as the ground truth answer and expects the systems to output a set of entities ordered by relevance; systems are evaluated using the mean reciprocal rank (Radev et al., 2003) of the answer entities with reference to the ground truth answer entity.

2. **List type**: This answer type expects the system to return an unordered list of entities as answer and evaluates them using a F-score based metric against a list of reference answer entities which can vary in number.

3. **Yes/No type**: This question type asks the systems to answer a given question with a binary output namely yes or no. The questions typically require reasoning and inference over the evidence snippets to be able to answer the questions correctly.

The dataset expected the participants to generate two types of answers, namely, exact and ideal answers. In ideal answers, the systems are expected to generate a well formed paragraph for each of the question types which explains the answer to the question. They call these answers 'ideal' because it is what a human would expect as an answer by a peer biomedical scientist. In the exact answers the systems are expected to generate "yes" or "no" in the case of yes/no questions, named entities in the case of factoid questions and list of named entities in the case of list questions.

4   Ideal Answers

This section describes our efforts to address the ideal answer category on BioASQ.

Our pipeline for ideal answers has three stages. The first stage involves pre-processing of answer snippets and ranking of answer sentences by various retrieval models described in the following sections. The retrieval model scores form the soft positional component introduced in the MMR algorithm. We perform sentence selection next, where we select the top 10 best sentences for generating an ideal answer. The third and final stage involves tiling together the selected sentences to generate a coherent, non redundant, ideal answer for the given question as mentioned in (Chandu et al., 2017). The subsequent subsections explain

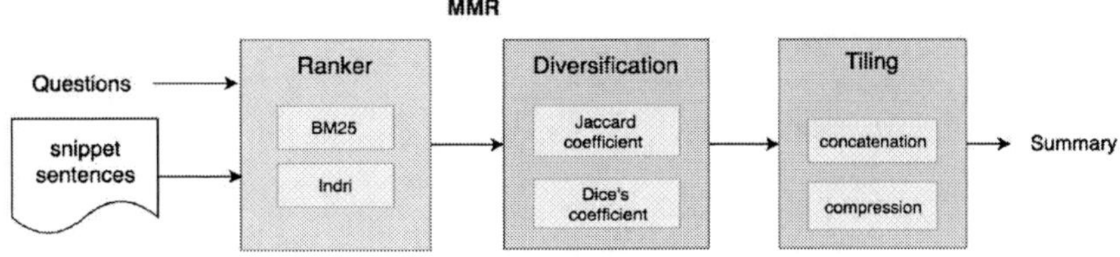

Figure 1: Pipeline for ideal answer generation

the pipeline for ideal answer type questions in detail (see Figure 1).

4.1 Question-Sentence Retrieval

In this section we describe various approaches which were adapted to improve the initial retrieval of candidate sentences. We used the standard BM25 algorithm with custom pre-processing of excluding medical entities from stop word removal.

4.1.1 Indri

Indri (Strohman et al., 2005) is a retrieval model based on the use of statistical language models and query likelihood. We employed a two-stage smoothing that considers characteristics of both the question and answer sentences.

The Indri score for a candidate sentence is estimated in a collection (C) of snippets as follows:

$$p(q_i|d) = (1 - \lambda)p_{mle}(q_i|d) + \lambda p_{mle}(q_i|C) \quad (1)$$

$$p_{mle}(q_i|d) = \frac{tf + \mu p_{mle}(q_i|C)}{length(d) + C} \quad (2)$$

$$p_{mle}(q_i|C) = \frac{ctf}{length(C)} \quad (3)$$

where, λ is the coefficient for linear interpolation based smoothing that accounts for question length smoothing and also compensates for differences in the word importance (gives idf-effects). Since the questions are of moderate length, after tuning, the best value of λ is attained at 0.75

In equation 2, μ is parameter for Bayesian smoothing using Dirichlet priors used for sentence length normalization, improving the estimates of the sentence sample.Since sentences of snippets can be of varying lengths, after tuning, the best value of μ is attained at 5000.

4.2 Sentence Selection

Once the top most relevant snippets have been chosen, we want to choose sentences from these snippets which are most relevant to a specific question. In this section we demonstrate how this selection is done.

4.2.1 MMR

We use the Maximum Marginal Relevance (MMR) algorithm (Forst et al., 2009) as the baseline for sentence selection. In contrast to the basic Jaccard similarity metric used in previous work (Chandu et al., 2017), we experimented with other similarity measures which consistently perform better than the Jaccard baseline. MMR ensures the selected set contains non-redundant yet complete information. The sentences are selected based on two aspects, the sentence's relevance to the question and how different it is to the already selected sentences. At each step we select a sentence to append to the ranking based on the equation below.

$$s_i = \arg\max_{s_j \in R \setminus S}(\lambda \cdot sim(q, s_i)$$
$$- (1 - \lambda) \cdot max_{s \in S}(sim_{sent}(s_i, s_j))) \quad (4)$$

We define a custom similarity metric between sentences which uses positional values of sentences from the initial ranking as follows:

$$sim_{sent}(s_i, s_j) = (1 - \beta) \cdot (1 - \frac{rank(d_i)}{n})$$
$$+ \beta \cdot sim(s_i, s_j) \quad (5)$$

Here, $sim_{sent}(s_i, s_j)$ is the sentence to sentence similarity, $sim(q, s_i)$ is the question - sentence similarity, $rank(d_i)$ is the rank of the snippet d_i, which contains the sentence s_i, S are Sentences already selected for summary i.e. which are ranked above this position. In the above equation, we tried various metrics to account for the sentence to sentence similarity. In cases where β is non-zero, equation 4 is identified as our SoftMMR which includes soft scoring based on sentence position.

β	Configuration	Rouge-2	Rouge-SU4
-	baseline	0.7064	0.6962
0.5	BM25, Jaccard	0.7175	0.7110
0.5	BM25, Dice	0.7193	0.7106
0.6	BM25, Dice	0.7133	0.7053
0.6	BM25, Jaccard	0.7133	0.7053
0.5	**Indri, Jaccard**	**0.7206**	**0.7135**
0.5	Indri, Dice	0.7113	0.7052

Table 1: ROUGE scores for different experiments on similarity metrics for extractive summarization

4.2.2 Dice's similarity Coefficient (DSC)

Dice's similarity Coefficient (DSC) (Srensen, 1948) is a quotient of similarity between two samples and ranges between 0 and 1 calculated as

$$dsc = (2 * n_t)/(n_x + n_y)$$

where n_t is the number of character bigrams found in both strings, n_x is number of bigrams in string x and n_y is the number of bigrams in string y. We used Dice coefficient as a similarity metric between two sentences in 5

4.3 Evaluation

The pipeline described above is primarily designed to improve the ROUGE evaluation metric (Lin, 2004). Although a higher ROUGE score does not necessarily reflect improved human readability, MMR can improve readability by reducing redundancy in generated answers. Results for ideal answers for Task 5 phase b are shown in Table 1. We also compare our results with other state of the art approaches in Table 4.

5 Exact answers

Exact answers represent the subset of the BioASQ task where the responses are not structured paragraphs, but instead either a single entity (*yes/no* types) or a combination of named entities (*factoid* or *list* types) that compose the correct reply to the given query. The main idea refers to evaluating if a response is able to capture the most important components of an answer. For factoid or list types of questions, we must return a list of the most likely entities to compose the answer. The main difference between them is that ground truth for *factoid* questions is composed of only one correct answer and the evaluation method is Mean Reciprocal Rank (MRR). However, the ground truth for

list is an actual list of correct answers with varying length, which uses F-measure as an evaluation metric. The BioASQ submission format allows everyone to submit 5 ranked answers for *factoid* and 1 to 10 answers for *list*. For *yes/no* questions, the ground truth is simply the yes or no label, using F-measure as an evaluation metric.

5.1 Yes/No type questions

Although yes/no questions require a simple binary response, calculating yes/no responses for the BioASQ question can be challenging:

1. There is an inherent class-bias towards the questions answered by yes in the dataset;

2. The dataset is quite small for training a complex semantic classifier;

3. An effective model must perform reasoning and inference using the limited information it has available, which is extremely difficult even for non-expert humans.

Due to the nature of the question type, these questions can not be simply classified by using word-level features. Learning the semantic relationship between the question and the sentences in the documents is quite elemental to solving this task. Hence, we present a Natural Language Inference (NLI)-based system that learns if the assertions made by the questions are true in the context of the documents. As a part of this system, we first generate assertions from questions and evaluate the entailment or contradiction of these assertions using a Recognizing Textual Entailment (RTE) model. We then use these entailment scores for all the sentences in the snippets or documents to heuristically evaluate if the answer to the yes/no question.

5.1.1 Assertion Extraction

The first step towards answering the question is to identify the assertions made by the question. For this, we use a statistical natural language parser to identify the syntactical structure in the question. We, then, heuristically generate assertions from the questions.

Consider the following example question:

Is the monoclonal antibody Trastuzumab (Herceptin) of potential use in the treatment of prostate cancer?

Upon parsing of this question, we have the phase constituents of the question. Almost all yes/no questions have a standard format that begins with an auxiliary verb followed by a noun phrase. In this example, we can toggle the question word with the first noun phrase to generate the assertion:

The monoclonal antibody Trastuzumab (Herceptin) is of potential use in the treatment of prostate cancer.

In a similar manner, we then create positive assertions for all *yes/no* questions. As a simple extension to this, we can also create negative assertions by using *not* along with the auxiliary verbs.

5.1.2 Recognizing Textual Entailment

The primary goal of our NLI module is to infer if any of the sentences among the answer snippets entails or contradicts the assertion posed by the question. We segmented the answer snippets for each question to produce a set of assertion-sentence pairs. To then evaluate if these assertions can be inferred or refuted from the sentences, we built a Recognizing Textual Entailment (RTE) model using the *InferSent* model (Conneau et al., 2017), which computes sentence embeddings for every sentence and has been shown to work well on NLI tasks. In training *InferSent*, we experienced two major challenges:

1. The number of assertion-sentence pairs in BioASQ is too few to train the textual entailment model effectively.

2. The models that are pre-trained on SNLI (Bowman et al., 2015) datasets use GLOVE (Pennington et al., 2014) embeddings that cannot be used for biomedical corpora which have quite different characteristics and vocabulary compared to the corpora that GLOVE was trained on.

However, we have pre-trained embeddings available that were trained on PubMed and PMC texts along with Wikipedia articles (Pyysalo et al., 2013). To leverage these embeddings, we implemented an embedding-transformation methodology to projecting the PubMed embeddings to GLOVE embedding space and then fine tune the pre-trained *InferSent* on the BioASQ dataset for textual entailment. The hypothesis is that, since both the embeddings had a significant fraction of documents in common (Wikipedia corpus), by transforming the embeddings from one space to another, the sentence embeddings from the model would still represent a lot of the semantic features of the input sentences that can subsequently used for classifying textual entailment. For this task, we explore both linear and non-linear methods of embedding transformation.

While simple, a linear projection of embeddings from one space to another has shown to be quite effective for a lot of multi-domain tasks. By imposing an orthogonality constraint on the project matrix, we model this problem as an orthogonal Procrustes problem:

Let d_p and d_g be the embedding dimensions of PubMed embeddings and GLOVE embeddings respectively. If E_p and E_g are the matrices of PubMed embeddings ($N \times d_p$) and their corresponding GLOVE embeddings ($N \times d_g$) for the words that both the embeddings have in common (N), the projection matrix ($d_g \times d_p$) can be computed as,

$$W^* = \arg\min_{W} \|W E_p^\mathsf{T} - E_g^\mathsf{T}\|$$

subject to the constraint that W is orthogonal. The solution to this optimization problem is given by using the singular value decomposition of $E_g^\mathsf{T} E_p$, i.e. $W^* = UV^\mathsf{T}$ where $E_g^\mathsf{T} E_p = U\Sigma V^\mathsf{T}$ With this simple linear transformation, we then computed the transformed embeddings for all the words in the PubMed embeddings that are not present in the GLOVE embeddings.

We also explore a non-linear transformation using a feed-forward neural network where the the objective is to learn function f such that, $f(e_p; \theta) = e_g$ where, e_p and e_g are PubMed and GLOVE embeddings respectively. We model f using a deep neural network with parameters θ, and train using the common words in both the embeddings.

The transformed embeddings from these models were used in conjunction with the pre-trained *InferSent* model to encode the semantic features of the biomedical sentences as sentence embeddings. Subsequently, we employ these sentence embeddings of the assertion-sentence pairs for a particular question to train a three-way neural classifier to predict if the relationship between the two is entailment, contradiction or neither.

It is worth noting here that the embedding transformation techniques that we implemented are not specific to the NLI tasks and, in fact, enable transfer learning of a much broader set of tasks on smaller datasets like BioASQ by using the pre-trained models on large datasets of other domains and fine-tuning on the smaller dataset.

5.1.3 Classification

As a final step, we use the textual entailment results for each assertion-sentence pair generated to heuristically classify the answer as *yes* or *no*. Since our system comprises multiple stages with the errors of each cascading to the final stage, we do not get perfect entailment results for the pairs. However, since we have a lot of pairs, we aggregate these entailment scores to compute the overall entailment or contraction scores to reduce the effect of accumulated errors for individual pairs on classification.

We used a simple unsupervised approach for classification by just comparing the overall entailment and contradiction scores, i.e. if the total number of snippet sentences that entail the assertion are N_e and the total number of snippet sentences that contradict are N_c, then,

$$\text{answer}_q = \begin{cases} \text{yes} & \text{if } N_e \geq N_c \\ \text{no} & \text{otherwise} \end{cases}$$

The end-to-end architecture of our system from the input questions and snippets to the answer is shown Figure 2.

5.1.4 Experimental Details

For parsing the questions, we used BLLIP reranking parser (Charniak and Johnson, 2005) (Charniak-Johnson parser) and used the model `GENIA+PubMed` for biomedical text. For training the textual entailment classifier using *InferSent*'s sentence embeddings, we used Stanford's SNLI dataset (Bowman et al., 2015) to achieve a test-set accuracy of 84.7%.

5.1.5 Results

The performance of the system on yes/no questions on the training set of phase 5b has been tabulated in table 2. While the accuracies are better than a random classifier, the task is far from being solved. Nonetheless, the classifier does handle the class bias in the training data and performance similarly on both the categories of answers. Moreover, this classifier achieved the second best test accuracy of 65.6% on phase 5 of BioASQ 5b (Table 4). While we implemented a simple heuristic based answer-classifier, we believe that a supervised classifier using the sentence embeddings as

Category	Accuracy (%)
Yes	56.5 (252/444)
No	58.9 (33/56)
Overall	57.0 (285/500)

Table 2: Class-wise accuracies on yes/no questions in training set of BioASQ Phase 5b

well as fine-tuning of the textual entailment classifier on BioASQ dataset would considerably enhance the overall performance of the system.

5.2 Factoid & List Type Questions

Most of the state-of-the-art models for this task involve training end-to-end deep neural architectures to identify a subset of entities (or phrases) from the relevant snippets that are most likely to answer the question. But, owing to the small size of the dataset, we cannot effectively train such models on the BioASQ dataset. Hence, we adopted a two-stage approach that first finds a set of entities that could potentially answer the question and a supervised classifier to rank the entities on the basis of their likelihood of answering the question.

For devising the model and evaluation, we primarily focused on factoid type questions since the methodology for the list-type question would be largely similar and different only in the number of top entities returned.

5.2.1 Candidate Selection

We found that the most critical step in the answer generation process is to identify the set of potential answer candidates that can be fed into a classifier or ranker to identify the best candidates. At first, in order to accomplish this, we used Named Entity Recognition (NER) taggers to form a set of candidate answers. The taggers that we used include Gram-CNN (Zhu et al., 2017), LingPipe(Carpenter, 2007) and PubTator (Wei et al., 2013). To analyze the effectiveness of these taggers, we performed an analysis on BioASQ training set 5b by evaluating the fraction of questions whose answers are included in the candidate entity set by the taggers.

Table 3 shows the relative performances of the three taggers, their union as well as intersection on train dataset of BioASQ 5b factoid type questions. A question is exactly answered if a tagger tags an entity that matches an answer exactly, and it is partially answered if there is a non-zero over-

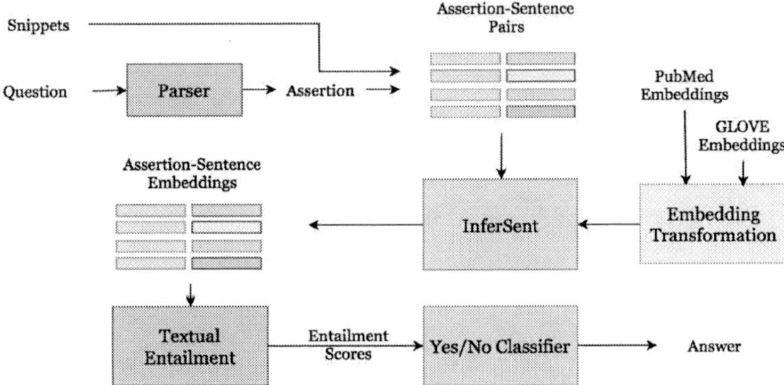

Figure 2: The complete system for yes/no answer classification using a question and relevant snippets

lap with an entity tagged and an answer for the question. We can notice that PubTator and Ling-Pipe have a good recall with relatively low precision, while Gram CNN has high recall but low precision. However, the final results with the Named Entity Taggers were not aligned with our expectations. This is mostly because the answers for BioASQ are usually a combination of BioNERs and complementary words, making it hard to define a pruning method that is able to yield satisfactory results. Surprisingly, a group of candidates formed of the 100 most frequent n-grams (n from 1 to 4) from the snippets' sentences were a better candidate group than the NER approach for our supervised ranking method (with NER taggers used as features instead of candidate entities).

5.2.2 Classification Features

Upon computing the set of candidate answers, we use the question q, set of relevant snippet sentences S and entity type t_i to devise a feature vector for each individual entity e_i that comprises the following features:

- BM25 Score: The BM25 scores for all the sentences are computed with the question as the query. Then, the scores of the sentence that contain the entity are aggregated to compute the BM25 score for the entity, i.e.

$$\text{Score}_{BM25}(e_i) = \sum_{s \in S} \text{Score}_{BM25}(e_i) \cdot \mathbb{1}(s, e_i)$$

where $\mathbb{1}(s, e_i)$ is 1 iff sentence s has entity e_i.
- Indri Score: Computed in the same manner as BM25 score in (i)
- Number of Sentences: Number of sentences $s \in S$ that contain the entity e_i

- NER Tagger: A multinomial feature that represents which tagger among PubTator, Ling-Pipe and GramCNN the entity was extracted with. This feature is included to identify the relative strengths of the different taggers.
- Tf Idf: The aggregate Tf-Idf scores of the entity with S as the set of documents
- Entity Type: Is a boolean feature that is 1 if the type of the entity (for example, *gene*) is present in the question, and 0 otherwise.
- Relative Frequency: The amount of times the entity appears on the snippets' sentences divided by the total appearance of all of the relevant entities.
- Query Presence: Is a boolean feature that is 1 if the query contains the entity completely and 0 otherwise.

NER Tags	% of questions		% of
	Exactly Answered	Partially Answered	tokens extracted
PubTator	32.05	72.15	52.27
Gram CNN	34.90	99.03	94.97
LingPipe	26.67	76.75	11.06
Union	49.04	99.65	99.25
Intersection	16.29	38.00	3.33

Table 3: Baseline recall of different NER Taggers measured by the fraction of questions that can be answered by an ideal classifier if the candidates are chosen using the tagger. We also measure precision as the fraction of total unique tokens from the documents that are tagged.

5.2.3 Unsupervised Ranking

As a baseline, we first present an unsupervised ranking system for the candidate answers. In this

Model	Exact Answers Yes/No type Accuracy (%)	Exact Answers Factoid type MRR	Exact Answers List type F1 score	Ideal Answers All types ROUGE-2
(Chandu et al., 2017)	-	-	-	0.653
(Peng et al., 2015)	**0.714**	0.272	0.187	-
(Wiese et al., 2017)	-	**0.392**	**0.361**	-
Sarrouti and Alaoui (2017)	0.461	0.207	0.243	0.577
BioAMA(Ours)	0.653	0.195	0.234	**0.721**

Table 4: Comparison of our model with other state of the art approaches

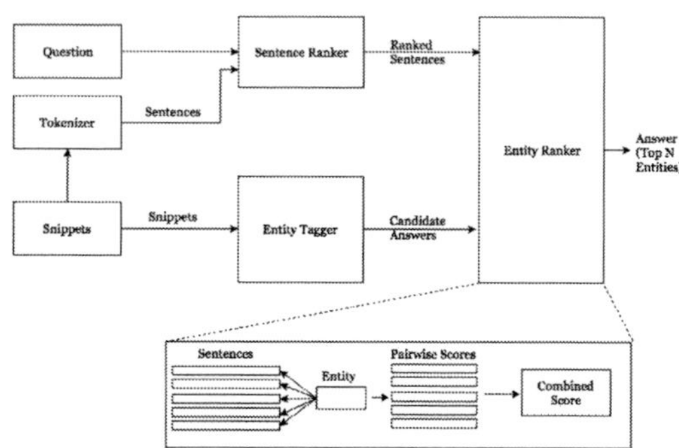

Figure 3: Unsupervised generation of factoid/list type answers using NER taggers and BM25 retrieval model

system, the snippet sentences are first ranked using the BM25 model. Then, for each entity, a score is computed by aggregating the BM25 scores of the sentences in which the entity is present. The rationale for this is that the entities in the top ranked sentences are more likely to be the answers. This entity score (which is equivalent to the BM25 score described in 5.2.2) is then used to rank the entities and return the top k entities as answers to the question. The overall unsupervised system is shown in Figure 3.

5.2.4 Learning To Rank

In order to rank the candidate entities in a supervised way, we use a ranking classifier based on the features described in 5.2.2. For ranking, we choose point-wise ranking classifiers over pairwise and list-wise, because it yields similar results to ranking methods with a less time-consuming and computationally expensive approach. We use a traditional SVM-Light (Joachims, 1998) implementation for point-wise ranking. The data for supervision is derived from the actual answers and candidate entities are ranked based on their over-

lap with the actual answers.

Once we rank the entities, we use a naive approach of merely taking top 5 entities as answers for factoid type and top 10 for list-type. One could, however, devise a separate model for identifying the number of top entities to return as answers for the list-type answers.

We found that using just the NER entities as the answer candidates, the classifier could achieve an MRR of 0.06 on factoid type questions and an F-measure of 0.18 for list type questions. However, by having all the n-grams ($n = 1, 2, 3, 4$) from the snippets as candidate answers and using NER tags as LeToR features, the performance was improved to an MRR of 0.195 on factoid type questions and an F1 score of 0.234 on list type questions. The results are summarized in Table 4.

6 Conclusion and Future Work

In this paper, we present a framework for tackling both ideal and exact answer type questions and obtain state of the art results on the ideal answer type questions on the BioASQ dataset. For exact answers, we incorporate neural entailment models along with a novel embedding transformation technique for answering yes/no questions, and employ LeToR ranking models to answer factoid/list based questions. For ideal answers, we improve the IR component of extractive summarization. Although this improves ROUGE scores considerably, the human readability aspect of the generated summary answer is not greatly improved. As future directions, we believe that effective abstractive summarization based approaches like Pointer Generator Networks (See et al., 2017) and Reinforcement Learning based techniques (Paulus et al., 2017) would improve the human readability of ideal answers. We aim to continue our research in this direction to achieve a good balance between ROUGE score and human readability.

References

Georgios Balikas, Anastasia Krithara, Ioannis Partalas, and George Paliouras. 2015. Bioasq: A challenge on large-scale biomedical semantic indexing and question answering. In *Revised Selected Papers from the First International Workshop on Multimodal Retrieval in the Medical Domain - Volume 9059*, pages 26–39, New York, NY, USA. Springer-Verlag New York, Inc.

Samuel R. Bowman, Gabor Angeli, Christopher Potts, and Christopher D. Manning. 2015. A large annotated corpus for learning natural language inference. In *Proceedings of the 2015 Conference on Empirical Methods in Natural Language Processing (EMNLP)*. Association for Computational Linguistics.

Bob Carpenter. 2007. Lingpipe for 99.99% recall of gene mentions. In *Proceedings of the Second BioCreative Challenge Evaluation Workshop*, volume 23, pages 307–309.

Khyathi Chandu, Aakanksha Naik, Aditya Chandrasekar, Zi Yang, Niloy Gupta, and Eric Nyberg. 2017. Tackling biomedical text summarization: Oaqa at bioasq 5b. In *BioNLP 2017*, pages 58–66. Association for Computational Linguistics.

Eugene Charniak and Mark Johnson. 2005. Coarse-to-fine n-best parsing and maxent discriminative reranking. In *ACL 2005, 43rd Annual Meeting of the Association for Computational Linguistics, Proceedings of the Conference, 25-30 June 2005, University of Michigan, USA*, pages 173–180.

Alexis Conneau, Douwe Kiela, Holger Schwenk, Loïc Barrault, and Antoine Bordes. 2017. Supervised learning of universal sentence representations from natural language inference data. *CoRR*, abs/1705.02364.

Jan Frederik Forst, Anastasios Tombros, and Thomas Roelleke. 2009. Less is more: Maximal marginal relevance as a summarisation feature. In *Advances in Information Retrieval Theory*, pages 350–353, Berlin, Heidelberg. Springer Berlin Heidelberg.

Thorsten Joachims. 1998. Text categorization with support vector machines: Learning with many relevant features. In *European conference on machine learning*, pages 137–142. Springer.

Chin-Yew Lin. 2004. Rouge: A package for automatic evaluation of summaries. In *Proc. ACL workshop on Text Summarization Branches Out*, page 10.

Romain Paulus, Caiming Xiong, and Richard Socher. 2017. A deep reinforced model for abstractive summarization. *CoRR*, abs/1705.04304.

Shengwen Peng, Ronghui You, Zhikai Xie, Beichen Wang, Yanchun Zhang, and Shanfeng Zhu. 2015. The fudan participation in the 2015 bioasq challenge: Large-scale biomedical semantic indexing and question answering. In *CLEF*.

Jeffrey Pennington, Richard Socher, and Christopher D. Manning. 2014. Glove: Global vectors for word representation. In *Empirical Methods in Natural Language Processing (EMNLP)*, pages 1532–1543.

S. Pyysalo, F. Ginter, H. Moen, T. Salakoski, and S. Ananiadou. 2013. Distributional semantics resources for biomedical text processing. In *Proceedings of LBM 2013*, pages 39–44.

Dragomir Radev, Y Hong Qi, Harris Wu, and Weiguo Fan. 2003. Evaluating web-based question answering systems.

Pranav Rajpurkar, Jian Zhang, Konstantin Lopyrev, and Percy Liang. 2016. Squad: 100, 000+ questions for machine comprehension of text. *CoRR*, abs/1606.05250.

Mourad Sarrouti and Said Ouatik El Alaoui. 2017. A biomedical question answering system in bioasq 2017. In *BioNLP 2017*, pages 296–301. Association for Computational Linguistics.

Abigail See, Peter J. Liu, and Christopher D. Manning. 2017. Get to the point: Summarization with pointer-generator networks. *CoRR*, abs/1704.04368.

Trevor Strohman, Donald Metzler, Howard Turtle, and W. Bruce Croft. 2005. Indri: a language-model based search engine for complex queries.

T. Srensen. 1948. A method of establishing groups of equal amplitude in plant sociology based on similarity of species and its application to analyses of the vegetation on danish commons. In *Kongelige Danske Videnskabernes Selskab*, pages 1–34.

Chih-Hsuan Wei, Hung-Yu Kao, and Zhiyong Lu. 2013. Pubtator: a web-based text mining tool for assisting biocuration. *Nucleic acids research*, 41(W1):W518–W522.

Dirk Weissenborn, Georg Wiese, and Laura Seiffe. 2017. Fastqa: A simple and efficient neural architecture for question answering. *CoRR*, abs/1703.04816.

Georg Wiese, Dirk Weissenborn, and Mariana L. Neves. 2017. Neural question answering at bioasq 5b. *CoRR*, abs/1706.08568.

Qile Zhu, Xiaolin Li, Ana Conesa, and Ccile Pereira. 2017. Gram-cnn: a deep learning approach with local context for named entity recognition in biomedical text. *Bioinformatics*, page btx815.

Phrase2VecGLM: Neural generalized language model–based semantic tagging for complex query reformulation in medical IR

Manirupa Das[†]**, Eric Fosler-Lussier**[†]**, Simon Lin**[‡]**, Soheil Moosavinasab**[‡]**,
David Chen**[‡]**, Steve Rust**[‡]**, Yungui Huang**[‡] **& Rajiv Ramnath**[†]
The Ohio State University[†] & Nationwide Children's Hospital[‡]
{das.65, fosler.1, ramnath.6}@osu.edu
{Simon.Lin, SeyedSoheil.Moosavinasab, David.Chen3,
Steve.Rust, Yungui.Huang}@nationwidechildrens.org

Abstract

In fact-based information retrieval, state-of-the-art performance is traditionally achieved by knowledge graphs driven by knowledge bases, as they can represent facts about and capture relationships between *entities* very well. However, in domains such as medical information retrieval, where addressing specific information needs of complex queries may require understanding query intent by capturing novel associations between potentially *latent concepts*, these systems can fall short. In this work, we develop a novel, completely unsupervised, neural language model–based ranking approach for semantic tagging of documents, using the document to be tagged as a query into the model to retrieve candidate phrases from top–ranked related documents, thus associating every document with *novel related concepts* extracted from the text. For this we extend the word embedding–based generalized language model (GLM) due to (Ganguly et al., 2015), to employ phrasal embeddings, and use the semantic tags thus obtained for downstream query expansion, both directly and in feedback loop settings. Our method, evaluated using the TREC 2016 clinical decision support challenge dataset, shows statistically significant improvement not only over various baselines that use standard MeSH terms and UMLS concepts for query expansion, but also over baselines using human expert–assigned concept tags for the queries, on top of a standard Okapi BM25–based document retrieval system.

1 Introduction

Existing state-of-the-art information retrieval (IR) systems such as knowledge graphs (Su et al., 2015; Sun et al., 2015), or information extraction techniques centered around entity relationships (Ritter et al., 2013), that often rely on some form of weak supervision from ontological or knowledgebase (KB) sources, tend to perform quite reliably on fact-based information retrieval and factoid question answering tasks. However, such systems may be limited in their ability to address the complex information needs of specific types of queries (Roberts et al., 2016; Diekema et al., 2003) in domains such as clinical decision support (Luo et al., 2008) or guided product search (Teo et al., 2016; McAuley and Yang, 2016), due to: 1) complex and subjective, or lengthy nature of the query containing multiple topics, 2) vocabulary mismatch between the query expression and knowledge representations in the document collection, and 3) lack of sufficiently complete knowledge bases of "related concepts", covering *all possible relations* between candidate concepts that may exist in a collection, essential for effectively addressing these types of queries (Hendrickx et al., 2009).

We hypothesize, that similar to human experts who can determine the *aboutness* of an unseen document by recalling meaningful concepts gleaned from similar past experiences via *shared contexts*, a completely unsupervised machine learning model could be trained to associate documents within a large collection with meaningful concepts *discovered* by fully leveraging *shared contexts* within and between documents, thus surfacing "related" concepts specific to the current context (Lin and Pantel, 2002; Halpin et al., 2007; Xu et al., 2014; Kholghi et al., 2015a; Turney and Pantel, 2010; Pantel et al., 2007; Bhagat and Hovy, 2013; Hendrickx et al., 2009). As a trivial example, ordinarily unrelated concepts (noun phrases,

Proceedings of the BioNLP 2018 workshop, pages 118–128
Melbourne, Australia, July 19, 2018. ©2018 Association for Computational Linguistics

in this work) such as "scarlet macaw" and "raccoon" occurring in separate documents d_1 and d_2 may become related by a novel context such as "exotic pets" that may occur as terms in a query or as a phrase in a document d_p which could be related to both d_1 and d_2. If by some means, documents d_1 and d_2 were semantically tagged with the phrase "exotic pets" via d_p, those documents would surface in the event of such a query (Hendrickx et al., 2009; Bhagat and Ravichandran, 2008). This could thus help to better close the vocabulary gap between potential user queries and the documents. To our knowledge, ours is the first work that employs word and phrase-level embeddings for local context analysis in a pseudo-relevance feedback setting (Xu and Croft, 2000), using a *language model-based document ranking framework*, to semantically tag documents with appropriate concepts for use in downstream retrieval tasks (Kholghi et al., 2015a; De Vine et al., 2014; Sordoni et al., 2014; Zhang et al., 2016; Zuccon et al., 2015; Tuarob et al., 2013).

The main contributions of our work, are as follows: 1) We present a novel use for a neural language modeling approach that leverages shared context between documents within a collection via phrase-based embeddings (1, 2, and 3-grams), finding the right trade-off between the local context around each term versus its global context within the collection, incorporating a local context analysis-based pseudo-relevance feedback mechanism(Xu and Croft, 2000) for concept extraction. 2) Our method is fully unsupervised, i.e. it includes no outside sources of knowledge in the training, leveraging instead the *shared contexts* within the document collection itself, via word and phrasal embeddings, mimicking a human that potentially reads through the documents in the collection and uses the seen information to make relevant concept tag judgments on unseen documents. 3) Our method presents a black-box approach for tagging any corpus of documents with meaningful concepts, treating it as a closed system. Thus the concept associations can be pre-computed offline or periodically, as new documents are added to the collection and can reside outside of the document retrieval system, allowing for it to be plugged into any such system, or for the underlying retrieval system to be changed. It is also in contrast to previous approaches to document categorization for retrieval, such as those based on cluster-

ing, e.g. clustering by committee (Lin and Pantel, 2002) or semantic class induction as in (Lin and Pantel, 2001b), LDA-based topic modeling (Blei et al., 2003; Griffiths and Steyvers, 2004; Tuarob et al., 2013) and supervised or active learning approaches (Kholghi et al., 2015a) for concept extraction in information retrieval.

2 Background and Motivation

The problem of *vocabulary mismatch* in information retrieval where *semantic overlap* may exist while there is no *lexical overlap*, can be greatly alleviated by the use of query expansion (QE) techniques; whereby a query is reformulated to improve retrieval performance and obtain *additional relevant documents* by expanding the original query with additional relevant terms, and re-weighting the terms in the expanded query (Xu and Croft, 2000; Rivas et al., 2014). This can also be done by learning *semantic classes* or *related candidate concepts* in the text and subsequently tagging documents or content with these semantic concept tags, that could then serve as a means for either query-document keyword matching, or for query expansion, to facilitate downstream retrieval or question answering tasks (Lin and Pantel, 2002; Xu and Croft, 2000; Lin and Pantel, 2001b; Xu et al., 2014; Bhagat and Ravichandran, 2008; Li et al., 2011; Tuarob et al., 2013; Halpin et al., 2007; Lin and Pantel, 2001a; McAuley and Yang, 2016). This is exactly the approach we adopt in order to achieve query expansion in an automated, fully unsupervised fashion, using a neural language model for local relevance feedback (Xu and Croft, 2000).

A major problem of approaches like LSA (Deerwester et al., 1990) and LDA–based topic modeling (Blei et al., 2003; Griffiths and Steyvers, 2004) is that they only consider word co-occurrences at the level of documents to model term associations, which may not always be reliable. Furthermore, these are parameterized approaches, where the number of topics K is fixed; and the final topics learnt are available as bags of words or n-grams from which topic labels must yet be inferred by an expert. In contrast, word and phrasal embeddings take into account *local co-occurrence information* of terms in the top ranked documents retrieved in response to a query (corresponding to the relevance feedback step in IR). This leads to a better modeling of query versus document term dependencies (Ganguly et al., 2015; Xu and Croft, 2000) lending itself to direct unsupervised extraction of meaningful terms related to a document, and eventually to the query.

Automatic query expansion techniques can be further categorized as either *global* or *local*. While global techniques rely on analysis of a whole collection to discover word relationships, local techniques emphasize analysis of the top-ranked documents retrieved for a query (Xu and Croft, 2000; Manning et al., 2009). Global methods include: (a) query expansion/reformulation with a thesaurus or ontology, e.g. WordNet, UMLS (b) query expansion via automatic thesaurus generation, and (c) techniques like spelling correction (Manning et al., 2009). Local methods adjust a query relative to the documents that initially appear to match the query, which is the basic idea behind our language modeling approach to semantic tagging. Basic local methods comprise: (a) relevance feedback, (b) pseudo-relevance feedback, (or blind relevance feedback), and (c) (global) indirect relevance feedback (Manning et al., 2009). *Pseudo-relevance feedback* automates the manual part of relevance feedback, so that the user gets improved retrieval performance without an extended interaction.

Here, we find an initial set of most relevant documents, then assuming that the top k ranked documents are relevant, relevance feedback is done as before under this assumption. Our proposed method tries to exactly mimic the human user behavior via pseudo-relevance feedback to semantically pre-tag documents that can later aid downstream novel retrieval for direct querying or refined querying. Thus, in our work we combine this local feedback approach with our neural language model, Phrase2VecGLM, as the query mechanism. Using a pseudo-document representation of *top-K TFIDF terms* for the document as a query into the GLM, we make novel use of Phrase2VecGLM, to semantically tag documents with phrases representative of latent concepts within those documents. This makes the collection more readily searchable by use of these tags for query expansion in downstream IR, particularly helpful in our specific use case of medical information retrieval (Luo et al., 2008; Kholghi et al., 2015b; De Vine et al., 2014; Halpin et al., 2007; Li et al., 2011; Zhang et al., 2016). Additionally, our method treats all queries in our dataset as unseen at test time, on which our actual results and gains are reported.

3 Dataset and Task

The TREC Clinical Decision Support (CDS) task track investigates techniques to evaluate biomedical literature retrieval systems for providing answers to generic clinical questions about patient cases (Roberts et al., 2016), with a goal toward making relevant biomedical information more discoverable for clinicians. For the 2016 TREC CDS challenge, actual electronic health records (EHR) of patients, in the form of case reports, typically describing a challenging medical case, as shown in Figure 1 are used. A case report is, for our purposes *a complex query* having a specific information need. There are 30 queries in the challenge dataset, corresponding to such case reports, divided into 3 topic types, at 3 levels of granularity **Note**, **Description** and **Summary** text.

The target document collection is the Open Access Subset of PubMed Central (PMC), containing 1.25 million articles consisting of *title*, *keywords*, *abstract* and *body* sections. In our work, we develop our query expansion method as a blackbox system using only a subset of 100K documents of the entire collection for which human judgments are made available by TREC. This allows us to derive "inferred measures" for *Normalized Discounted Cumulative Gain* (NDCG) and *Precision at 10* (P@10) scores for our evaluation (Voorhees, 2014). However, we evaluate our method on the entire collection of 1.25 million PMC articles on a separate search engine setup using an ElasticSearch (Gormley and Tong, 2015) instance, that indexes this entire set of articles on all available fields. Our unsupervised document tagging method as outlined in Section 4 employs only the *abstract* field of the 100K PMC articles, for developing the Phrase2VecGLM language model–based document ranking subsequently used in query expansion.

4 Methodology

In our work, a *concept* is defined as a "candidate term" or "noun phrase" scored by a chosen metric e.g. top-K TFIDF, for downstream use in our algorithm (see Section 4.1 & Algorithm 1). They are used in both, training, as building blocks for unsupervised model creation by first learning a phrasal embedding space on the document collection and

```
▼<topic number="17" type="test">
  ▼<note>
    This is a 76-year-old female with pmh of diastolic CHF, atrial fibrillation on coumadin, presenting with Hct 16.9 and shortness of breath. She had
    routine labs drawn yesterday at her PCP's office. Once her hematocrit came she was called and instructed to come to the ED. She is also reporting
    progressive shortness of breath worse with exertion over the past two weeks. She denies fevers, chills, chest pain, palpitaitons, cough, abdominal
    pain, constipation or diahrrea, melena, blood in her stool, dysuria, rash. She reports orthopnea. In the ED: vitals were 98.4 131/49, 60 24 100% 2L.
    ekg with NSR, twi in V1, no significant change from previous. Repeat CBC showed Hct 16.1 with haptoglobin < 20, and elevated LDH to 315. In addition,
    her guaiac was reported as being positive. Past medical history: Hypertension Atrial flutter/fibrillation, s/p cardioversion [**2797-1-27**] Diastolic
    heart failure Hysterectomy Bilateral hip replacements Social History: Married for 53 years with four children. She is retired from the airport. She
    does not smoke or drink. Occupation: retired from airport Drugs: denies Tobacco: denies any history Alcohol: denies
  </note>
  ▼<description>
    This is a 76-year-old female with personal history of diastolic congestive heart failure, atrial fibrillation on Coumadin, presenting with low
    hematocrit and shortness of breath. Her hematocrit dropped from 28 to 16.9 over the past 6 weeks with progressive shortness of breath, worse with
    exertion over the past two weeks. She reports orthopnea. She denies fevers, chills, chest pain, palpitaitons, cough, abdominal pain, constipation or
    diahrrea, melena, blood in her stool, dysuria or rash. Her electrocardiogram present no significant change from previous. Her Guaiac was reported as
    being positive.
  </description>
  ▼<summary>
    76-year-old female with personal history of diastolic congestive heart failure, atrial fibrillation on Coumadin, presenting with low hematocrit and
    dyspnea.
  </summary>
</topic>
```

Figure 1: Sample query from the TREC 2016 challenge dataset, representing a clinical note with patient history, at **Note**, **Description** and **Summary** granularity levels.

subsequent construction of the GLM (Section 4.2), and at inference, for semantically concept–tagging documents. At the time of evaluation, concepts refer to either query terms representing a query document (Yang et al., 2009), or concept tags for target documents. Thus our concepts, predominantly noun phrases, vary from a single unigram term to consisting of up to three terms as employed by our phrase-embedding based language model (Section 4.1). Word embedding techniques use the information around the local context of each word to derive the embeddings. We therefore hypothesize that using these embeddings within a language model (LM) could help to derive terms or concepts that may be closely associated with a given document (Ganguly et al., 2015). Then further extending the model to use embeddings of candidate noun phrases, we could leverage such shared contexts for query expansion, despite no lexical overlap between the query and a given document. This could potentially help both: 1) the global context analysis for IR leading to better downstream retrieval performance from direct query expansion, and, 2) the local context analysis from top-ranked documents aiding query refinement for complex query reformulation within a relevance feedback loop (Su et al., 2015; Xu and Croft, 2000).

Thus, using our phrasal embedding based general language model, Phrase2VecGLM, described in Section 4.2 we generate top-ranked document sets for each document in the collection, treating each document as a query. We subsequently select concepts to tag query documents with, from the top-ranked documents sets for each query. We apply our language model-based concept discovery to query expansion (QE) both *directly* on the challenge dataset queries, as well as via *relevance feedback*, using the concept tags for the top-ranked documents as QE terms. We evaluate the expanded queries on a separate ElasticSearch–based search engine setup, showing improvement in both methods of query expansion (Gormley and Tong, 2015; Chen et al., 2016).

4.1 Pre-processing corpus for Phrasal GLM

We first pre-process the documents in our collection by lower-casing the text, removing most punctuation, like commas, periods, ampersands etc. keeping however, the hyphens, in order to retain hyphenated unigrams, also keeping semi-colons and colons for context. We use regular expressions to retain periods that occur within a decimal value replacing these with the string *decimal* that then gets its own vector representation.

Since we implement both *unigram* and *phrasal embedding–based* GLMs, we process the same document collection accordingly, for each. For the unigram model, our tokens are single or hyphenated words in the corpus. For the phrasal model, we do an additional step of iterating through each document in the corpus, extracting the noun phrases in each using the *textblob* (Loria, 2014) toolkit. This at times gave phrases of up to a length of six, so we only admit ones of size up to three which may include some hyphenated words, to avoid tiny frequency counts. We then plug these extracted phrases back into the documents to obtain a "phrase-based corpus" for training, that has both unigrams and variable-length phrases upto 3-grams, with no tokens repeated for the n-gram processed corpus.

We then pre-compute various document and collection level statistics such as raw counts, term frequencies (phrase frequencies for phrasal corpus), IDF and TF-IDF (Sparck Jones, 1972) for the terms and phrases. Following this, we proceed to generate various embedding models (Mikolov et al., 2013) for both our unigram and phrasal corpora having different length vector representations and context windows using the *gensim* (Řehůřek and Sojka, 2010) package, using the processed text. In particular we generate word embeddings trained with the skip-gram model with negative sampling (Mikolov et al., 2013) with vector length settings of 50 with a context window of 4, and also length 100 with a context window of 5. We also train with the CBOW learning model with negative sampling (Mikolov et al., 2013) for generating embeddings of length 200 with a context window of 7. But we report all of our results on experiments run off the models having an embedding length of 50. Our method is outlined in detail, in the pseudocode shown in Algorithm 1, and assumes that the document and collection statistics as well as the embedding models are already computed and available. We now describe how the processed corpus and the collection and document-level statistics are employed as building blocks to construct our phrasal embedding-based generalized language model, Phrase2VecGLM.

4.2 Phrasal Embedding-based GLM

Standard Jelinek–Mercer smoothing–based language models used for query–document matching can lead to poor probability estimation when query terms *do not* appear in the document due to a key *independence* assumption in these models, wherein query terms are sampled *independently* from *either* the document or the collection (Zhai and Lafferty, 2004). Thus given our goal of alleviating vocabulary mismatch to reformulate complex queries, we find that the word-embedding based generalized language model due to Ganguly et al. (2015), that models *term dependencies* using vector embeddings of terms, lends itself exactly for this purpose as it *relaxes* this independence assumption to incorporate term similarities via vector embeddings. This leads to better probability estimations in the event of semantic overlap between query terms and documents while no lexical overlap by proposing a generative process in which a "noisy channel" may *transform* a term t sampled from a document d or the collection C, with probabilities α and β respectively, into a

query term q'. Thus, by this model we have:

$$
\begin{aligned}
\prod_{q' \in q} P(q'|d) = \prod_{q' \in q} [& \lambda P(q'|d) \\
& + \alpha \sum_{t \in d} \hat{P}_{sim\_doc}(q', t|d) \\
& + \beta \sum_{t \in d} \hat{P}_{sim\_Coll}(q', t|d) \\
& + (1 - \lambda - \alpha - \beta) P(q'|C)]
\end{aligned}
\tag{1}
$$

Here $P(q'|d)$ and $P(q'|C)$ are the same as direct term sampling without transformation, from either the document d or collection C, by a regular Jeliner-Mercer smoothing-based LM as in Equation (2), when $t = q'$:

$$
\begin{aligned}
P(d|q) &= \prod_{q' \in q} \lambda.\hat{P}(q'|d) + (1 - \lambda).\hat{P}(q'|C) \\
&= \prod_{q' \in q} \lambda \frac{tf(q', d)}{|d|} + (1 - \lambda).\frac{cf(q')}{|C|}
\end{aligned}
\tag{2}
$$

However, when $t \neq q'$ we may sample the term t either from document d or collection C where the term t is *transformed* to q'. When t is sampled from d, since the probability of selecting a query term q', given the sampled term t, is *proportional* to the *similarity* of q' with t, where $sim(q', t)$ is the cosine similarity between the *vector representations* of q' and t, and $\sum(d)$ is the sum of the similarity values between *all term pairs* occurring in document d, the document term transformation probability can be estimated as:

$$
\hat{P}_{sim\_doc}(q', t|d) = \frac{sim(q', t)}{\sum(d)} . \frac{tf(t, d)}{|d|}
\tag{3}
$$

Similarly when t is sampled from C, where for the normalization constant, instead of considering all (q', t) pairs in C, we restrict to a small neighbourhood of say 3 terms around the query term q', i.e. $N_{q'}$, to reduce the effect of noisy terms, then the collection term transformation probability can be estimated as:

$$
\hat{P}_{sim\_Coll}(q', t|d) = \frac{sim(q', t)}{\sum N_{q'}} . \frac{cf(t)}{|C|}
\tag{4}
$$

Equation 1 combines all these term transformation events by denoting the probability of observing a query term q' without transformation (standard LM) as λ, that of document sampling–based transformation as α and the probability of collection sampling–based transformation as β.

Thus, per Equations (2) and (1), deriving the posterior probabilities $P(d|q)$ for ranking documents with respect to a query involves maximizing the conditional log likelihood of the query terms in a query q given the document d, as shown:

$$P(d|q) = - \sum_{q' \in q} [log(P(q'|d))] \qquad (5)$$

We use their original word (uni-gram) embedding–based model as a *baseline* in our work. Our model, Phrase2VecGLM, further augments the original model using *variable length noun-phrases* in the vocabulary prior to learning the embedding space for the GLM. While the model by Ganguly et al, is designed as an IR matching function, we extend this model in our work to incorporate embeddings of *candidate noun phrases* from the collection, and re-purpose the model to be used as a *pseudo-relevance feedback function* to select new query expansion terms (Xu and Croft, 2000). Thus, working with the hypothesis that *concepts* in the form of "candidate noun-phrases" provide more *support for meaning*, we update the vocabulary to include noun-phrases of up to a length of three, extracted from the text. The vocabulary terms now consist of phrases, introducing more contextually meaningful terms into the set used in term similarity calculations (Equation 3). This improves concept matching, giving additional coverage toward final query term expansion via LM–based document ranking.

5 Algorithm

Our algorithm (Algorithm 1) works by intrinsically using the Phrase2VecGLM model (Section 4.2) for query expansion, to discover concepts that are similar in the shared local contexts that they occur in, within documents ranked as top-K relevant to a *query document*, and using one of two options for specified threshold criteria to tag the document, as described below. Thus our algorithm consists of two main parts: 1) A document scoring and ranking module applying directly the phrasal embeddings–based general language model described in sections 4.2, 5.1 & algorithm 1, and, 2) A concept selection module to tag the query document with, coming from the set of top ranked matching documents to a query document from step 1. There are a couple of different variations implemented for the concept selection scheme: (i) Selecting the top

TF-IDF term from each of the top-K matching documents as the set of diverse concepts, representative of the query document, and (ii) Selecting the top-similar concept terms matching each of the representative query document terms, using word2vec/Phrase2Vec similarities on the top-ranked set of documents (Mikolov et al., 2013). The code for the corpus pre-processing, model building and inference (semantically tagging documents) is made available online [1] and the dataset is available publicly [2].

5.1 Implementation Details

In the pseudocode given by Algorithm 1, $< docStats >$ represents a set of tuples containing various pre-computed document level frequency and similarity statistics, having elements like $docTermsFreqsRawCounts$, $docTermsTFIDFs$, $docTermPairSimilaritySums$. $< collStats >$ represents a similar set for collection level frequency and similarity measures with elements like $collTermsFreqRawCountsIDFs$ and $collTermPairSimilaritySums$. The procedure also assumes available, the precomputed hashtable $dqTerms$, holding the top TF-IDF terms for each document d, used for querying into the GLM. We have excluded the implementation details for the methods $selectConceptsEmbeddingsModel$, $selectConceptsTFIDF$ and also the GLM method (which essentially computes Equations (1) and (5) for the query document to be tagged with concepts.

6 Experimental Setup

We run two different sets of experiments: (1) *Direct* query expansion of the 30 queries in the TREC dataset, using UMLS concepts (Manual, 2008) for our augmented baselines, and, (2) *Feedback loop–based* query expansion where we use the concept tags for a subset of the top returned articles for the Summary Text–based queries ran against an ElasticSearch index, as query expansion terms, (here MeSH terms-based QE (Adams and Bedrick, 2014) is an augmented baseline), and evaluate both types of runs against our Elastic-Search (ES) index setup described in Section 6.2.

---

[1] https://github.com/manirupa/Phrase2VecGLM
[2] http://www.trec-cds.org/2016.html#documents

Algorithm 1 Document Ranking and Concept Selection by Phrase2VecGLM

Initialize hashtables $rankedListBestMatchedDocs, word2vecConcepts, TFIDFConcepts$;   ▷
These hold ranked document matches and selected concept tags for documents $d \in C$;

  1:   **procedure** GENERATEDOCUMENTRANKINGSCONCEPTS($queryDocs, vectorEmbeddingsModel, <$
        $docStats >, < collStats >, lambda, alpha, beta, query\_length, K$)
  2:       **for** $d \in queryDocs$ **do**
  3:           rankedListBestMatchedDocs$[d]$ = Phrase2VecGLM($dqTerms[d], query\_length, lambda, alpha, beta,$
  4:           $< docStats >, < collStats >$)
  5:           word2vecConcepts$[d]$ =
  6:           selectConceptsEmbeddingsModel($dqTerms[d], < docStats >$
  7:           $rankedListBestMatchedDocs[d], vectorEmbeddingsModel, K$)
  8:           TFIDFConcepts$[d]$ =
  9:           selectConceptsTFIDF($dqTerms[d], < docStats >,$
 10:          $rankedListBestMatchedDocs[d], K$)
 11:       **end for**
 12: **end procedure**

For direct query expansion we take all granularity levels of query topics described in Section 3, i.e. Summary, Description and Notes text, and feed these into our GLMs obtaining the top-K ranked documents for each query and drawing our query expansion concept tags from this set according to the algorithm described in Section 5. For our augmented query baselines, we use UMLS terms within the above query texts generated from the UMLS Java Metamap API that is quite effective in finding optimal phrase boundaries (Bodenreider, 2004; Chen et al., 2016).

For the relevance feedback–based query expansion, we take the top 10-15 documents returned by our ES index setup for each of the Summary Text queries and use the concept tags assigned to each of these top returned documents by our unigram and phrasal GLMs as the concept tags for query expansion for the original query. We then re-run these expanded queries through the ES search engine to record the retrieval performance. The MeSH terms used for the augmented baseline for the feedback loop case, are directly available for a majority of the PMC articles from the TREC dataset itself. Section 4.1 outlines the details of how the dataset was processed to generate the vocabulary and various elements of the GLM.

6.1   Human–Judged Query Annotation

Additionally, to evaluate our feedback loop method against a human judgments–based baseline, we use Expert Term annotations for the query topics available from a 2016 submission to TREC CDS, where 3 physicians were invited to partic-ipate in a manual query expansion experiment. Each physician was assigned 10 out of the 30 query topics from the 2016 challenge. Based on the clinical note, each physician provided a list of 2 to 4 key-phrases. The key-phrases did not have to be part of the note, but could be derived from the physician's knowledge after reading the note (Chen et al., 2016). The search keywords for the query topics thus manually provided by these *domain experts*, were used to retrieve corresponding matching PMC article IDs from the PubMed domain. The expert then spot-checked the top–ranked articles to see if these were mostly relevant. If so, they finalized the keywords assigned. Otherwise, they kept fine-tuning the keywords, until they got a desired set of results, simulating exactly the adaptive decision support (relevance feedback loop) in IR. We also develop an interpolated model with a coefficient γ that interpolates between the unigram and phrasal models, which gets performance comparable to the phrasal model, but does not outperform the other models by itself, hence we do not report those results here. Because the challenge data provides relevance judgments only on a subset of documents (which Phrase2VeGLM is trained on), we report our results using the *inferred measures* (Voorhees, 2014), for "normalized discounted cumulative gain" (NDCG) and "Precision at 10" (P@10). Although the TREC CDS 2016 query set is categorized into three topic types for Diagnosis, Tests and Treatment, we do not divide our evaluation runs into three corresponding sets, evaluating our method's performance on the entire TREC query data set instead.

6.2 Evaluation on ElasticSearch (BM25)

For the search engine–based evaluation of our proposed method, we replicated an ElasticSearch (ES) instance setup with similar settings used in a 2016 challenge submission (Chen et al., 2016). Among the different algorithms available, BM25 (with parameters k1=3 and b=0.75) was selected as the ranking algorithm in our setup due to slightly better performance observed than others, with a logical OR querying model implemented, and the *minimum percentage match* criterion in ES, for search queries, set at 15% of the keywords matched for a document. Since our GLM outlined in Section 4.2 uses the *abstract* field of the article for query expansion, we boosted the *abstract* field 4 times and the *title* field 2 times in our ES search index setup.

6.3 Results and Discussion

Table 1 outlines our results obtained with the various experimental runs described in Section 6. The hyper–parameters for our best performing models were empirically determined and set to be at $(\lambda, \alpha, \beta) = (0.2, 0.3, 0.2)$ for the word embedding–based GLM and $(\lambda, \alpha, \beta) = (0.2, 0.4, 0.2)$ for the phrasal embedding–based GLM, similar to those reported by Ganguly et al., (2015). All models were evaluated for statistical significance against the respective baselines using a two-sided Wilcoxon signed rank test, for $p \ll 0.01$, indicated by bold face value, if found to be significant.

As seen from the results, our unigram and phrasal GLM–based methods for query expansion appear quite promising for both direct query expansion and feedback loop based decision support. For both methods, our *trivial baseline* is the BM25 algorithm of ElasticSearch itself, that uses only the Summary text from the clinical note as the query, with no expanded set of terms.

We summarize our key findings as follows: We run two additional baselines for generation of QE terms: (i) a vanilla language model using standard Jelinek-Mercer smoothing, equivalent to Phrase2VecGLM with settings $(\lambda, \alpha, \beta) = (0.5, 0.0, 0.0)$ such that the embedding space is *not* used to derive term similarities, and (ii) the standard Phrase2vec embedding space model itself (De Vine et al., 2014) prior to deriving the GLM. Both these baselines actually perform worse than the trivial BM25 baseline for QE on the Summary

text, in both direct and relevance feedback settings.

For direct query expansion, UMLS concepts found within the Summary, Description and Notes text of the query itself, were used as augmented baselines. Of these, the Notes UMLS–based expansion worked rather poorly (we attribute this to extra noise concepts in the lengthy Notes text). Though Description text–based UMLS terms did worse than our vanilla Summary text baseline, the Description UMLS terms run through the unigram GLM to get expanded terms did significantly better than Description UMLS terms indicating that our method helps improve term expansion. For direct query expansion, the biggest gain against the baseline was observed for the Summary text UMLS terms run through the unigram GLM to get expanded terms, with a P@10 value of 0.2817. The phrasal model did comparably to the unigram model, however did not beat it, for the direct setting of query expansion.

For the feedback loop based query expansion method, we had two separate human judgment–based baselines, one using the MeSH terms available from PMC for the top 15 documents returned in a first round of querying the ES index with Summary text, and the other based on the expert annotations of the 30 query topics as described in Section 6. The MeSH terms baseline got a P@10 of 0.2294, even less than our vanilla Summary Text baseline with no expanded terms, while our Expert Terms baseline beat this baseline significantly. One reason for the lower performance of the MeSH terms model, we believe, is lack of MeSH term coverage for all the documents chosen. Our unigram GLM–based expanded terms from the top–15 documents returned by Summary Text beat the Expert Terms baseline quite significantly with P@10 of **0.2792**. This was outperformed by the phrasal GLM–based expanded terms model with P@10 of **0.2872**.

Finally our combined model using the unigram + phrasal GLM terms from the top–15 off of the Summary text, beat our vanilla baseline, and was outperformed by our very best combined terms model which generated unigram + phrasal GLM–based terms for the top–15 documents for each query, off of the **Summary + Summary UMLS concepts**, getting a P@10 of **0.3091**. As an example to illustrate, a set of concept tags learned by our unigramGLM model may look like:

Query Expansion Method	Metric		
	Query Text	NDCG **	P@10 **
Direct setting:			
BM25+Standard LM (Jelinek-Mercer sm.) QE Terms (**baseline**)	Summary	0.0475	0.1172
BM25+Phrase2Vec (without GLM) QE Terms (**baseline**)	Summary	0.0932	**0.2267**
BM25+DescUMLS QE Terms (**augmented baseline**)	Summary	0.1070	**0.2299**
BM25+DescUMLS+unigramGLM QE Terms (**model**)	Summary	0.1010	**0.2414**
BM25+None (**baseline**)	Summary	0.1060	**0.2489**
BM25+SumUMLS QE Terms (**augmented baseline**)	Summary	0.1466	**0.2644**
BM25+SumUMLS+unigramGLM QE Terms (model)	Summary	0.1387	**0.2817**
Feedback Loop setting:			
BM25+Standard LM (Jelinek-Mercer sm.) QE Terms (**baseline**)	Summary	0.0265	0.0867
BM25+Phrase2Vec (without GLM) QE Terms (**baseline**)	Summary	0.0662	**0.1318**
BM25+MeSH QE Terms (**baseline**)	Summary	0.0970	**0.2294**
BM25+None (**baseline**)	Summary	0.1060	**0.2489**
BM25+**Human Expert** QE Terms (**augmented baseline**)	Summary	0.1029	**0.2511**
BM25+unigramGLM QE Terms (**model**)	Summary	0.1173	**0.2792** *
BM25+Phrase2VecGLM QE Terms (model)	Summary	0.1159	**0.2872** *
Feedback Loop Combined Models			
BM25+unigramGLM Terms+Phrase2VeclGLM Terms (**baseline**)	Summary	0.1057	0.2756
BM25+SumUMLS+unigramGLM Terms+Phrase2VecGLM QE Terms (model)	Summary	**0.1206**	**0.3091** *

Table 1: Results for IR after Query Expansion (QE) by different methods using unigram and phrasal GLM–generated QE terms, in **direct** and **feedback loop** settings. Bold face values indicate statistical significance at $p << 0.01$ over the previous result or baseline. Single asterisks indicate our best performing models. Double asterisks indicate *inferred* measures (Voorhees, 2014). Numbers are from evaluation of ranking results based on document relevance judgments available for all 30 queries in the dataset.

<*'query_doc':(4315343, ['dementia', 'cognitive', 'bp']), 'concept_tags': ['alzheimers', 'diabetes', 'behavioral']* >, and for the phrasalGLM model we may have: <*'query_doc':(3088738, ['albendazole', 'eosinophilic ascites', 'parasitic infection']), 'concept_tags': ['corticosteroid therapy', 'case hyperinfection', 'strongyloides stercoralis']* >.

7 Conclusions and Future Work

In this work, we demonstrate that our proposed method of semantic tagging for query expansion, via word and phrasal GLM–based document ranking for pseudo-relevance feedback, can prove an effective means to serve complex, specific information needs such as clinical queries in medical information retrieval that require adaptive decision support, performing better in some cases than even human expert–provided query expansion terms. This is especially helpful to solve the problem of *lack of keyword coverage* for all documents in any collection, e.g. MeSH terms for PMC articles. In future we hope to leverage end-to-end recurrent neural architectures such as LSTMs, possibly with attention mechanisms (Rocktäschel et al., 2015; Bahdanau et al., 2014) to improve our current method of semantic tagging for complex querying in medical IR.

Acknowledgments

The authors would like to thank Alan Ritter, whose invaluable feedback helped to significantly improve portions of evaluation and presentation of this work, and our collaborators at Nationwide Children's Hospital whose valuable time, support and resources made this work possible. We also thank our anonymous reviewers for their feedback.

References

Joel Adams and Steven Bedrick. 2014. Automatic classification of pubmed abstracts with latent semantic indexing: Working notes. In *CLEF (Working Notes)*, pages 1275–1282. Citeseer.

Dzmitry Bahdanau, Kyunghyun Cho, and Yoshua Bengio. 2014. Neural machine translation by jointly learning to align and translate. *arXiv preprint arXiv:1409.0473*.

Rahul Bhagat and Eduard Hovy. 2013. What is a paraphrase? *Computational Linguistics*, 39(3):463–472.

Rahul Bhagat and Deepak Ravichandran. 2008. Large scale acquisition of paraphrases for learning surface patterns. In *ACL*, volume 8, pages 674–682.

David M Blei, Andrew Y Ng, and Michael I Jordan. 2003. Latent dirichlet allocation. *Journal of machine Learning research*, 3(Jan):993–1022.

Olivier Bodenreider. 2004. The unified medical language system (umls): integrating biomedical terminology. *Nucleic acids research*, 32(suppl_1):D267–D270.

Wei Chen, Soheil Moosavinasab, Anna Zemke, Ariana Prinzbach, Steve Rust, Yungui Huang, and Simon Lin. 2016. Evaluation of a machine learning method to rank pubmed central articles for clinical relevancy: Nch at trec 2016 cds. *TREC 2016 Clinical Decision Support Track*.

Lance De Vine, Guido Zuccon, Bevan Koopman, Laurianne Sitbon, and Peter Bruza. 2014. Medical semantic similarity with a neural language model. In *Proceedings of the 23rd ACM International Conference on Conference on Information and Knowledge Management*, pages 1819–1822. ACM.

Scott Deerwester, Susan T Dumais, George W Furnas, Thomas K Landauer, and Richard Harshman. 1990. Indexing by latent semantic analysis. *Journal of the American society for information science*, 41(6):391.

Anne Diekema, Ozgur Yilmazel, Jiangping Chen, Sarah Harwell, Lan He, and Elizabeth D Liddy. 2003. What do you mean? finding answers to complex questions. In *New Directions in Question Answering*, pages 87–93.

Debasis Ganguly, Dwaipayan Roy, Mandar Mitra, and Gareth JF Jones. 2015. A word embedding based generalized language model for information retrieval. In *Proceedings of the 38th International ACM SIGIR Conference on Research and Development in Information Retrieval*, pages 795–798. ACM.

Clinton Gormley and Zachary Tong. 2015. *Elasticsearch: The Definitive Guide.* " O'Reilly Media, Inc.".

Thomas L Griffiths and Mark Steyvers. 2004. Finding scientific topics. *Proceedings of the National academy of Sciences*, 101(suppl 1):5228–5235.

Harry Halpin, Valentin Robu, and Hana Shepherd. 2007. The complex dynamics of collaborative tagging. In *Proceedings of the 16th International Conference on World Wide Web*, pages 211–220. ACM.

Iris Hendrickx, Su Nam Kim, Zornitsa Kozareva, Preslav Nakov, Diarmuid Ó Séaghdha, Sebastian Padó, Marco Pennacchiotti, Lorenza Romano, and Stan Szpakowicz. 2009. Semeval-2010 task 8: Multi-way classification of semantic relations between pairs of nominals. In *Proceedings of the Workshop on Semantic Evaluations: Recent Achievements and Future Directions*, pages 94–99. Association for Computational Linguistics.

Mahnoosh Kholghi, Laurianne Sitbon, Guido Zuccon, and Anthony Nguyen. 2015a. Active learning: a step towards automating medical concept extraction. *Journal of the American Medical Informatics Association*, 23(2):289–296.

Mahnoosh Kholghi, Laurianne Sitbon, Guido Zuccon, and Anthony Nguyen. 2015b. Active learning: a step towards automating medical concept extraction. *Journal of the American Medical Informatics Association*, 23(2):289–296.

Chenliang Li, Anwitaman Datta, and Aixin Sun. 2011. Semantic tag recommendation using concept model. In *Proceedings of the 34th international ACM SIGIR conference on Research and development in Information Retrieval*, pages 1159–1160. ACM.

Dekang Lin and Patrick Pantel. 2001a. Discovery of inference rules for question-answering. *Natural Language Engineering*, 7(4):343–360.

Dekang Lin and Patrick Pantel. 2001b. Induction of semantic classes from natural language text. In *Proceedings of the Seventh ACM SIGKDD International Conference on Knowledge Discovery and Data Mining*, pages 317–322. ACM.

Dekang Lin and Patrick Pantel. 2002. Concept discovery from text. In *Proceedings of the 19th International Conference on Computational Linguistics-Volume 1*, pages 1–7. Association for Computational Linguistics.

Steven Loria. 2014. Textblob: simplified text processing. *Secondary TextBlob: Simplified Text Processing*.

Gang Luo, Chunqiang Tang, Hao Yang, and Xing Wei. 2008. Medsearch: a specialized search engine for medical information retrieval. In *Proceedings of the 17th ACM conference on Information and knowledge management*, pages 143–152. ACM.

Christopher D Manning, Prabhakar Raghavan, and Hinrich Schütze. 2009. Introduction to information retrieval. *An Introduction To Information Retrieval*, 151(177):5.

NLM UMLS Knowledge Sources Manual. 2008. National library of medicine. *Bethesda, Maryland.*

Julian McAuley and Alex Yang. 2016. Addressing complex and subjective product-related queries with customer reviews. In *Proceedings of the 25th International Conference on World Wide Web*, pages 625–635. International World Wide Web Conferences Steering Committee.

Tomas Mikolov, Ilya Sutskever, Kai Chen, Greg S Corrado, and Jeff Dean. 2013. Distributed representations of words and phrases and their compositionality. In *Advances in Neural Information Processing Systems*, pages 3111–3119.

Patrick Pantel, Rahul Bhagat, Bonaventura Coppola, Timothy Chklovski, and Eduard H Hovy. 2007. Isp: Learning inferential selectional preferences. In *HLT-NAACL*, pages 564–571.

Radim Řehůřek and Petr Sojka. 2010. Software Framework for Topic Modelling with Large Corpora. In *Proceedings of the LREC 2010 Workshop on New Challenges for NLP Frameworks*, pages 45–50, Valletta, Malta. ELRA. http://is.muni.cz/publication/884893/en.

Alan Ritter, Luke Zettlemoyer, Oren Etzioni, et al. 2013. Modeling missing data in distant supervision for information extraction. *Transactions of the Association for Computational Linguistics*, 1:367–378.

Andreia Rodrıguez Rivas, Eva Lorenzo Iglesias, and L Borrajo. 2014. Study of query expansion techniques and their application in the biomedical information retrieval. *The Scientific World Journal*, 2014.

Kirk Roberts, Matthew S Simpson, Ellen M Voorhees, and William R Hersh. 2016. Overview of the trec 2015 clinical decision support track. In *TREC*.

Tim Rocktäschel, Edward Grefenstette, Karl Moritz Hermann, Tomáš Kočiskỳ, and Phil Blunsom. 2015. Reasoning about entailment with neural attention. *arXiv preprint arXiv:1509.06664.*

Alessandro Sordoni, Yoshua Bengio, and Jian-Yun Nie. 2014. Learning concept embeddings for query expansion by quantum entropy minimization. In *AAAI*, volume 14, pages 1586–1592.

Karen Sparck Jones. 1972. A statistical interpretation of term specificity and its application in retrieval. *Journal of documentation*, 28(1):11–21.

Yu Su, Shengqi Yang, Huan Sun, Mudhakar Srivatsa, Sue Kase, Michelle Vanni, and Xifeng Yan. 2015. Exploiting relevance feedback in knowledge graph search. In *Proceedings of the 21th ACM SIGKDD International Conference on Knowledge Discovery and Data Mining*, pages 1135–1144. ACM.

Huan Sun, Hao Ma, Wen-tau Yih, Chen-Tse Tsai, Jingjing Liu, and Ming-Wei Chang. 2015. Open domain question answering via semantic enrichment.

In *Proceedings of the 24th International Conference on World Wide Web*, pages 1045–1055. ACM.

Choon Hui Teo, Houssam Nassif, Daniel Hill, Sriram Srinivasan, Mitchell Goodman, Vijai Mohan, and SVN Vishwanathan. 2016. Adaptive, personalized diversity for visual discovery. In *Proceedings of the 10th ACM Conference on Recommender Systems*, pages 35–38. ACM.

Suppawong Tuarob, Line C Pouchard, and C Lee Giles. 2013. Automatic tag recommendation for metadata annotation using probabilistic topic modeling. In *Proceedings of the 13th ACM/IEEE-CS joint conference on Digital libraries*, pages 239–248. ACM.

Peter D Turney and Patrick Pantel. 2010. From frequency to meaning: Vector space models of semantics. *Journal of artificial intelligence research*, 37:141–188.

Ellen M Voorhees. 2014. The effect of sampling strategy on inferred measures. In *Proceedings of the 37th international ACM SIGIR conference on Research & development in information retrieval*, pages 1119–1122. ACM.

Jinxi Xu and W Bruce Croft. 2000. Improving the effectiveness of information retrieval with local context analysis. *ACM Transactions on Information Systems (TOIS)*, 18(1):79–112.

Wei Xu, Alan Ritter, Chris Callison-Burch, William B Dolan, and Yangfeng Ji. 2014. Extracting lexically divergent paraphrases from twitter. *Transactions of the Association for Computational Linguistics*, 2:435–448.

Yin Yang, Nilesh Bansal, Wisam Dakka, Panagiotis Ipeirotis, Nick Koudas, and Dimitris Papadias. 2009. Query by document. In *Proceedings of the Second ACM International Conference on Web Search and Data Mining*, pages 34–43. ACM.

Chengxiang Zhai and John Lafferty. 2004. A study of smoothing methods for language models applied to information retrieval. *ACM Transactions on Information Systems (TOIS)*, 22(2):179–214.

Ye Zhang, Md Mustafizur Rahman, Alex Braylan, Brandon Dang, Heng-Lu Chang, Henna Kim, Quinten McNamara, Aaron Angert, Edward Banner, Vivek Khetan, et al. 2016. Neural information retrieval: A literature review. *arXiv preprint arXiv:1611.06792.*

Guido Zuccon, Bevan Koopman, Peter Bruza, and Leif Azzopardi. 2015. Integrating and evaluating neural word embeddings in information retrieval. In *Proceedings of the 20th Australasian Document Computing Symposium*, page 12. ACM.

Convolutional neural networks for chemical-disease relation extraction are improved with character-based word embeddings

Dat Quoc Nguyen and **Karin Verspoor**
School of Computing and Information Systems
The University of Melbourne, Australia
{dqnguyen, karin.verspoor}@unimelb.edu.au

Abstract

We investigate the incorporation of character-based word representations into a standard CNN-based relation extraction model. We experiment with two common neural architectures, CNN and LSTM, to learn word vector representations from character embeddings. Through a task on the BioCreative-V CDR corpus, extracting relationships between chemicals and diseases, we show that models exploiting the character-based word representations improve on models that do not use this information, obtaining state-of-the-art result relative to previous neural approaches.

1 Introduction

Relation extraction, the task of extracting semantic relations between named entities mentioned in text, has become a key research topic in natural language processing (NLP) with a variety of practical applications (Bach and Badaskar, 2007). Traditional approaches for relation extraction are feature-based and kernel-based supervised learning approaches which utilize various lexical and syntactic features as well as knowledge base resources; see the comprehensive survey of these traditional approaches in Pawar et al. (2017). Recent research has shown that neural network (NN) models for relation extraction obtain state-of-the-art performance. Two major neural architectures for the task include the convolutional neural networks, CNNs, (Zeng et al., 2014; Nguyen and Grishman, 2015; Zeng et al., 2015; Lin et al., 2016; Jiang et al., 2016; Zeng et al., 2017; Huang and Wang, 2017) and long short-term memory networks, LSTMs (Miwa and Bansal, 2016; Zhang et al., 2017; Katiyar and Cardie, 2017; Ammar et al., 2017). We also find combinations of those two architectures (Nguyen and Grishman, 2016; Raj et al., 2017).

Relation extraction has attracted particular attention in the high-value biomedical domain. Scientific publications are the primary repository of biomedical knowledge, and given their increasing numbers, there is tremendous value in automating extraction of key discoveries (de Bruijn and Martin, 2002). Here, we focus on the task of understanding relations between chemicals and diseases, which has applications in many areas of biomedical research and healthcare including toxicology studies, drug discovery and drug safety surveillance (Wei et al., 2015). The importance of chemical-induced disease (CID) relation extraction is also evident from the fact that chemicals, diseases and their relations are among the most searched topics by PubMed users (Islamaj Dogan et al., 2009). In the CID relation extraction task formulation (Wei et al., 2015, 2016), CID relations are typically determined at document level, meaning that relations can be expressed across sentence boundaries; they can extend over distances of hundreds of word tokens. As LSTM models can be difficult to apply to very long word sequences (Bradbury et al., 2017), CNN models may be better suited for this task.

New domain-specific terms arise frequently in biomedical text data, requiring the capture of unknown words in practical relation extraction applications in this context. Recent research has shown that character-based word embeddings enable capture of unknown words, helping to improve performance on many NLP tasks (dos Santos and Gatti, 2014; Ma and Hovy, 2016; Lample et al., 2016; Plank et al., 2016; Nguyen et al., 2017). This may be particularly relevant for terms such as gene or chemical names, which often have identifiable morphological structure (Krallinger et al., 2017).

We investigate the value of character-based word embeddings in a standard CNN model for relation extraction (Zeng et al., 2014; Nguyen and Grishman, 2015). To the best of our knowledge,

Proceedings of the BioNLP 2018 workshop, pages 129–136
Melbourne, Australia, July 19, 2018. ©2018 Association for Computational Linguistics

there is no prior work addressing this.

We experiment with two common neural architectures of CNN and LSTM for learning the character-based embeddings, and evaluate the models on the benchmark BioCreative-V CDR corpus for chemical-induced disease relation extraction (Li et al., 2016a), obtaining state-of-the-art results.

2 Our modeling approach

This section describes our relation extraction models. They can be viewed as an extension of the well-known CNN model for relation extraction (Nguyen and Grishman, 2015), where we incorporate character-level representations of words.

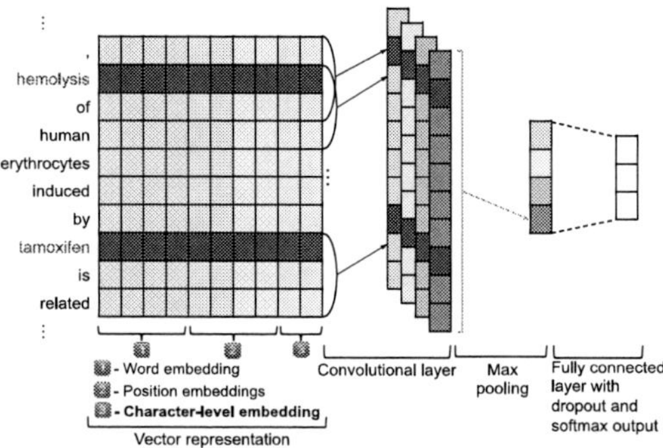

Figure 1: Our model architecture. Given the input relation mention marked with two entities "hemolysis" and "tamoxifen", the convolutional layer uses the window size $k = 3$ and the number of filters $m = 4$.

Figure 1 presents our model architecture. Given an input fixed-length sequence (i.e. a *relation mention*) of n word tokens $w_1, w_2, w_3, ..., w_n,$[1] marked with two entity mentions, the vector representation layer encodes each ith word in the input relation mention by a real-valued vector representation $v_i \in \mathbb{R}^d$. The convolutional layer takes the input matrix $S = [v_1, v_2, ..., v_n]^\mathsf{T}$ to extract high level features. These high level features are then fed into the max pooling layer to capture the most important features for generating a feature vector of the input relation mention. Finally, the feature vector is fed into a fully-connected neural network with softmax output to produce a probability distribution over relation types. For convenience, we detail the vector representation layer in Section 2.2 while the remaining layers appear in Section 2.1.

---

[1] We set n to be the length of the longest sequence and pad shorter sequences with a special "PAD" token.

2.1 CNN layers for relation extraction

Convolutional layer: This layer uses different filters to extract features from the input matrix $S = [v_1, v_2, ..., v_n]^\mathsf{T} \in \mathbb{R}^{n \times d}$ by performing convolution operations. Given a window size k, a filter can be formalized as a weight matrix $F = [f_1, f_2, ..., f_k]^\mathsf{T} \in \mathbb{R}^{k \times d}$. For each filter F, the convolution operation is performed to generate a feature map $x = [x_1, x_2, ..., x_{n-k+1}] \in \mathbb{R}^{n-k+1}$:

$$x_j = g\big(\textstyle\sum_{h=1}^{k} f_h v_{j+h-1} + b\big)$$

where $g(.)$ is some non-linear activation function and $b \in \mathbb{R}$ is a bias term.

Assume that we use m different weight matrix filters $F^{(1)}, F^{(2)}, ..., F^{(m)} \in \mathbb{R}^{k \times d}$, the process above is then repeated m times, resulting in m feature maps $x^{(1)}, x^{(2)}, ..., x^{(m)} \in \mathbb{R}^{n-k+1}$.

Max pooling layer: This layer aims to capture the most relevant features from each feature map x by applying the popular max-over-time pooling operation: $\hat{x} = \max\{x\} = \max\{x_1, x_2, ..., x_{n-k+1}\}$. From m feature maps, the corresponding outputs are concatenated into a feature vector $z = [\hat{x}^{(1)}, \hat{x}^{(2)}, ..., \hat{x}^{(m)}] \in \mathbb{R}^m$ to represent the input relation mention.

Softmax output: The feature vector z is then fed into a fully connected NN followed by a softmax layer for relation type classification. In addition, following Kim (2014), for regularization we apply dropout on z only during training. The softmax output procedure can be formalized as:

$$p = \mathsf{softmax}\big(\mathbf{W}_1(z * r) + b_1\big)$$

where $p \in \mathbb{R}^t$ is the final output of the network in which t is the number of relation types, and $\mathbf{W}_1 \in \mathbb{R}^{t \times m}$ and $b_1 \in \mathbb{R}^t$ are a transformation weight matrix and a bias vector, respectively. In addition, $*$ denotes an element-wise product and $r \in \mathbb{R}^m$ is a vector of independent Bernoulli random variables, each with probability ρ of being 0 (Srivastava et al., 2014).

2.2 Input vector representation

This section presents the vector representation $v_i \in \mathbb{R}^d$ for each ith word token in the input relation mention $w_1, w_2, w_3, ..., w_n$. Let word tokens w_{i_1} and w_{i_2} be two entity mentions in the input.[2] We obtain v_i by concatenating word embeddings $e_{w_i} \in \mathbb{R}^{d_1}$, position embeddings $e_{i-i_1}^{(p1)}$

---

[2] If an entity spans over multiple tokens, we take only the last token in the entity into account (Nguyen et al., 2016).

and $e_{i-i_2}^{(p2)} \in \mathbb{R}^{d_2}$, and character-level embeddings $e_{w_i}^{(c)} \in \mathbb{R}^{d_3}$ (so, $d = d_1 + 2 \times d_2 + d_3$):

$$v_i = e_{w_i} \circ e_{i-i_1}^{(p1)} \circ e_{i-i_2}^{(p2)} \circ e_{w_i}^{(c)}$$

Word embeddings: Each word type w in the training data is represented by a real-valued word embedding $e_w \in \mathbb{R}^{d_1}$.

Position embeddings: In relation extraction, we focus on assigning relation types to entity pairs. Words close to target entities are usually informative for identifying a relationship between them. Following Zeng et al. (2014), to specify entity pairs, we use position embeddings $e_{i-i_1}^{(p1)}$ and $e_{i-i_2}^{(p2)} \in \mathbb{R}^{d_2}$ to encode the relative distances $i - i_1$ and $i - i_2$ from each word w_i to entity mentions w_{i_1} and w_{i_2}, respectively.

Character-level embeddings: Given a word type w consisting of l characters $w = c_1 c_2 ... c_l$ where each jth character in w is represented by a character embedding $c_j \in \mathbb{R}^{d_4}$, we investigate two approaches for learning character-based word embedding $e_w^{(c)} \in \mathbb{R}^{d_3}$ from input $c_{1:l} = [c_1, c_2, ..., c_l]^\mathsf{T}$ as follows:

(1) Using **CNN** (dos Santos and Gatti, 2014; Ma and Hovy, 2016): This CNN contains a convolutional layer to generate d_3 feature maps from the input $c_{1:l}$, and a max pooling layer to produce a final vector $e_w^{(c)}$ from those feature maps for representing the word w.

(2) Using a sequence BiLSTM (**BiLSTM$_{\text{seq}}$**) (Lample et al., 2016): In the BiLSTM$_{\text{seq}}$, the input is the sequence of l character embeddings $c_{1:l}$, and the output is a concatenation of outputs of a forward LSTM (LSTM$_\text{f}$) reading the input in its regular order and a reverse LSTM (LSTM$_\text{r}$) reading the input in reverse:

$$e_w^{(c)} = \text{BiLSTM}_{\text{seq}}(c_{1:l}) = \text{LSTM}_\text{f}(c_{1:l}) \circ \text{LSTM}_\text{r}(c_{l:1})$$

2.3 Model training

The baseline CNN model for relation extraction (Nguyen and Grishman, 2015) is denoted here as **CNN**. The extensions incorporating CNN and BiLSTM character-based word embeddings are **CNN+CNNchar** and **CNN+LSTMchar**, respectively. The model parameters, including word, position, and character embeddings, weight matrices and biases, are learned during training to minimize the model negative log likelihood (i.e. cross-entropy loss) with L_2 regularization.

3 Experiments

3.1 Experimental setup

We evaluate our models using the BC5CDR corpus (Li et al., 2016a) which is the benchmark dataset for the chemical-induced disease (CID) relation extraction task (Wei et al., 2015, 2016).[3] The corpus consists of 1500 PubMed abstracts: 500 for each of training, development and test. The training set is used to learn model parameters, the development set to select optimal hyperparameters, and the test set to report final results. We make use of gold entity annotations in each case. For evaluation results, we measure the CID relation extraction performance with F1 score. More details of the dataset, evaluation protocol, and implementation are in the Appendix.

3.2 Main results

Table 1 compares the CID relation extraction results of our models to prior work. The first 11 rows report the performance of models that use the same experimental setup, without using additional training data or various features extracted from external knowledge base (KB) resources. The last 6 rows report results of models exploiting various kinds of features based on external relational KBs of chemicals and diseases, in which the last 4 SVM-based models are trained using both training and development sets.

The models exploiting more training data and external KB features obtained the best F1 scores. Panyam et al. (2016) and Xu et al. (2016) have shown that without KB features, their model performances (61.7% and 67.2%) are decreased by 5 and 11 points of F1 score, respectively.[4] Hence we find that external KB features are essential; we plan to extend our models to incorporate such KB features in future work.

In terms of models *not* exploiting external data or KB features (i.e. the first 11 rows in Table 1), our CNN+CNNchar and CNN+LSTMchar obtain the highest F1 scores; with 1+% absolute F1 improvements to the baseline CNN (p-value < 0.05).[5] In addition, our models obtain 2+% higher

---

[3] http://www.biocreative.org/tasks/
biocreative-v/track-3-cdr/

[4] Pons et al. (2016) and Peng et al. (2016) did not provide results without using the KB-based features. Xu et al. (2016) and Pons et al. (2016) did not provide results in using only the training set for learning models.

[5] Improvements are significant with p-value < 0.05 for a bootstrap significance test.

Model	P	R	F1
MaxEnt (Gu et al., 2016)	62.0	55.1	58.3
Pattern rule-based (Lowe et al., 2016)	59.3	62.3	60.8
LSTM-based (Zhou et al., 2016)	64.9	49.3	56.0
LSTM-based & PP (Zhou et al., 2016)	55.6	68.4	61.3
CNN-based (Gu et al., 2017)	60.9	59.5	60.2
CNN-based & PP (Gu et al., 2017)	55.7	68.1	61.3
BRAN (Verga et al., 2017)	55.6	70.8	**62.1**
SVM+APG (Panyam et al., 2018)	53.2	69.7	60.3
CNN	54.8	69.0	61.1
CNN+CNNchar	57.0	68.6	**62.3**
CNN+LSTMchar	56.8	68.8	62.2
Linear+TK (Panyam et al., 2016)	63.6	59.8	61.7
SVM (Peng et al., 2016)	62.1	64.2	63.1
SVM (+dev.) (Peng et al., 2016)	68.2	66.0	67.1
SVM (+dev.+18K) (Peng et al., 2016)	71.1	72.6	**71.8**
SVM (+dev.) (Xu et al., 2016)	65.8	68.6	67.2
SVM (+dev.) (Pons et al., 2016)	73.1	67.6	70.2

Table 1: Precision (P), Recall (R) and F1 scores (in %). "& PP" refers to the use of additional post-processing heuristic rules. "BRAN" denotes bi-affine relation attention networks. "SVM+APG" denotes a model using SVM with All Path Graph kernel. "Linear+TK" denotes a model combining linear and tree kernel classifiers. "+dev." denotes the use of both training and development sets for learning models. Note that Peng et al. (2016) also used an extra training corpus of 18K weakly-annotated PubMed articles.

F1 score than the traditional feature-based models MaxEnt (Gu et al., 2016) and SVM+APG (Panyam et al., 2018). We also achieve 2+% higher F1 score than the LSTM- and CNN-based methods (Zhou et al., 2016; Gu et al., 2017) which exploit LSTM and CNN to learn relation mention representations from dependency tree-based paths.[6] Dependency trees have been actively used in traditional feature-based and kernel-based methods for relation extraction (Culotta and Sorensen, 2004; Bunescu and Mooney, 2005; GuoDong et al., 2005; Mooney and Bunescu, 2006; Mintz et al., 2009) as well as in the biomedical domain (Fundel et al., 2007; Panyam et al., 2016, 2018; Quirk and Poon, 2017). Although we obtain better results, we believe dependency tree-based feature representations still have strong potential value. Note that to obtain dependency trees, previous work on CID relation extraction used the Stanford depen-

---

[6]Zhou et al. (2016) and Gu et al. (2017) used the same post-processing heuristics to handle cases where models could not identify any CID relation between chemicals and diseases in an article, resulting in final F1 scores at 61.3%.

dency parser (Chen and Manning, 2014). However, this dependency parser was trained on the Penn Treebank (in the newswire domain) (Marcus et al., 1993); training on a domain-specific treebank such as CRAFT (Bada et al., 2012) should help to improve results (Verspoor et al., 2012).

We also achieve slightly better scores than the more complex model BRAN (Verga et al., 2017), the Biaffine Relation Attention Network, based on the Transformer self-attention model (Vaswani et al., 2017). BRAN additionally uses byte pair encoding (Gage, 1994) to construct a vocabulary of subword units for tokenization. Using subword tokens to capture rare or unknown words has been demonstrated to be useful in machine translation (Sennrich et al., 2016) and likely captures similar information to character embeddings. However, Verga et al. (2017) do not provide comparative results using only original word tokens. Therefore, it is difficult to assess the usefulness specifically of using byte-pair encoded subword tokens in the CID relation extraction task, as compared to the impact of the full model architecture. We also plan to explore the usefulness of subword tokens in the baseline CNN for future work, to enable comparison with the improvement when using the character-based word embeddings.

It is worth noting that both CNN+CNNchar and CNN+LSTMchar return similar F1 scores, showing that in this case, using either CNN or BiLSTM to learn character-based word embeddings produces a similar improvement to the baseline. There does not appear to be any reason to prefer one of these in our relation extraction application.

4 Conclusion

In this paper, we have explored the value of integrating character-based word representations into a baseline CNN model for relation extraction. In particular, we investigate the use of two well-known neural architectures, CNN and LSTM, for learning character-based word representations. Experimental results on a benchmark chemical-disease relation extraction corpus show that the character-based representations help improve the baseline to attain state-of-the-art performance. Our models are suitable candidates to serve as future baselines for more complex models in the relation extraction task.

Acknowledgment: This work was supported by the ARC Discovery Project DP150101550.

References

Martin Abadi, Paul Barham, Jianmin Chen, Zhifeng Chen, Andy Davis, Jeffrey Dean, Matthieu Devin, Sanjay Ghemawat, Geoffrey Irving, Michael Isard, Manjunath Kudlur, Josh Levenberg, Rajat Monga, Sherry Moore, Derek G. Murray, Benoit Steiner, Paul Tucker, Vijay Vasudevan, Pete Warden, Martin Wicke, Yuan Yu, and Xiaoqiang Zheng. 2016. TensorFlow: A system for large-scale machine learning. In *Proceedings of the 12th USENIX Symposium on Operating Systems Design and Implementation.* pages 265–283.

Waleed Ammar, Matthew Peters, Chandra Bhagavatula, and Russell Power. 2017. The AI2 system at SemEval-2017 Task 10 (ScienceIE): semi-supervised end-to-end entity and relation extraction. In *Proceedings of the 11th International Workshop on Semantic Evaluation.* pages 592–596.

Nguyen Bach and Sameer Badaskar. 2007. A Review of Relation Extraction. Technical report, Carnegie Mellon University.

Michael Bada, Miriam Eckert, Donald Evans, Kristin Garcia, Krista Shipley, Dmitry Sitnikov, William A Baumgartner, K Bretonnel Cohen, Karin Verspoor, Judith A Blake, et al. 2012. Concept annotation in the CRAFT corpus. *BMC bioinformatics* 13(1):161.

James Bradbury, Stephen Merity, Caiming Xiong, and Richard Socher. 2017. Quasi-Recurrent Neural Networks. In *Proceedings of the 5th International Conference on Learning Representations.*

Razvan Bunescu and Raymond Mooney. 2005. A Shortest Path Dependency Kernel for Relation Extraction. In *Proceedings of Human Language Technology Conference and Conference on Empirical Methods in Natural Language Processing.* pages 724–731.

Danqi Chen and Christopher Manning. 2014. A Fast and Accurate Dependency Parser using Neural Networks. In *Proceedings of the 2014 Conference on Empirical Methods in Natural Language Processing.* pages 740–750.

Billy Chiu, Gamal Crichton, Anna Korhonen, and Sampo Pyysalo. 2016. How to Train good Word Embeddings for Biomedical NLP. In *Proceedings of the 15th Workshop on Biomedical Natural Language Processing.* pages 166–174.

François Chollet et al. 2015. Keras. https://github.com/fchollet/keras.

Aron Culotta and Jeffrey Sorensen. 2004. Dependency Tree Kernels for Relation Extraction. In *Proceedings of the 42nd Meeting of the Association for Computational Linguistics.* pages 423–429.

Berry de Bruijn and Joel Martin. 2002. Getting to the (c)ore of knowledge: mining biomedical literature. *International Journal of Medical Informatics* 67(1):7 – 18.

Cicero dos Santos and Maira Gatti. 2014. Deep Convolutional Neural Networks for Sentiment Analysis of Short Texts. In *Proceedings of COLING 2014, the 25th International Conference on Computational Linguistics: Technical Papers.* pages 69–78.

Timothy Dozat. 2016. Incorporating Nesterov Momentum into Adam. In *Proceedings of the ICLR 2016 Workshop Track.*

Katrin Fundel, Robert Kffner, and Ralf Zimmer. 2007. RelExRelation extraction using dependency parse trees. *Bioinformatics* 23(3):365–371.

Philip Gage. 1994. A New Algorithm for Data Compression. *The C Users Journal* 12(2):23–38.

Jinghang Gu, Longhua Qian, and Guodong Zhou. 2016. Chemical-induced disease relation extraction with various linguistic features. *Database* 2016:baw042.

Jinghang Gu, Fuqing Sun, Longhua Qian, and Guodong Zhou. 2017. Chemical-induced disease relation extraction via convolutional neural network. *Database* 2017:bax024.

Zhou GuoDong, Su Jian, Zhang Jie, and Zhang Min. 2005. Exploring Various Knowledge in Relation Extraction. In *Proceedings of the 43rd Annual Meeting on Association for Computational Linguistics.* pages 427–434.

YiYao Huang and William Yang Wang. 2017. Deep Residual Learning for Weakly-Supervised Relation Extraction. In *Proceedings of the 2017 Conference on Empirical Methods in Natural Language Processing.* pages 1803–1807.

Rezarta Islamaj Dogan, G. Craig Murray, Aurlie Nvol, and Zhiyong Lu. 2009. Understanding PubMed user search behavior through log analysis. *Database* 2009.

Xiaotian Jiang, Quan Wang, Peng Li, and Bin Wang. 2016. Relation Extraction with Multi-instance Multi-label Convolutional Neural Networks. In *Proceedings of COLING 2016, the 26th International Conference on Computational Linguistics: Technical Papers.* pages 1471–1480.

Arzoo Katiyar and Claire Cardie. 2017. Going out on a limb: Joint Extraction of Entity Mentions and Relations without Dependency Trees. In *Proceedings of the 55th Annual Meeting of the Association for Computational Linguistics (Volume 1: Long Papers).* pages 917–928.

Yoon Kim. 2014. Convolutional Neural Networks for Sentence Classification. In *Proceedings of the 2014 Conference on Empirical Methods in Natural Language Processing.* pages 1746–1751.

Eliyahu Kiperwasser and Yoav Goldberg. 2016. Simple and Accurate Dependency Parsing Using Bidirectional LSTM Feature Representations. *Transactions of the Association for Computational Linguistics* 4:313–327.

Martin Krallinger, Obdulia Rabal, Anlia Loureno, Julen Oyarzabal, and Alfonso Valencia. 2017. Information retrieval and text mining technologies for chemistry. *Chemical reviews* 117(12):7673–7761.

Guillaume Lample, Miguel Ballesteros, Sandeep Subramanian, Kazuya Kawakami, and Chris Dyer. 2016. Neural Architectures for Named Entity Recognition. In *Proceedings of the 2016 Conference of the North American Chapter of the Association for Computational Linguistics: Human Language Technologies.* pages 260–270.

Jiao Li, Yueping Sun, Robin J. Johnson, Daniela Sciaky, Chih-Hsuan Wei, Robert Leaman, Allan Peter Davis, Carolyn J. Mattingly, Thomas C. Wiegers, and Zhiyong Lu. 2016a. BioCreative V CDR task corpus: a resource for chemical disease relation extraction. *Database* 2016:baw068.

Zhiheng Li, Zhihao Yang, Hongfei Lin, Jian Wang, Yingyi Gui, Yin Zhang, and Lei Wang. 2016b. CIDExtractor: A chemical-induced disease relation extraction system for biomedical literature. In *Proceedings of the 2016 IEEE International Conference on Bioinformatics and Biomedicine.* pages 994–1001.

Yankai Lin, Shiqi Shen, Zhiyuan Liu, Huanbo Luan, and Maosong Sun. 2016. Neural Relation Extraction with Selective Attention over Instances. In *Proceedings of the 54th Annual Meeting of the Association for Computational Linguistics (Volume 1: Long Papers).* pages 2124–2133.

Carolyn E. Lipscomb. 2000. Medical Subject Headings (MeSH). *Bulletin of the Medical Library Association* 88(3):265–266.

Daniel M. Lowe, Noel M. OBoyle, and Roger A. Sayle. 2016. Efficient chemical-disease identification and relationship extraction using Wikipedia to improve recall. *Database* 2016:baw039.

Xuezhe Ma and Eduard Hovy. 2016. End-to-end Sequence Labeling via Bi-directional LSTM-CNNs-CRF. In *Proceedings of the 54th Annual Meeting of the Association for Computational Linguistics (Volume 1: Long Papers).* pages 1064–1074.

Mitchell P Marcus, Mary Ann Marcinkiewicz, and Beatrice Santorini. 1993. Building a large annotated corpus of English: The Penn Treebank. *Computational linguistics* 19(2):313–330.

Mike Mintz, Steven Bills, Rion Snow, and Daniel Jurafsky. 2009. Distant supervision for relation extraction without labeled data. In *Proceedings of the Joint Conference of the 47th Annual Meeting of the ACL and the 4th International Joint Conference on Natural Language Processing of the AFNLP.* pages 1003–1011.

Makoto Miwa and Mohit Bansal. 2016. End-to-End Relation Extraction using LSTMs on Sequences and Tree Structures. In *Proceedings of the 54th Annual Meeting of the Association for Computational Linguistics (Volume 1: Long Papers).* pages 1105–1116.

Raymond J. Mooney and Razvan C. Bunescu. 2006. Subsequence Kernels for Relation Extraction. In *Advances in Neural Information Processing Systems 18*, pages 171–178.

Dat Quoc Nguyen, Mark Dras, and Mark Johnson. 2017. A Novel Neural Network Model for Joint POS Tagging and Graph-based Dependency Parsing. In *Proceedings of the CoNLL 2017 Shared Task: Multilingual Parsing from Raw Text to Universal Dependencies.* pages 134–142.

Thien Huu Nguyen, Kyunghyun Cho, and Ralph Grishman. 2016. Joint Event Extraction via Recurrent Neural Networks. In *Proceedings of the 2016 Conference of the North American Chapter of the Association for Computational Linguistics: Human Language Technologies.* pages 300–309.

Thien Huu Nguyen and Ralph Grishman. 2015. Relation Extraction: Perspective from Convolutional Neural Networks. In *Proceedings of the 1st Workshop on Vector Space Modeling for Natural Language Processing.* pages 39–48.

Thien Huu Nguyen and Ralph Grishman. 2016. Combining Neural Networks and Log-linear Models to Improve Relation Extraction. In *Proceedings of IJCAI Workshop on Deep Learning for Artificial Intelligence.*

Nagesh C. Panyam, Karin Verspoor, Trevor Cohn, and Kotagiri Ramamohanarao. 2018. Exploiting graph kernels for high performance biomedical relation extraction. *Journal of Biomedical Semantics* 9(1):7.

Nagesh C. Panyam, Karin M. Verspoor, Trevor Cohn, and Kotagiri Ramamohanarao. 2016. Exploiting Tree Kernels for High Performance Chemical Induced Disease Relation Extraction. In *Proceedings of the 7th International Symposium on Semantic Mining in Biomedicine.* pages 42–47.

Sachin Pawar, Girish K. Palshikar, and Pushpak Bhattacharyya. 2017. Relation Extraction: A Survey. *arXiv preprint* arXiv:1712.05191.

Yifan Peng, Chih-Hsuan Wei, and Zhiyong Lu. 2016. Improving chemical disease relation extraction with rich features and weakly labeled data. *Journal of Cheminformatics* 8(1):53.

Barbara Plank, Anders Søgaard, and Yoav Goldberg. 2016. Multilingual Part-of-Speech Tagging with Bidirectional Long Short-Term Memory Models and Auxiliary Loss. In *Proceedings of the 54th Annual Meeting of the Association for Computational Linguistics (Volume 2: Short Papers).* pages 412–418.

Ewoud Pons, Benedikt F.H. Becker, Saber A. Akhondi, Zubair Afzal, Erik M. van Mulligen, and Jan A. Kors. 2016. Extraction of chemical-induced diseases using prior knowledge and textual information. *Database* 2016:baw046.

Chris Quirk and Hoifung Poon. 2017. Distant Supervision for Relation Extraction beyond the Sentence Boundary. In *Proceedings of the 15th Conference of the European Chapter of the Association for Computational Linguistics: Volume 1, Long Papers*. pages 1171–1182.

Desh Raj, Sunil Kumar Sahu, and Ashish Anand. 2017. Learning local and global contexts using a convolutional recurrent network model for relation classification in biomedical text. In *Proceedings of the 21st Conference on Computational Natural Language Learning*. pages 311–321.

Rico Sennrich, Barry Haddow, and Alexandra Birch. 2016. Neural Machine Translation of Rare Words with Subword Units. In *Proceedings of the 54th Annual Meeting of the Association for Computational Linguistics (Volume 1: Long Papers)*. pages 1715–1725.

Nitish Srivastava, Geoffrey Hinton, Alex Krizhevsky, Ilya Sutskever, and Ruslan Salakhutdinov. 2014. Dropout: A Simple Way to Prevent Neural Networks from Overfitting. *Journal of Machine Learning Research* 15:1929–1958.

Ashish Vaswani, Noam Shazeer, Niki Parmar, Jakob Uszkoreit, Llion Jones, Aidan N Gomez, Ł ukasz Kaiser, and Illia Polosukhin. 2017. Attention is All you Need. In *Advances in Neural Information Processing Systems 30*, pages 5998–6008.

Patrick Verga, Emma Strubell, Ofer Shai, and Andrew McCallum. 2017. Attending to All Mention Pairs for Full Abstract Biological Relation Extraction. In *Proceedings of the 6th Workshop on Automated Knowledge Base Construction*.

Karin Verspoor, Kevin Bretonnel Cohen, Arrick Lanfranchi, Colin Warner, Helen L Johnson, Christophe Roeder, Jinho D Choi, Christopher Funk, Yuriy Malenkiy, Miriam Eckert, et al. 2012. A corpus of full-text journal articles is a robust evaluation tool for revealing differences in performance of biomedical natural language processing tools. *BMC bioinformatics* 13(1):207.

Chih-Hsuan Wei, Yifan Peng, Robert Leaman, Allan Peter Davis, Carolyn J. Mattingly, Jiao Li, Thomas C. Wiegers, and Zhiyong Lu. 2015. Overview of the BioCreative V Chemical Disease Relation (CDR) Task. In *Proceedings of the Fifth BioCreative Challenge Evaluation Workshop*. pages 154–166.

Chih-Hsuan Wei, Yifan Peng, Robert Leaman, Allan Peter Davis, Carolyn J. Mattingly, Jiao Li, Thomas C. Wiegers, and Zhiyong Lu. 2016. Assessing the state of the art in biomedical relation extraction: overview of the BioCreative V chemical-disease relation (CDR) task. *Database* 2016:baw032.

Jun Xu, Yonghui Wu, Yaoyun Zhang, Jingqi Wang, Hee-Jin Lee, and Hua Xu. 2016. CD-REST: a system for extracting chemical-induced disease relation in literature. *Database* 2016:baw036.

Daojian Zeng, Kang Liu, Yubo Chen, and Jun Zhao. 2015. Distant Supervision for Relation Extraction via Piecewise Convolutional Neural Networks. In *Proceedings of the 2015 Conference on Empirical Methods in Natural Language Processing*. pages 1753–1762.

Daojian Zeng, Kang Liu, Siwei Lai, Guangyou Zhou, and Jun Zhao. 2014. Relation Classification via Convolutional Deep Neural Network. In *Proceedings of COLING 2014, the 25th International Conference on Computational Linguistics: Technical Papers*. pages 2335–2344.

Wenyuan Zeng, Yankai Lin, Zhiyuan Liu, and Maosong Sun. 2017. Incorporating Relation Paths in Neural Relation Extraction. In *Proceedings of the 2017 Conference on Empirical Methods in Natural Language Processing*. pages 1768–1777.

Meishan Zhang, Yue Zhang, and Guohong Fu. 2017. End-to-End Neural Relation Extraction with Global Optimization. In *Proceedings of the 2017 Conference on Empirical Methods in Natural Language Processing*. pages 1730–1740.

Huiwei Zhou, Huijie Deng, Long Chen, Yunlong Yang, Chen Jia, and Degen Huang. 2016. Exploiting syntactic and semantics information for chemicaldisease relation extraction. *Database* 2016:baw048.

Appendix

Dataset and evaluation protocol: We evaluate our models using the BC5CDR corpus (Li et al., 2016a), which is the benchmark dataset for the BioCreative-V shared task on chemical-induced disease (CID) relation extraction (Wei et al., 2015, 2016).[7] The BC5CDR corpus consists of 1500 PubMed abstracts: 500 each for training, development and test set. In all articles, chemical and disease entities were manually annotated using the Medical Subject Headings (MeSH) concept identifiers (Lipscomb, 2000).

CID relations were manually annotated for each relevant pair of chemical and disease concept identifiers at the *document level* rather than for each pair of entity mentions (i.e. the relation annotations are not tied to specific mention annotations). Figure 2 shows examples of CID relations. We follow Gu et al. (2016) (see relation instance construction and hypernym filtering sections) and Gu

---

[7] http://www.biocreative.org/tasks/
biocreative-v/track-3-cdr/

```
1601297|t|Electrocardiographic evidence of myocardial injury in psychiatrically
hospitalized cocaine abusers.
1601297|a|The electrocardiograms (ECG) of 99 cocaine-abusing patients were
compared with the ECGs of 50 schizophrenic controls. Eleven of the cocaine
abusers and none of the controls had ECG evidence of significant myocardial
injury defined as myocardial infarction, ischemia, and bundle branch block.
1601297 33       50        myocardial injury    Disease  D009202
1601297 83       90        cocaine   Chemical D003042
1601297 135      142       cocaine   Chemical D003042
1601297 194      207       schizophrenic        Disease  D012559
1601297 232      239       cocaine   Chemical D003042
1601297 305      322       myocardial injury    Disease  D009202
1601297 334      355       myocardial infarction            Disease  D009203
1601297 357      365       ischemia Disease  D007511
1601297 371      390       bundle branch blockDisease  D002037
1601297 CID      D003042 D009203
1601297 CID      D003042 D002037
```

Figure 2: A part of an annotated PubMed article.

et al. (2017) to transfer these annotations to *mention level* relation annotations.

In the evaluation phase, mention-level classification decisions must be transferred to the document level. Following Gu et al. (2016), Li et al. (2016b) and Gu et al. (2017), these are derived from either (i) a pair of entity mentions that has been positively classified to form a CID relation based on the document or (ii) a pair of entity mentions that co-occurs in the document, and that has been annotated as having a CID relation in a document in the training set.

In an article, a pair of chemical and disease concept identifiers may have multiple entity mention pairs, expressed in different relation mentions.

The longest relation mention has about 400 word tokens; the longest word has 37 characters.

We use the training set to learn model parameters, the development set to select optimal hyperparameters, and the test to report final results using gold entity annotations. For evaluation results, we measure the CID relation extraction performance using F1 score.

Implementation details: We implement CNN, CNN+CNNchar, CNN+LSTMchar using Keras (Chollet et al., 2015) with a TensorFlow backend (Abadi et al., 2016), and use a fixed random seed. For both CNN+CNNchar and CNN+LSTMchar, character embeddings are randomly initialized with 25 dimensions, i.e. $d_4 = 25$. For CNNchar, the window size is 5 and the number of filters at 50, resulting in $d_3 = 50$. For LSTMchar, we set the number of LSTM units at 25, also resulting in $d_3 = 50$.

For all three models, position embeddings are randomly initialized with 50 dimensions, i.e. $d_2 = 50$. Word embeddings are initialized by using 200-dimensional pre-trained word vectors from Chiu

et al. (2016), i.e. $d_1 = 200$; and word types (including a special "UNK" word token representing unknown words), which are not in the embedding list, are initialized randomly. Following Kiperwasser and Goldberg (2016), the "UNK" word embedding is learned during training by replacing each word token w appearing n_w times in the training set with "UNK" with probability $\mathrm{p}_{unk}(w) = \frac{0.25}{0.25+n_w}$ (this procedure only involves the word embedding part in the input vector representation layer). We use ReLU for the activation function g, and fix the window size k at 5 and the L_2 regularization value at 0.001.

We train the models with Stochastic gradient descent using Nadam (Dozat, 2016). For training, we run for 50 epochs. We perform a grid search to select the optimal hyperparameters by monitoring the F1 score after each training epoch on the development set. Here, we select the initial Nadam learning rate $\lambda \in \{5\text{e-}06, 1\text{e-}05, 5\text{e-}05, 1\text{e-}04, 5\text{e-}04\}$, the number of filters $m \in \{100, 200, 300, 400, 500\}$ and the dropout probability $\rho \in \{0.25, 0.5\}$. We choose the model with highest F1 on the development set, which is then applied to the test set for the evaluation phase.

Domain Adaptation for Disease Phrase Matching
with Adversarial Networks

Miaofeng Liu♦*, **Jialong Han**♠, **Haisong Zhang**♠, and **Yan Song**♠
♦University of Science and Technology of China
♠Tencent AI Lab
{water3er, jialonghan}@gmail.com, {hansonzhang, clksong}@tencent.com

Abstract

With the development of medical information management, numerous medical data are being classified, indexed, and searched in various systems. Disease phrase matching, *i.e.,* deciding whether two given disease phrases interpret each other, is a basic but crucial preprocessing step for the above tasks. Being capable of relieving the scarceness of annotations, domain adaptation is generally considered useful in medical systems. However, efforts on applying it to phrase matching remain limited. This paper presents a domain-adaptive matching network for disease phrases. Our network achieves domain adaptation by adversarial training, *i.e.,* preferring features indicating whether the two phrases match, rather than which domain they come from. Experiments suggest that our model has the best performance among the very few non-adaptive or adaptive methods that can benefit from out-of-domain annotations.

1 Introduction

In recent years, hospitals depend more on information systems to store and retrieve medical data for diagnosis and treatment. To facilitate reliable and efficient processing of medical data, *disease phrase matching* has been identified as a crucial task in those medical systems. Given two disease phrases, this task requires identifying whether they are able to interpret each other.

Owing to complicated medical terminologies, overlapping words or similar syntactic structures are not reliable cues for disease phrase matching. Table 1 shows two matching candidates for *"Latent syphilis, specified as early or late"* (Phrase

Phrase 1	Phrase 2	Label
Latent syphilis, specified as early or late	*Syphilis latent*	Yes
Latent syphilis, specified as early or late	*Late syphilis, specified*	No

Table 1: Examples of disease phrase matching.

1). In the first one, the absent participial modifier and the different word order do not prevent the two phrases from matching. The second one is, however, a false match, though it shares more words and similar syntactic structures with Phrase 1.

Given the variability of human languages, supervised phrase or sentence matching is widely applied in information identification (Madnani et al., 2012; Yin et al., 2016), textual entailment (Marelli et al., 2014), web search (Li et al., 2014), entity linking (Traylor et al., 2017), and disease inference (Nie et al., 2015). As deep learning drew attentions on various tasks (Lecun et al., 2015), dedicated neural matching models are also designed in two types of structures. 1) **Siamese-based networks** (Neculoiu et al., 2016; Mueller and Thyagarajan, 2016): the input phrases are first encoded by the same network; the encoded vectors are then used to compute similarities by metrics like Cosine. 2) **Matching-aggregating networks**: fine-grained units of the two phrases are represented and matched in word-by-word (Rocktäschel et al., 2015), one-direction (Wang and Jiang, 2016), or bilateral-multi-perspective (Wang et al., 2017) manners to produce matching features; the features are aggregated into a vector, based on which the matching label is predicted.

Despite encouraging results in other areas, neural matching models still face specific challenges on medical data. Different medical subfields like physiology and urology may adopt diverse terminologies. Due to their professional nature, it is hard to obtain human annotations at scale for a single subfield. This causes systems on a partic-

*Work was done during the internship at Tencent AI Lab.

Proceedings of the BioNLP 2018 workshop, pages 137–141
Melbourne, Australia, July 19, 2018. ©2018 Association for Computational Linguistics

ular target subfield or domain to have too few annotations to learn a complicated neural model. It may be tempting to involve annotations from one or more source domains for more training data. But since all above models assume in-domain annotations, the effect of source-domain annotations remains uncertain on the trained models.

This paper takes a perspective that is orthogonal to works on designing sophisticated matching networks. We employ domain adaptation in disease phrase matching to effectively exploit source annotations. Based on Bilateral Multi-Perspective Matching (BiMPM) (Wang et al., 2017), we propose a Domain-Adaptive BiMPM (DA-BiMPM) model. Inspired by domain-adversarial training (Ganin et al., 2016) on text classification (Liu et al., 2017), relation extraction (Fu et al., 2017), and paraphrase identification (Yu et al., 2018), we introduce a domain discriminator in addition to the matching predictor in BiMPM. With such a discriminator, DA-BiMPM is encouraged to learn features predictive of the matching labels, while being least discriminative of which domain the data comes from. In doing so, it is expected that the learned models distill domain-insensitive knowledge from source annotations. On two medical datasets from different subfields, we set up non-adaptive baselines fed with or without source-domain annotations, as well as an adaptive one. Experimental results show that, when trivially involving source-domain data, only the strongest baseline BiMPM can achieve a slight gain. Compared with the adaptive approach, DA-BiMPM is capable of making more improvement on BiMPM.

2 Preliminaries

Before going into details of DA-BiMPM, we start with introducing the BiMPM model (Wang et al., 2017), which is illustrated by components outside the dotted box in Figure 1. Its encoding, matching, and aggregation layers are described as follows.

Phrase Encoder. Given a disease phrase $P = (p_1, \ldots, p_n)$ with n words, BiMPM encode it as follows. First, it transforms P in to a vector sequence $P = (p_1, \ldots, p_n)$. Each word is represented by concatenating a pre-trained GloVe (Pennington et al., 2014) vector and a character-BiLSTM-encoded vector. A BiLSTM is then applied on P to represent context in both directions:

$$\overleftarrow{\mathbf{H}}^P = (\overleftarrow{\mathbf{h}}_1^P, \overleftarrow{\mathbf{h}}_2^P, \ldots, \overleftarrow{\mathbf{h}}_n^P) = \overleftarrow{\text{LSTM}}(P) \quad (1)$$

$$\overrightarrow{\mathbf{H}}^P = (\overrightarrow{\mathbf{h}}_1^P, \overrightarrow{\mathbf{h}}_2^P, \ldots, \overrightarrow{\mathbf{h}}_n^P) = \overrightarrow{\text{LSTM}}(P) \quad (2)$$

Phrase Matcher. Given context representations

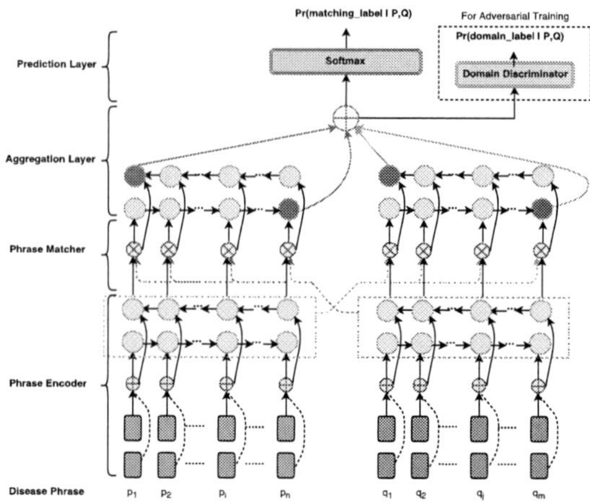

Figure 1: The architecture of (DA-)BiMPM.

of P and Q, a *phrase matcher* compares them with each time step of one against all of the other's in both directions. For example, when comparing word p_i with Q, we generated a *matching vector*

$$\mathbf{m}_i^P = (\overrightarrow{\mathbf{h}}_i^P \otimes \overrightarrow{\mathbf{H}}^Q, \overleftarrow{\mathbf{h}}_i^P \otimes \overleftarrow{\mathbf{H}}^Q) \quad (3)$$

Here $\otimes$ denotes the *multi-perspective matching* operation defined in (Wang et al., 2017). We refer readers to this paper for details.

Aggregation Layer. Given all matching vectors $\mathbf{M}^P = (\mathbf{m}_1^P, \ldots, \mathbf{m}_n^P)$ by comparing P to Q, and $\mathbf{M}^Q$ vice versa, we apply another BiLSTM layer to aggregate both of them, respectively. Formally,

$$(\overrightarrow{\mathbf{A}}^P, \overleftarrow{\mathbf{A}}^P) = \text{BiLSTM}(\mathbf{M}^P) \quad (4)$$

$$(\overrightarrow{\mathbf{A}}^Q, \overleftarrow{\mathbf{A}}^Q) = \text{BiLSTM}(\mathbf{M}^Q) \quad (5)$$

Finally, we concatenate the four ending hidden vectors of the BiLSTM layer, *i.e.*, $\overrightarrow{\mathbf{a}}_n^P, \overleftarrow{\mathbf{a}}_1^P, \overrightarrow{\mathbf{a}}_m^Q$, and $\overleftarrow{\mathbf{a}}_1^Q$, as the *matching features* $\mathbf{F}$.

To decide whether P and Q match, we apply a fully connected softmax layer on $\mathbf{F}$ to produce the prediction $y(\mathbf{F})$. Denoting all parameters of the feature extraction layers by ϕ_f, and the prediction layer ϕ_y, for ground truth $y^{(k)}$ of the k-th phrase pair, the instance-level matching loss is

$$l^{(k)}(\phi_f, \phi_y) = l(y(\mathbf{F}^{(k)}), y^{(k)}) \quad (6)$$

3 Domain-Adversarial Training

Given the configurations of BiMPM, the network parameters $\{\phi_f, \phi_y\}$ are optimized to minimize the gap between predicted and ground-truth matching labels. When source-domain training data is involved, due to the large parameter space of ϕ_f, the model may be satisfied with fitting labels in each domain separately instead of finding a unified explanation. This limitation thus causes the model to miss potential benefits of learning domain-independent matching features.

To fully utilize source-domain annotations, we apply domain-adversarial training (Ganin et al., 2016) on BiMPM. As illustrated by the dotted box in Figure 1, we add a domain discriminator $d(.)$ on $\mathbf{F}$, *i.e.*, the matching features. The discriminator is configured with the same fully-connected and softmax layers as the matching prediction layer. Given the domain $d^{(k)}$ where the k-th phrase pair is from, the domain loss is similarly given as

$$l_d^{(k)}(\phi_f, \phi_d) = l_d(d(\mathbf{F}^{(k)}), d^{(k)}) \tag{7}$$

Different from minimizing the matching loss $l^{(k)}$, we optimize the domain loss $l_d^{(k)}$ in the contrary direction. In other words, we prefer $\{\phi_f, \phi_d\}$ that preserve little domain-specific information.

Formally, given training phrase pairs in the target domain with indices $k \in T$, and source-domain data with indices $k \in S$, our joint objective function is given as follows by interpolating both the matching and the domain losses:

$$L(\phi_f, \phi_y, \phi_d) = \frac{1}{|S \cup T|} \sum_{k \in S \cup T} l^{(k)}(\phi_f, \phi_y)$$
$$- \lambda \Big[\frac{1}{|S|} \sum_{k \in S} l_d^{(k)}(\phi_f, \phi_d) + \frac{1}{|T|} \sum_{k \in T} l_d^{(k)}(\phi_f, \phi_d) \Big] \tag{8}$$

When optimizing the objective function, we seek for a saddle $\{\hat{\phi}_f, \hat{\phi}_y, \hat{\phi}_d\}$ such that:

$$\hat{\phi}_f, \hat{\phi}_y = \operatorname*{arg\,min}_{\phi_f, \phi_y} L(\phi_f, \phi_y, \hat{\phi}_d) \tag{9}$$

$$\hat{\phi}_d = \operatorname*{arg\,max}_{\phi_d} L(\hat{\phi}_f, \hat{\phi}_y, \phi_d) \tag{10}$$

By considering domain adaptation and matching label prediction in the joint objective, the training process pursuits a balance between both aspects. Interactions between the matching loss and the domain loss will force their shared parameters, *i.e.*, $\hat{\phi}_f$, to be generalizable across domains.

4 Experiments

4.1 Datasets and Baselines

We employ ICD10DATA[1] and MIMIC (Johnson et al., 2016) as the source and target domain datasets, respectively. ICD10DATA consists of diverse disease names from multiple medical subfields[2] and their approximate synonyms. MIMIC is a public dataset on computational physiology. The used phrase pairs are composed of terminology co-reference pairs of disease entities. Because both datasets consist of only positive pairs, we have to generate negative pairs. For each positive pair $\langle P, Q \rangle$, we corrupt Q with a random phrase

Dataset	# of Pairs	Subfield	Domain
ICD10DATA	29,783	Mixed	Source
MIMIC	22,504	Physiology	Target

Table 2: Statistics of source and target datasets.

from all other pairs containing neither P nor Q. We summarize both datasets in Table 2.

We adopt a training/validation/testing split of 3:1:1 on the target dataset, and conduct 5-fold cross validation. Average results on the five testing sets are reported. When involving the source dataset to help train better classifiers for the target domain, we use all annotations for training. We compare DA-BiMPM with five baselines:

Cosine: Phrases are represented by summing their GloVe (Pennington et al., 2014) word vectors. Their similarities are measured by Cosine scores.

Support Vector Machine (SVM): An SVM classifier is trained and applied on the concatenation of the phrase pairs' GloVe vectors.

Random Forest: Instead of SVM, this baseline applies random forest to train matching classifiers.

Siamese-LSTM: We use an existing implementation[3] of Mueller and Thyagarajan (2016).

BiMPM: This is the matching-aggregating network (Wang et al., 2017) described in Section 2.

In DA-BiMPM, we adopt the same configuration with that of BiMPM. We empirically set λ in Equation 8 to 0.5 throughout the experiments.

4.2 Preliminary Results

Figure 2 demonstrates the changes of the three losses in Equations 6, 7, and 8, respectively. We observe that, as training proceeds to about 100 iterations, all losses tend to decrease and then converge. Readers may notice that the domain loss follows a decreasing trend, which seems inconsistent with its negative coefficient in Equation 8. Note that the matching and domain losses are both functions of the feature extraction parameters ϕ_f, thus are correlated. As the matching loss decreases, ϕ_f may inevitably capture domain-dependent information. Therefore, the trade-off between minimizing the matching loss and maximizing the domain loss cannot achieve both objectives in positive directions. It can only prevent the latter loss from decreasing too much. The same figure also shows that, after 20 iterations, the validation accuracy grows quickly and then converges to 96.04%, yielding a testing accuracy of 96.96%.

---

[1] http://www.icd10data.com/. We only used the ICD-10-CM (diagnosis) subset.

[2] We uniformly treat them as from one source domain.

[3] https://github.com/dhwajraj/deep-siamese-text-similarity

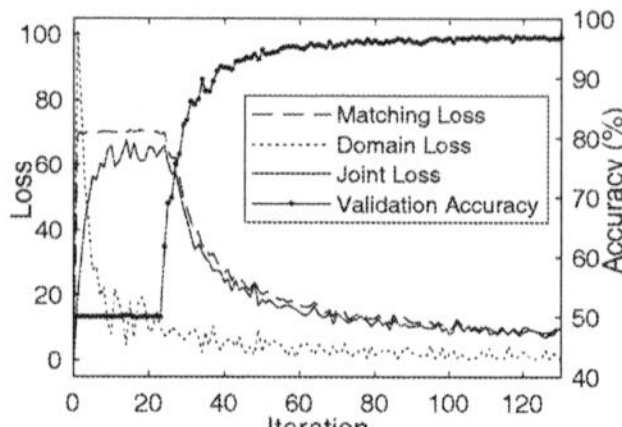

Figure 2: Training losses and validation accuracy.

Table 3: Testing accuracy (%) w/ or w/o source annotations.

Model	S. + T.	T. Only
Cosine	48.22	53.73
SVM	78.54	80.04
Random Forest	83.61	86.15
Siamese-LSTM	90.75	90.97
BiMPM	91.27	91.06
DA-BiMPM	96.96	N/A

Table 4: (DA-)BiMPM's testing accuracy (%) w.r.t. different settings.

Setting	Accuracy
BiMPM (S. Only)	90.74
BiMPM (T. Only)	91.06
BiMPM (S. + T.)	91.27
BiMPM (DDC variant)	92.39
DA-BiMPM (unsupervised)	96.12
DA-BiMPM (supervised)	96.96

4.3 Comparative Studies

In Table 3, we report the performance of all approaches. For each baseline, we train the model by aggregating both the entire source-domain dataset and the training set of the target domain. For comparison, we also trained them without the source dataset. We have the following observations.

First, when given the combined training set, the performance of the five baselines increases by the order they are presented. Specifically, the simplest Cosine approach is close to random guesses. Supervised methods like SVM and Random Forest, on the other hand, produce much better results. Neural network approaches, including Siamese-LSTM and matching-aggregating-based BiMPM, have the best performance among all baselines.

Moreover, including the source-domain dataset for training have different effects on the baselines. For the first four baselines, this dataset harms the training process and results in inferior performance. In contrast, BiMPM achieves slightly better accuracy by involving source-domain annotations. We note that such different effects may be due to a different model complexity. As a complicated model, BiMPM is able and tends to benefit from larger training data, even if they are from different domains. In summary, if exploited in a straight-forward manner, source-domain annotations cannot always guarantee better performance.

Finally, DA-BiMPM achieves more than five points of performance gain on top of BiMPM. Note that BiMPM has already taken advantage of source-domain annotations. Compared with BiMPM, DA-BiMPM only accepts domain labels as additional training information. The matching classifier trained by DA-BiMPM has the same structure, and requires the same input to make predictions, with that of BiMPM. This indicates that DA-BiMPM is making domain-adaptive exploitation of source-domain data from the feature level.

In Table 4, we evaluate DA-BiMPM in the unsupervised setting, *i.e.*, considering only source annotations in matching loss. This is done by not involving any target data when updating the prediction layer. We compete with Deep Domain Confusion (DDC) (Tzeng et al., 2014), where an adaptation layer based on Maximum Mean Discrepancy (Borgwardt et al., 2006) is applied after the phrase matcher. We also include (DA-)BiMPM's results in other relevant settings for comparison. It is observed that approaches with more information achieve better accuracy. Specifically, with access to the source data and distribution of the target training set, the unsupervised DA-BiMPM outperforms DDC-based BiMPM by nearly four points.

4.4 A Case Study

To further examine the impact of domain adaptation, we study a phrase pair *"bleed"* and *"gun shot wound to the head"* in the target set. When involving only target data, BiMPM correctly judged the pair as a mismatch. We find that, if involved in a pair, *"bleed"* on both sides tends to suggest a match. The numbers of instances for and against this feature are 1,401 and 747, respectively.

After trivially accessing source data, BiMPM achieved a slight gain. However, the above statistics are both 41 on the source set, implying a different data distribution. BiMPM was mislead on the above pair, and gave a false positive label. Meanwhile, DA-BiMPM overcomes the domain difference, and corrected the label to negative.

5 Conclusion

We present DA-BiMPM, a domain-adversarial network for disease phrase matching. It outperforms the base model BiMPM as well as four other baselines, with or without source annotations. Experiments also demonstrate that, when trivially combined with target-domain training data, source-domain data does not always make positive impacts. However, DA-BiMPM can better exploit the source-domain data, even if BiMPM or its DDC variant have taken advantage of it.

References

Karsten M Borgwardt, Arthur Gretton, Malte J Rasch, Hans-Peter Kriegel, Bernhard Schölkopf, and Alex J Smola. 2006. Integrating structured biological data by kernel maximum mean discrepancy. *Bioinformatics* 22(14):e49–e57.

Lisheng Fu, Thien Huu Nguyen, Bonan Min, and Ralph Grishman. 2017. Domain adaptation for relation extraction with domain adversarial neural network. In *Proceedings of the Eighth International Joint Conference on Natural Language Processing (Volume 2: Short Papers)*. volume 2, pages 425–429.

Yaroslav Ganin, Evgeniya Ustinova, Hana Ajakan, Pascal Germain, Hugo Larochelle, François Laviolette, Mario Marchand, and Victor Lempitsky. 2016. Domain-adversarial training of neural networks. *Journal of Machine Learning Research* 17(59):1–35.

Alistair EW Johnson, Tom J Pollard, Lu Shen, Liwei H Lehman, Mengling Feng, Mohammad Ghassemi, Benjamin Moody, Peter Szolovits, Leo Anthony Celi, and Roger G Mark. 2016. Mimic-iii, a freely accessible critical care database. *Scientific data* 3.

Yann Lecun, Yoshua Bengio, and Geoffrey Hinton. 2015. Deep learning. *Nature* 521(7553):436–444.

Hang Li, Jun Xu, et al. 2014. Semantic matching in search. *Foundations and Trends® in Information Retrieval* 7(5):343–469.

Pengfei Liu, Xipeng Qiu, and Xuanjing Huang. 2017. Adversarial multi-task learning for text classification. In *Proceedings of the 55th Annual Meeting of the Association for Computational Linguistics (Volume 1: Long Papers)*. volume 1, pages 1–10.

Nitin Madnani, Joel Tetreault, and Martin Chodorow. 2012. Re-examining machine translation metrics for paraphrase identification. In *Proceedings of the 2012 Conference of the North American Chapter of the Association for Computational Linguistics: Human Language Technologies*. Association for Computational Linguistics, pages 182–190.

Marco Marelli, Luisa Bentivogli, Marco Baroni, Raffaella Bernardi, Stefano Menini, and Roberto Zamparelli. 2014. Semeval-2014 task 1: Evaluation of compositional distributional semantic models on full sentences through semantic relatedness and textual entailment. In *SemEval@ COLING*. pages 1–8.

Jonas Mueller and Aditya Thyagarajan. 2016. Siamese recurrent architectures for learning sentence similarity. In *Thirtieth AAAI Conference on Artificial Intelligence*. pages 2786–2792.

Paul Neculoiu, Maarten Versteegh, and Mihai Rotaru. 2016. Learning text similarity with siamese recurrent networks. In *Repl4nlp Workshop at ACL*.

Liqiang Nie, Meng Wang, Luming Zhang, Shuicheng Yan, Bo Zhang, and Tat-Seng Chua. 2015. Disease inference from health-related questions via sparse deep learning. *IEEE Transactions on Knowledge and Data Engineering* 27(8):2107–2119.

Jeffrey Pennington, Richard Socher, and Christopher Manning. 2014. Glove: Global vectors for word representation. In *Proceedings of the 2014 conference on empirical methods in natural language processing (EMNLP)*. pages 1532–1543.

Tim Rocktäschel, Edward Grefenstette, Karl Moritz Hermann, Tomáš Kočiský, and Phil Blunsom. 2015. Reasoning about entailment with neural attention. *arXiv preprint arXiv:1509.06664* .

Aaron Traylor, Nicholas Monath, Rajarshi Das, and Andrew McCallum. 2017. Learning string alignments for entity aliases. In *6th Workshop on Automated Knowledge Base Construction (AKBC)*.

Eric Tzeng, Judy Hoffman, Ning Zhang, Kate Saenko, and Trevor Darrell. 2014. Deep domain confusion: Maximizing for domain invariance. *arXiv preprint arXiv:1412.3474* .

Shuohang Wang and Jing Jiang. 2016. Learning natural language inference with lstm. In *Proceedings of NAACL-HLT*. pages 1442–1451.

Zhiguo Wang, Wael Hamza, and Radu Florian. 2017. Bilateral multi-perspective matching for natural language sentences. *arXiv preprint arXiv:1702.03814* .

Wenpeng Yin, Hinrich Schütze, Bing Xiang, and Bowen Zhou. 2016. ABCNN: attention-based convolutional neural network for modeling sentence pairs. *TACL* 4:259–272.

Jianfei Yu, Minghui Qiu, Jing Jiang, Jun Huang, Shuangyong Song, Wei Chu, and Haiqing Chen. 2018. Modelling domain relationships for transfer learning on retrieval-based question answering systems in e-commerce. In *Proceedings of the Eleventh ACM International Conference on Web Search and Data Mining*. ACM, pages 682–690.

Predicting Discharge Disposition Using Patient Complaint Notes in Electronic Medical Records

Mohamad Salimi
Department of Computer Science
Queens College, CUNY
Krasnoff Quality Management Institute
Northwell Health
`mohamad.salimi45@qmail.cuny.edu`

Alla Rozovskaya
Department of Computer Science
Queens College, CUNY
`arozovskaya@qc.cuny.edu`

Abstract

Overcrowding in emergency rooms is a major challenge faced by hospitals across the United States. Overcrowding can result in longer wait times, which, in turn, has been shown to adversely affect patient satisfaction, clinical outcomes, and procedure reimbursements. This paper presents research that aims to automatically predict discharge disposition of patients who received medical treatment in an emergency department. We make use of a corpus that consists of notes containing patient complaints, diagnosis information, and disposition, entered by health care providers. We use this corpus to develop a model that uses the complaint and diagnosis information to predict patient disposition. We show that the proposed model substantially outperforms the baseline of predicting the most common disposition type. The long-term goal of this research is to build a model that can be implemented as a real-time service in an application to predict disposition as patients arrive.

1 Introduction

Studies show that wait times not only affect patient satisfaction, but also the perception of providers and quality of care (Chandra et al., 1981). Furthermore, the Center for Medicare and Medicaid Services is tying reimbursements to the Hospital Consumer Assessment of Healthcare Providers and Systems (HCAHPS) scores. As a financial and patient experience priority, hospitals are focused on addressing issues that affect patient satisfaction. One common issue is long wait times in the emergency rooms, that are due to high volume and overcrowding. Another issue is that of bed management and its effect on wait times. If no in-patient beds are available, admitted patients are kept in the emergency department until beds open. This is commonly referred to as patient boarding and has been shown to negatively affect outcomes and wait times. (Felton et al., 2011).

Improved bed management and resource utilization are necessary to achieve shorter wait times. This paper describes a first attempt at an experimental model which aims to predict discharge disposition based on chief complaint (i.e. symptoms description) and diagnosis information contained in clinical notes. The corpus that we use contains approximately 260,000 annotated emergency department records. The records contain free text of a complaint and admit diagnosis, and are labeled with the disposition information. The disposition, which is the destination after medical treatment, can be classified as *Admit, Discharge, Observation, Expire, Left Against Medical Advice (AMA), Asthma Observation Unit (AOU), Eloped,* or *Transfer.*

A model that predicts disposition type could be realized as an informational alert system integrated with electronic medical record software. Patient complaints are made available before discharge dispositions, allowing for an immediate prediction of disposition. In some cases the complaint is available hours before the discharge diagnosis or the disposition. Thus, such a model could provide integrated real-time forecasting on potential discharges and in-patient admissions.

The rest of the paper is organized as follows. Sec. 2 presents related work. Sec. 3 describes the corpus. Sec. 4 presents the experimental setup. Results are reported in Sec. 5. We present error analysis in Sec. 6 and conclude in Sec. 7.

Proceedings of the BioNLP 2018 workshop, pages 142–146
Melbourne, Australia, July 19, 2018. ©2018 Association for Computational Linguistics

2   Related Work

The Academic Emergency Medicine journal published preliminary results that attempt to predict emergency department in-patient admissions to improve same-day patient flow (Peck et al., 2012). They used three methods – expert opinion, Naive Bayes, and a generalized linear regression model – to analyze two months worth of emergency department data from the Boston VA healthcare System. However, Peck et al. (2012) focused strictly on predicting *admit* dispositions only, while we aim to predict all possible outcomes. Furthermore, their results focus strictly on structured fields, such as urgency level, age, sex, chief complaint, and the provider seen, while we work with free text in clinical notes. Another issue with the above model is that by including the provider seen to predict admission, they are tightly coupling the model to the Boston VA health care System.

Previous work has been done which aims to predict patient outcomes using unstructured text. Yamashita et al. (2016) analyzed admission records of 1,222 patients who had a clinical pathway of cerebral infraction. The goal was to develop a method for automatically performing clinical evaluations and to identify early interventions for cases that may have clinically important outcomes.

There has been a lot of other related work in the NLP area on unstructured electronic medical records and, in particular, in the clinical domain. For example, Jonnagaddala et al. (2015) developed a model to automatically identify smoking status using a SVM model. Jung et al. (2011) extracted events from clinical notes and used this information to construct a timeline of patient medical history. Both of the above mentioned works also used unstructured clinical notes, but focused on identifying patient history information. Cogley et al. (2012) used machine learning to determine whether a patient experienced a particular medical condition. However, while Cogley et al. (2012) looked at patient history and physical examination reports, we wish to predict disposition from complaint and admitting diagnosis alone.

3   Data

The data used in this project is provided by the Krasnoff Quality Management Institute of Northwell Health. Northwell Health is a not-for-profit healthcare network that includes 22 hospitals and

Disposition	Percentage (%)
Admit	30.88
AMA	0.89
AOU	0.05
Discharge	63.66
Eloped	0.27
Expired	0.08
Observation	3.56
Transfer	0.56

Table 1:   Distribution of disposition labels in the training corpus.

over 500 medical offices. Krasnoff Quality Management Institute provides analytics support for the many facilities across Northwell Health. The dataset contains de-identified emergency department records from several facilities across the system. Each record includes information about the patient complaint, diagnosis, and the resulting disposition, filled out by clerical staff or nurses. We use a subset of the entire dataset in the present study, approximately 260,000 records. 215,000 are used to train the model, and 45,000 are used for testing.

There are eight possible values for the disposition outcome. Table 1 shows the distribution of the values in the corpus. Note that the outcome types are not evenly distributed. The most common disposition type, *discharge*, accounts for over 63% of all disposition types, and the two most frequent types, *discharge* and *admission* to the hospital, account for over 94% of all disposition types. The *observation* unit, which is an area in some emergency rooms which allows for extended evaluation for patients whose stays will likely be less than one day, follows as the third most common disposition (3.56%). Left against medical advice *(AMA)*, asthma observation unit *(AOU)*, left without notice *(eloped)*, death in the ER *(expired)*, and *transfer* to a different facility all account for less than 1% of total number of records.

Example Records Each instance in the dataset contains information about the symptoms, the diagnosis, and is annotated with its final disposition. The notes in the dataset do not contain information related to the treatment of the patient. Below we show several complaint instances from the corpus. As expected, since this information was entered by clinical staff, the text is quite noisy, contains a lot of specific medical abbreviations ("pt"), incomplete sentences, and typos ("cant").

Data Point	Value
Complaint	flu-like symptoms
Admit diag.	fever
Discharge diag.	viral illness
Complaint	Abdominal pain & heart burn
Admit diag.	NULL
Discharge diag.	enteritis

Table 2: Complaint, admit diagnosis, and discharge diagnosis examples.

- "pt called EMS 'I cant see' pt says she cant open her eyes"
- "bite, animal pain in limb puncture wound of left thigh, initial encounter, observation"
- "head injury car passenger injured in collision with two- or three-wheeled motor vehicle in traffic accident, initial encounter mvc (motor vehicle collision)"

The records also contain admitting diagnosis and discharge diagnosis. The admitting diagnosis is entered shortly after the complaint and may be updated by staff. The discharge diagnosis is entered once the patient's visit is complete. Table 2 shows two examples.

4 Experiments

Our aim is to create a prototype model that will be able to make predictions with the complaint and admit diagnosis extracted from clinical notes. Our model is trained with the Averaged Perceptron (Freund and Schapire, 1999) algorithm implemented with Learning Based Java (Rizzolo, 2011). While classical Perceptron comes with generalization bound related to the margin of the data, Averaged Perceptron also comes with a PAC-like generalization bound (Freund and Schapire, 1999). This linear learning algorithm is known, both theoretically and experimentally, to be among the best linear learning approaches and is competitive with SVM and Logistic Regression, while being more efficient in training. We do not use neural network approaches in this work both due to the moderate size of the dataset (neural models have been shown to have a steep learning curve (Koehn and Knowles, 2017) and also because our goal is to develop a model that would be as efficient as possible. We train the classifier on the training partition of the corpus and report results on the test partition. All the data was normalized by removing special characters, lowercased, and POS

Disposition	Rel. freq. (%)	Accuracy (%)
Discharge	63.4	75.7
Admit	31.1	77.9
Observation	3.6	96.3
Eloped	0.3	0.0
AMA	0.7	0.0
Transfer	0.6	0.0
AOU	0.1	0.0
Expire	0.1	99.9
Total	-	75.7

Table 3: Accuracy results by disposition type.

tagged with the NLTK tagger (Bird, 2006).

4.1 Features

The features include bag-of-word unigrams and bigrams, and collocations. To control for the vocabulary size, we only include the top unigrams and bigrams occurring in the training data. 75 unigram and bigram features are included.

The collocation features are based on a list of keywords extracted from the top 50 words occurring in the training data. Each collocation feature is a conjunction of the keyword, word tokens and part-of-speech tags occurring in the 2-word window around the keyword. Sample collocation features are shown below:

- W_{i-2}, POS_{i-2}, W_{i-1}, POS_{i-1}, Infection, W_{i+1}, POS_{i+1}, W_{i+2}, POS_{i+2}
- W_{i-2}, POS_{i-2}, W_{i-1}, POS_{i-1}, Pain, W_{i+1}, POS_{i+1}, W_{i+2}, POS_{i+2}

5 Results

We evaluate the model using both accuracy and F-score. Table 3 shows accuracy results by disposition type. We note that the most frequent class baseline that corresponds to selecting the discharge disposition, is 63.4. This is substantially lower than the overall accuracy of 75.7. The accuracy for the *discharge* class is 75.7%, while the accuracy for the second most frequent class, *admit*, is 77.9% (recall from Table 1 that the two labels account for over 94% of all instances in the training data). The performance on the least common disposition labels is poor, with the exception of *expire* (this is further discussed in the next section).

We further evaluate by computing precision, recall, and F-score for each class (Table 4). In general, the performance is higher for more frequent

Disposition	Precision (%)	Recall (%)	F-score (%)
Discharge	75.2	92.0	82.8
Admit	70.4	50.2	58.6
Observation	100.0	0.1	0.2
Eloped	0.0	0.0	0.0
AMA	0.0	0.0	0.0
Transfer	0.0	0.0	0.0
AOU	0.0	0.0	0.0
Expire	64.1	68.0	66.0
Total	73.2	74	70.8

Table 4: Precision, recall, and f-score results by disposition type.

classes, and very poor for the least common labels. The best F-score of 82.8% is achieved for the most frequent class, *discharge*. Again, one interesting exception here is the *expire* class.

6   Error Analysis

We analyze several cases on which the classifier's predictions were incorrect. The first instance (shown below) had a prediction for *discharge* but the correct label was *admit*.

- abdominal pain pleural effusion in other conditions classified elsewhere pleural effusion associated with hepatic disorder

"Pleural effusions", which is a condition in which excess fluid buildup is present around the lungs, is a potentially serious condition. In our corpus, pleural effusion cases were over five times more likely to be admitted than discharged. In this case, the addition of "abdominal pain" feature resulted in the classifier considering it a *discharge* record.

The next record was a *discharge* which was predicted to be an *admit*. This may be due to the presence of the word "bleeding".

- abdominal pain diverticulitis of large intestine without perforation or abscess without bleeding

Finally, some notes are extremely short, such as the complaint "chest pain", which was labeled as *admit* but the model classified it as *discharge*, due to it being the most common disposition for "chest pain".

It is clear that this task is challenging, given the brief and noisy nature of the clinical notes, which contributes to data sparseness, and ambiguity of features that may indicate multiple likely disposition outcomes.

Lastly, some dispositions were not classifiable by the model. In particular, we conjecture that leaving against medical advice (*AMA*) may be tied to factors not seen in symptoms such as social determinants. *Observation* and *transfer* classification may be improved with features that better target those dispositions. Clinical experts will need to be engaged for this task to better understand the feasibility of predicting those dispositions.

7   Conclusion

We presented a model for predicting emergency room disposition from clinical notes. We used a corpus of emergency room records that contains information on symptoms, diagnosis, and disposition labels, entered by medical staff. We showed that the proposed model significantly outperforms the baseline approach of selecting the most frequent class. The nature of the corpus is such that two most common classes account for over 94% of all cases. Although most machine learning problems have to do with label imbalance, we believe that our task is unique in that the imbalance is extreme. The performance of the model is better than the baseline in the most prevalent dispositions, as well as one very rare disposition of *expire*. The other least frequent classes are not classifiable by the model. We hypothesize that some dispositions may be tied to factors not reflected in symptoms, such as social determinants.

Although the results are promising, more work is needed to reach the level where such a model can be utilized in real-time applications. For example, text correction and text normalization of the clinical data might be helpful, given that the notes contain a lot of noise. However, we believe that the proposed experiment is an important step towards building a real-time system that can provide predictions as complaints come into emergency departments. Such a system can be utilized to assist clinical leadership in staffing and operational decisions.

Acknowledgments

We thank the anonymous reviewers for their helpful comments.

References

Steven Bird. 2006. Nltk: The natural language toolkit. In *Proceedings of the COLING/ACL 2006 Interactive Presentation Sessions*, pages 69–72. Association for Computational Linguistics.

Ashok K. Chandra, Dexter C. Kozen, and Larry J. Stockmeyer. 1981. Alternation. *Journal of the Association for Computing Machinery*, 28(1):114–133.

James Cogley, Nicola Stokes, Joe Carthy, and John Dunnion. 2012. Analyzing patient records to establish if and when a patient suffered from a medical condition. In *Proceedings of the 2012 Workshop on Biomedical Natural Language Processing*, BioNLP '12, pages 38–46, Stroudsburg, PA, USA. Association for Computational Linguistics.

Brent M. Felton, Earl J. Reisdorff, Christopher N. Krone, and Gus A. Laskaris. 2011. Emergency department overcrowding and inpatient boarding: A statewide glimpse in time.

Yoav Freund and Robert E. Schapire. 1999. Large margin classification using the perceptron algorithm. *Machine Learning*.

Jitendra Jonnagaddala, Hong-Jie Dai, Pradeep Ray, and Siaw-Teng Liaw. 2015. A preliminary study on automatic identification of patient smoking status in unstructured electronic health records. In *Proceedings of BioNLP 15*, pages 147–151, Beijing, China. Association for Computational Linguistics.

Hyuckchul Jung, James Allen, Nate Blaylock, William de Beaumont, Lucian Galescu, and Mary Swift. 2011. Building timelines from narrative clinical records: Initial results based-on deep natural language understanding. In *Proceedings of BioNLP 2011 Workshop*, pages 146–154, Portland, Oregon, USA. Association for Computational Linguistics.

P. Koehn and R. Knowles. 2017. Six challenges for neural machine translation. In *Proceedings of the First Workshop on Neural Machine Translation*. Association for Computational Linguistics.

J S Peck, J C Benneyan, D J Nightingale, and S A Gaehde. 2012. Predicting emergency department inpatient admissions to improve same-day patient flow.

N. Rizzolo. 2011. Learning based programming.

Takanori Yamashita, Yoshifumi Wakata, Hidehisa Soejima, Naoki Nakashima, and Sachio Hirokawa. 2016. Prediction of key patient outcome from sentence and word of medical text records. In *Proceedings of the Clinical Natural Language Processing Workshop (ClinicalNLP)*, pages 86–90, Osaka, Japan. The COLING 2016 Organizing Committee.

Bacteria and Biotope Entity Recognition Using A Dictionary-Enhanced Neural Network Model

Qiuyue Wang and Xiaofeng Meng
School of Information, Renmin University of China
Beijing 100872, China
{qiuyuew, xfmeng}@ruc.edu.cn

Abstract

Automatic recognition of biomedical entities in text is the crucial initial step in biomedical text mining. In this paper, we investigate employing modern neural network models for recognizing biomedical entities. To compensate for the small amount of training data in biomedical domain, we propose to integrate dictionaries into the neural model. Our experiments on BB3 data sets demonstrate that state-of-the-art neural network model is promising in recognizing biomedical entities even with very little training data. When integrated with dictionaries, its performance could be greatly improved, achieving the competitive performance compared with the best dictionary-based system on the entities with specific terminology, and much higher performance on the entities with more general terminology.

1 Introduction

In the microbial community, knowledge about habitats of bacteria is crucial for the study, e.g. metagenomics. To extract such information from the biomedical literature, the very first step is to accurately recognize bacteria and habitat entities in text. State-of-the-art systems mainly have taken two approaches: dictionary-based and feature-based.

Dictionary-based approach looks for all the possible names in one or more dictionaries (or ontologies, or databases, or gazetteers) of entities. The performance depends on the quality and comprehensiveness of the dictionaries built for each entity type, which require a lot of expert knowledge and maintenance costs. It is well suited for entities with closely defined vocabularies of specific names, such as species and diseases, but fails to accurately recognize entities with names consisting of more common words, e.g. habitat entities. TagIt (Cook et al., 2016) is a dictionary-based system participating BioNLP Shared Task

2016, which yielded the best performance in recognizing bacteria entities, however could not compete with other machine learning systems on recognizing habitat entities.

Feature-based machine learning systems are currently more widely used in biomedical entity recognition. When properly trained, a machine learning model can potentially recognize new entity names and new spelling variations of an entity name. Traditional machine learning approaches, are feature-rich supervised learning classifiers, requiring significant domain-specific feature engineering. Recently neural network models gain increasingly more research attention as they could automatically learn useful features from raw data. Compared with the work on NER in general domain (Lample et al., 2016, Chiu and Nichols, 2016, Ma and Hovy, 2016), there is little published work on employing modern neural network models for BioNER. It is probably due to the small sizes of human-annotated corpora in biomedical domain, which makes it very hard to train non-trivial neural network models.

In this paper, we investigate employing state-of-the-art neural network models to recognize biomedical entities. Our experiments on BB3 data sets show that even with very little training data, modern neural network model is promising in recognizing bacteria and habitat entities. To compensate for the shortage of annotated training data, we propose to utilize dictionaries or ontologies, which is abundant in biomedical domain, to enhance the neural models. The experiment results demonstrate that our dictionary-enhanced neural model yielded better performance than the currently best systems, especially on habitat entities.

2 Dictionary-Enhanced BiLSTM-CRF Model

Following the most state-of-the-art neural network models for general domain NER, we design a similar BiLSTM-CRF model as shown in Figure 1 for recognizing bacteria and habitats in text.

Proceedings of the BioNLP 2018 workshop, pages 147–150
Melbourne, Australia, July 19, 2018. ©2018 Association for Computational Linguistics

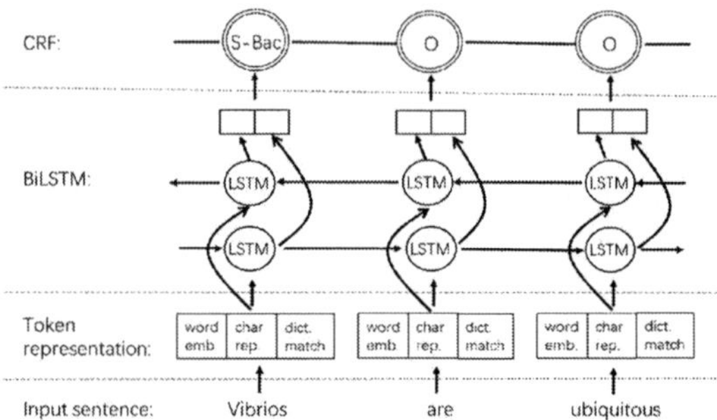

Figure 1: The BiLSTM-CRF model for entity recognition.

When receiving a sentence of tokens as input, e.g. *"Vibrios are ubiquitous"*, the system first forms a representation for each token, which is the concatenation of its word embedding, character-based representation and dictionary-matching representation of the token.

Next, the vector representations of tokens are fed into a bidirectional LSTM. The hidden state for each token position in BiLSTM is the concatenation of the hidden states from the forward and backward LSTMs. As a result, it contains both the left and right context information useful to make prediction for this token.

Finally, a Conditional Random Field (CRF) layer, modeling the dependencies between successive labels, is added on top of the BiLSTM network to find the most likely sequence of labels as the final output.

2.1 Word Embedding

There are various word embedding techniques, e.g. word2vec, Glove and fasttext. They address different types of semantic similarities, and thus perform differently for different NLP tasks. We tested Glove and fasttext for our task and found that fasttext performed better. Thus, we used the fasttext method to train word embeddings. Word embedding dimension is set to 100.

We downloaded PubMed 2017 baseline, extracted all the titles and abstracts, segmented them into tokens using different strategies:

- using a segmentation model for general English text or a model specially trained on biomedical text.
- removing punctuations or not.
- converting all characters into lower case and all digits to "0" or not.

We compared the performance of all the above strategies in the experiments, and found that re-

moving punctuations, lowercasing all characters and converting all digits to "0" did not result in better performed embeddings. So, we generated embeddings without removing punctuations and any other conversions.

2.2 Character-Based Representation

Although the word embeddings capture the semantic similarities between tokens, they ignore the character-level regularities of the token, like suffixes or prefixes, which are proven to be effective in NER tasks. We generate a character-based representation for each token using a LSTM model like that proposed in Lample et al., 2016. The dimensions of the character embedding and the hidden states of the BiLSTM are both set to be 25, so the dimension of the final character-level representation is 50.

2.3 Dictionary-Matching Representation

To train a non-trivial neural network model without overfitting it, a huge amount of annotated data are needed, which are much costlier to obtain in biomedical domain than in general domain since expert domain knowledge is required for annotating data. On the other hand, dictionaries, ontologies and databases are abundant in biomedical domain. We propose to make better use of such available knowledge in neural network models to compensate for the small sizes of annotated data.

In this paper, we incorporate dictionaries into the neural network model by adding a third part to the token representation: dictionary-matching representation. For each given dictionary, a dictionary matching feature is assigned to each input token. The matching feature indicates whether a word sequence formed by the token and its consecutive neighbors is in the dictionary. The maximal length of the word sequence is set to 6. When there are multiple overlapping matches, longer matches are preferred over shorter matches, and earlier matches in the sentence are preferred over later matches. The matching feature can take one of the five values: 'B', 'I', 'O', 'E', 'S', which means 'Begin', 'Inside', 'Outside', 'End' and 'Single' respectively, indicating the position of the token in the matched word sequence. Figure 2 shows an example sentence and the dictionary matching feature for each of its tokens. There are two types of entities to be recognized: bacteria and habitats, and two dictionaries are applied, one for each entity type.

	Vibrios	*are*	*ubiquitous*	*to*	*oceans*	*,*	*coastal*	*waters*	*,*	*and*	*estuaries*	*.*
Bacteria	S	O	O	O	O	O	O	O	O	O	O	O
Habitat	O	O	O	O	S	O	B	E	O	O	S	O

Figure 2: Dictionary matchings of an example sentence.

To generate the dictionary-matching representation for the token, we embed the matching feature for each dictionary into a 5-dimensional real-valued vector and then concatenate the vectors for all the dictionaries. As in Figure 2, the dictionary-matching representation of a token will be a 10-dimensional vector representing the matching features of this token in two dictionaries.

3 Experiments and Results

We implemented our models based on the open source code of NeuroNER[1] (Dernoncourt et al., 2017) and evaluated their performance using the dataset provided by the Bacteria Biotope task in the BioNLP Shared Task 2016 (BB3).

The BB3 task has no separate task for named entity recognition. It is jointly evaluated with downstream applications such as categorization or event extraction. Only in the BB3-cat+ner subtask, the official BB3 evaluation service additionally outputs the boundaries scoring about the system's ability to predict entity boundaries, in terms of SER (Slot Error Rate), Precision and Recall. For this reason, we primarily focus on the BB3-cat+ner subtask. We use the SER, Precision and Recall, output by the official BB3 evaluation service, as the evaluation metrics for our experiments. According to the official evaluation (Deléger et al., 2016), TagIt system achieved the best performance on detecting bacteria boundaries (SER: 0.236, recall: 0.772, precision: 0.954), while LIMSI system worked best on habitat entities (SER: 0.597, recall: 0.504, precision: 0.728). Bacteria are easier to recognize than habitats because bacteria names are mainly specific terms from a closely defined vocabulary, i.e. NCBI Taxonomy, with little variations, while habitat names usually consist of common English nouns and adjectives, e.g. "egg", "water", "fish" and expressed in various ways.

3.1 Dataset and Preprocessing

The dataset of the BB3-cat+ner subtask consists of 161 documents, split into training, development and test sets, which include 71, 36 and 54 documents and 1122, 698, 1022 entity occurrences respectively.

Entities occurring in the training or development documents are annotated in BRAT format. We preprocessed the data by first segmenting all the text into sentences of tokens using spaCy[2], and then tagging each token with a label in BIOES labelling scheme. For example, "B-Bacteria" means the token is the beginning word of a bacteria entity mention, and "S-Habitat" means the token is by itself the mention for a habitat entity.

3.2 Word Embeddings

For segmenting text to train word embeddings, we could use a segmentation model for general English text, or alternatively a model specifically trained on biomedical text. For the general model, we used spaCy, and for the specific model, we applied OpenNLP with its specially trained model on the GENIA corpus.

As shown by the first two lines in Table 1, using a specific model trained on domain text gained higher precision while lower recall than using a general English model. It also shows that the state-of-the-art BiLSTM-CRF model is a promising approach for recognizing biomedical entities, even with very little training data like in BB3 task.

3.3 Integration of Dictionaries

In general, performance of neural models could get far improved by using more training data. However, it is costly to collect a large amount of training data in biomedical domain. Recently, more and more research work focused on finding ways to compensate for the shortage of training data, e.g. using semi-supervised learning or multi-task learning techniques. In this paper, we exploited the way of integrating dictionaries or ontologies into the neural network model to improve performance. For detecting bacteria and habitats, we use the most recent comprehensive dictionaries[3] specially built for these two types of entities by TagIt. We tested two strategies of matching with dictionary entries: case-sensitive and case-

---

[1] http://neuroner.com/

[2] https://spacy.io/

[3] https://github.com/bitmask/BioNLP-BB3

Systems	Overall				Bacteria boundaries			Habitat boundaries		
	SER	Recall	Prec.	F1	SER	Recall	Prec.	SER	Recall	Prec.
spaCy	0.487	0.596	0.789	0.678	0.415	0.693	0.814	0.519	0.544	0.775
OpenNLP	0.493	0.549	0.830	0.661	0.376	0.656	0.891	0.558	0.490	0.785
spaCy (B+H)	0.435	0.624	0.828	0.712	0.324	0.701	0.919	0.503	0.578	0.768
spaCy (B+H, lower)	0.429	0.617	0.852	0.715	0.318	0.710	0.918	0.499	0.556	0.801
OpenNLP (B+H)	0.442	0.578	**0.876**	0.697	0.330	0.684	0.925	0.514	0.511	**0.835**
OpenNLP (B+H, lower)	**0.415**	0.617	0.867	**0.721**	0.301	0.707	**0.938**	0.483	0.563	0.816
TagIt (Cook et al., 2016)	-	-	-	-	0.236	0.772	0.954	0.599	0.476	0.675
LIMSI (Grouin, 2016)	-	-	-	-	0.277	0.751	0.903	0.597	0.504	0.728

Table 1: Experiment results.

insensitive matching. In Table 1, "B+H" represents for using bacteria and habitat dictionaries, and "lower" means case-insensitive matching with the dictionary.

From Table 1, we can have the following observations:

(1) By comparing the "B+H" lines with the first two lines, we can observe that integrating dictionaries into neural models can significantly improve the performance. For example, the overall SER is reduced by 12%-16%.

(2) By comparing the "B+H" lines with "B+H, lower" lines, we see that case-insensitive matching with dictionary is more effective than case-sensitive matching.

(3) Compared with the existing two best systems using traditional dictionary-based (TagIt) or feature-based (LIMSI) approaches, our best model "OpenNLP (B+H, lower)" can perform competitively on recognizing bacteria entities and much better on recognizing habitat entities.

4   Conclusions and Future Work

To the best of our knowledge, this is the first work of applying state-of-the-art neural network models in recognizing bacteria and biotope entities. The experiment results on BB3 task show that it is promising even with very small sized training data. Its performance can be much improved by integrating dictionaries, achieving competitive performance on bacteria entities and much better performance on habitat entities compared with the best traditional methods.

As for future work, we intend to (1) test our model on more types of biomedical entities; (2) investigate other ways of integrating dictionaries or ontologies with neural networks; (3) extend our model to deal with the embedded entities and discontinuous entities, which are special challenges for BioNER.

Acknowledgments

This work was supported by the National Natural Science Foundation of China (No. 61532010), the National Key Research and Development Program of China (No. 2016YFB000603), and the Opening Project of State Key Laboratory of Digital Publishing Technology.

References

Jason P.C. Chiu and Eric Nichols. 2016. Named Entity Recognition with Bidirectional LSTM-CNNs. *TACL*, vol. 4, pp. 357–370.

Helen V Cook, Evangelos Pafilis, Lars J. Jensen. 2016. A dictionary- and rule-based system for identification of bacteria and habitats in text. *BioNLP 2016*.

Louise Deléger, et al. 2016. Overview of the Bacteria Biotope task at BioNLP Shared Task 2016. *BioNLP 2016*.

Franck Dernoncourt, Ji Young Lee, Peter Szolovits. 2017. NeuroNER: an easy-to-use program for named-entity recognition based on neural networks. *EMNLP 2017*.

Cyril Grouin, 2016. Identification of Mentions and Relations between Bacteria and Biotope from PubMed Abstracts. *BioNLP 2016*.

Guillaume Lample, et al. 2016. Neural architectures for named entity recognition. *NAACL-HLT 2016*, pages 260–270.

Xuezhe Ma and Eduard Hovy. 2016. End-to-end Sequence Labeling via Bi-directional LSTM-CNNs-CRF. *ACL 2016*, pages 1064–1074.

SingleCite: Towards an improved Single Citation Search in PubMed

Lana Yeganova, Donald C Comeau, Won Kim,
W John Wilbur, Zhiyong Lu
National Center for Biotechnology Information, NLM, NIH, Bethesda, MD, USA
{yeganova, comeau, wonkim, wilbur, luzh}@mail.nih.gov

Abstract

A search that is targeted at finding a specific document in databases is called a Single Citation search. Single citation searches are particularly important for scholarly databases, such as PubMed®, because users are frequently searching for a specific publication. In this work we describe SingleCite, a single citation matching system designed to facilitate user's search for a specific document. We report on the progress that has been achieved towards building that functionality.

1 Introduction

PubMed, a search engine that works on MED-LINE®, processes on average 3 million queries a day and is recognized as a primary tool for scholars in the biomedical field (Falagas, Pitsouni, Malietzis, & Pappas, 2008; Lu, 2011; Wildgaard & Lund, 2016). Given the significance of Pub-Med, improving query understanding offers tremendous opportunities for providing better search results. In this work we present SingleCite, a single citational matching tool designed with the goal to improve current single citation searching functionality in PubMed.

PubMed queries are generally being classified as informational or navigational. Informational queries, also known as topical searches, such as *colon cancer*, or *familial Mediterranean fever*, are intended to satisfy information needs on a search topic. They tend to retrieve many documents, the information need is typically not satisfied with just one result, and the user does not know in advance which document will be the most useful. Navigational queries, also called known-item queries (Ogilvie & Callan, 2003), such as *Katanaev AND Cell 2005, 120(1):111-22*, are intended to retrieve a specific publication. Processing navigational queries requires techniques rather different from those used for information searches, and includes access to structured citation data, syntactic parsers, and intelligent metadata (volume, issue, page, date fields) parsers. Parsing and managing citations is a critical task of digital libraries and has been studied extensively (Anzaroot & McCallum, 2013; Kim, Le, & Thoma, 2008; Zhang, Cao, & Yu, 2011). Addressing navigational queries is particularly important for scholarly citation databases, including PubMed, where navigational searches constitute about half of all queries (Islamaj, Murray, Névéol, & Lu, 2009; Yeganova, Kim, Comeau, Wilbur, & Lu, 2018), unlike general search domain where they represent a significantly smaller portion (Jansen, Booth, & Spink, 2007). Moreover, because of specificity of the expected response, retrieving the correct document is of great importance.

Users that have a specific document in mind, frequently enter a query they believe uniquely identifies that document. A specific document may be accessed in various ways. Author name(s) queries and title queries are two most frequent navigational search patterns (Yeganova et al., 2018). Other search patterns include combinations of author(s), year, key words, journal, volume, issue, page and date fields. Not all navigational queries lead to retrieving a single citation. Author name queries may retrieve several PMIDs written by the same person. Similarly, title queries targeted at retrieving a document with a particular title, may be interpreted as key words and retrieve multiple matching documents. Our single citation matching tool is not intended to handle such queries. It is designed for queries that provide enough information to establish a high probability match between a query and a single correct document. When such a document is found, PubMed redirects a user directly to that document, instead of a summary page which generally contains many retrieved results.

151

Proceedings of the BioNLP 2018 workshop, pages 151–155
Melbourne, Australia, July 19, 2018. ©2018 Association for Computational Linguistics

Here we present SingleCite, a single citation matching algorithm designed to retrieve a high probability match for a navigational query targeting a unique document. The algorithm establishes a query-document mapping by building a regression function to predict the probability of a retrieved document being the target based on three variables: the score of the highest scoring retrieved document, the difference in score between the two top retrieved documents, and the fraction of a query matched by the candidate citation. We demonstrate the advantage of our method by comparing it with the currently existing single citation matching scheme in PubMed and manually annotating a random sample of 1,000 queries on which the two methods disagreed. We also apply SingleCite on 1 million zero-hit PubMed queries and recover a single citation match for 3.3% of them.

2   Methods

To create the mapping between a query and a candidate PubMed document we propose an algorithm that predicts the probability of a retrieved document being the target given a query. We propose three variables to measure the success of match between a query and a PubMed record: the log odds score of the top scoring pmid, the difference between log odds scores of the two top scoring pmids, and the fraction of alpha-numeric query characters that match the record. In the next subsection we address the details of how we compute the log odds score between a query and a PubMed record. Then we describe how we build the regression function that takes as input the three variables and predicts the probability of a retrieved document being the target. In this work we also propose techniques to create artificial queries, where each query is created from a known document. This query set is essential for training the regression functions in the absence of manually annotated data.

2.1 Computing the query-document score

We represent PubMed documents by their bibliographic data including article title, author name(s), journal title, volume, issue, page, and date as features. Features from abstracts are not used as they are generally not as specific and less likely to be the source of a user's query terminology for a single citation. The seven fields of interest will be referred to as citation fields. We index the elements of citation fields by including all non-stop word single tokens and capitalized stop words, that are then lower cased. We also index all token pairs with the following exceptions: do not include first name or initials alone, do not include the last page of a page range alone, do not include the issue, except as paired with the volume.

The features are then weighted with the IDF weights approximating naïve Bayesian weights, and the resultant weighted features are added up for each element of a document matching the query. Using these IDF weights we compute the log odds score that the matching document is what the user was seeking.

To produce log odds scoring that is as close to the truth as possible we make some modifications to the weighting. The first problem is that IDF weighting is used for both word pairs and single words. To correct for this dependency, we modify the IDF weights of pairs as follows:

$$modIDF_{(w1,w2)} = IDF_{(w1,w2)} - IDF_{w1}$$

We also adjust the IDF weights to correct for the unevenness in the amount of dependency within fields in the bibliographic record. The unevenness is caused by terms in some fields being more independent then in others. For example, the terms in the author, page, volume, issue and date fields tend to be independent of each other. On the other hand, in fields such as article titles and journal titles terms are more dependent. Intuitively, it is significantly more difficult to predict author first name given the last name, or to predict issue given the volume, then to predict a word following another word in a title.

Query: "Strategies for assessing and fostering hope Penrod.J.& Morse.J.M"

Clicked Article: "Penrod, J., & Morse, J. M. (1997). Strategies for assessing and fostering hope: The Hope Assessment."

Derived Query Parse: Strategies for assessing and fostering hope [Title] Penrod.J. [Author] & Morse.J.M [Author].

Figure 1: Query annotation based on clicked article.

We use a machine annotated training set to optimize the weight modification. The machine annotated queries are created from NCBI PubMed logs by sampling navigational queries that are followed by a user clicked document. Given a query and a clicked pmid, we interpret the parts of the query by mapping them to citation fields of the clicked document (title, author, journal, volume, issue, page and date). This approach allows us to obtain an unlimited amount of citation query–pmid pairs. Figure 1 presents an example of such annotation. Using the machine annotated queries as a training set we now modify the IDF weights to improve the matching between the query elements and PubMed citation. To correct for the dependencies within the title fields, we upweight the IDF weights for terms coming from all the remaining fields by the factor of 1.4. The factor of 1.4 is empirically determined using a grid search.

Given a query, we can now score all PubMed records and retrieve top ten ranks. As users frequently submit queries with misspelled words (Behnert & Lewandowski, 2017) we have incorporated spell checking limited to a single edit correction per term into our processing. This is implemented by retrieving the top ten scoring records based on the original query and then applying spelling correction to the query one term at a time. This may increase the match score between the query and a record. If we have increased the difference between the top score and the next best score, the revised query is accepted as the preferred result. Otherwise the original scores are retained.

Now that we can compute the scores between a query and candidate pmids, the next step is how to interpret and combine the scores. To address that question we build regression functions to map the log odds scores and the fraction of the query matched, to the probability the top scoring document being the target. Since the training of a regression function requires labeled query-pmid pairs, we propose methods for producing artificial queries.

2.1 Artificial Queries

We propose techniques for creating an artificial dataset of annotated citation queries modelled upon user's actual queries. Simulating test collections for evaluating retrieval quality has been explored in the literature (Azzopardi & de Rijke, 2006; Azzopardi, de Rijke, & Balog, 2007) as it offers a viable alternative to manually annotating queries. Constructing simulated known-item queries present a particularly well-defined task; the retrieval goal is the document from which a query is constructed.

We have already shown how to get an unlimited supply of query-document pairs. From each such pair, we can take the annotated query as a model describing the fields from which the query is composed and the length of each such piece. We then randomly sample a PubMed document. Using the pattern of the annotated query, we generate a synthetic query from the reference PubMed document mimicking the structure of the annotated query. For example, if the annotated query contains an author name, we extract an author name from the document that is closest in length to the author name element in the model query. The same technique holds for all the fields found in the model query. A second technique randomly selects a PubMed document, creates its citation as a text string, and then randomly splits it into two strings. Each of these strings then simulates a cut-and-paste query.

The advantage of such queries is that we know the target document the query is intended to retrieve. However, we have no guarantee the query will retrieve the document on which it is based. Using these two techniques, we created a set of one million queries.

2.2 Training the Regression Functions using Artificial Queries

Based on the synthetic queries which have known target documents in PubMed, the goal is to build a regression model for estimating the relationship between the three dependent variables and the predictor. Predictor in this model is label of a query document pair, 1 if the document is identified correctly, and 0 otherwise. For each query we carry out retrieval using our system and record the top scoring documents from PubMed; x and y represent values determined by the retrieval as

$$x = score1; \; y = (score1\text{-}score2)\,/\,score1.$$

To be kept, a score had to be greater than a certain lower bound and we only record at most scores for the top three documents. The first stage of our computation is to estimate the probability $p(t \in PubMed|x, y)$, where t represents the target document of the query. We construct the first regression function which estimates that probability given x and y. The second stage of the computation is to estimate $p(d_1 = t|x, y, t \in PubMed)$, the probability that the document at rank 1 is the target document. Again, we use all the artificial queries and their retrieved documents as long as at least two scores were above the threshold to directly estimate $p(d_1 = t|x, y, t \in PM)$. This is obtained by a straightforward application of the two-dimensional isotonic regression algorithm (Spouge, Wan, & Wilbur, 2003). Consequently, we can combine this probability with the previously estimated $p(t \in PubMed|x, y)$ and obtain:

$$p(d_1 = t|x, y) =$$
$$p(d_1 = t|x, y, t \in PM)p(t \in PM|x, y).$$

Given retrieval results from our system for a query, $p(d_1 = t|x, y)$ provides an estimate of how likely the user was looking for document d_1.

The final step in the model constructs a third regression by taking as input $p(d_1 = t|x, y)$ and the query fraction matched. We hypothesize that if the query is a sufficiently good match to a PubMed record and there is a reasonable gap to the next best score, the top scoring record may be of interest even if not exactly what the user was seeking. We conjecture this to depend on the quality of match and how much of the query is involved in the match. The difficulty however is that we do not have a way to simulate this problem with known answers. Instead, we compare our system output to the output of a legacy system (known to have high precision) possessed and currently used by NCBI to processes single citation queries. A total of 343,731 unique queries were collected from PubMed logs on October 12, 2016. These were the queries that triggered the single citation matching system in PubMed. The existing system produced a presumed high-quality answer for 58,375 queries. SingleCite produced probabilistic output of variable quality for 232,256 of these queries. For the 51,472 queries where the existing and the new system both made predictions, we counted predictions as correct when the two systems agreed on the retrieved pmid (45,713) and incorrect otherwise (5,759). Using this data, we build the regression function that combines the probability of top scoring document being the target obtained from previous step and the fraction of the query matched for the 51,472 queries. We empirically chose a threshold of 0.98 and accept predictions from the third regression function that are above or at that value.

3 Evaluation

We ran SingleCite on the 343,731 query set mentioned above, and predicted high probability answers on 26,892 queries (with the 0.98 threshold) where the legacy system made no predictions. To evaluate the accuracy of our algorithm, we randomly sampled 500 queries from the set where we alone made predictions and examined the quality of the answers. We found 7 answers clearly wrong and 5 probably wrong but potentially useful. Wrong answers were mostly seen with the shorter queries. These results are consistent with a 98% accuracy level. We further randomly sampled 200 queries from the set of 11,688 queries where the legacy system alone made the prediction. There we found 22% of answers clearly wrong and 8% probably wrong, but potentially useful. The remaining 70% of queries produced a single citation match that we thought was correct. On close examination of queries missed by SingleCite, we identified a few opportunities for improvement, including enriching the index with journal name abbreviations (currently index contains only full journal names), and better handling of hyphenated last names (for example, query containing *Shiloh* did not retrieve the target document containing *Shiloh-Malawsky* as an author).

As a second experiment, we ran SingleCite on one million queries randomly sampled from queries submitted to PubMed in 2017 that produced no results using the legacy system. We found a single citation match for 3.34% of them.

4 Conclusion

Here we present our preliminary work on the single citation matching tool aimed to facilitate user's search for a specific document in PubMed. The method depends on good feature engineering combined with novel approaches for adjusting feature weights when combining elements from different fields. We also describe how we create one million synthetic queries, each along with the PMID of the document used as the source. SingleCite shows promising results compared to the existing system for finding single citations. The tool can also be used as part of NLP pipeline for identifying citations in text, abstract or full text, and mapping them to corresponding PMIDs. The tool can further be useful for citation management systems and portfolio analysis.

Acknowledgments

This research was supported by the Intramural Research Program of the NIH, National Library of Medicine.

References

Anzaroot, S., & McCallum, A. (2013). *A New Dataset for Fine-Grained Citation Field Extraction*. Paper presented at the Proceedings of the 30 th International Conference on Machine Learning, Atlanta, Georgia, USA, 2013. JMLR: W&CP volume 28. Copyright 2013 by the author(s).

Azzopardi, L., & de Rijke, M. (2006). *Automatic construction of known-item finding test beds*. Paper presented at the SIGIR '06.

Azzopardi, L., de Rijke, M., & Balog, K. (2007). *Building Simulated Queries for Known-Item Topics*. Paper presented at the SIGIR'07, Amsterdam, The Netherlands.

Behnert, C., & Lewandowski, D. (2017). Known-item searches resulting in zero hits: Considerations for discovery systems. *The Journal of Academic Librarianship, 43*(2), 128-134.

Falagas, M., Pitsouni, E., Malietzis, G., & Pappas, G. (2008). Comparison of PubMed, Scopus, Web of Science, and Google Scholar: strengths and weaknesses. *The FASEB Journal, 22*(2), 338-342.

Islamaj, R., Murray, C., Névéol, A., & Lu, Z. (2009). Understanding PubMed user search behavior through log analysis. *Database*.

Jansen, B. J., Booth, D. L., & Spink, A. (2007). *Determining the User Intent of Web Search Engine Queries*. Paper presented at the WWW 2007, Banff, Alberta, Canada.

Kim, J., Le, D., & Thoma, G. (2008). *Naive Bayes Classifier for Extracting Bibliographic Information From Biomedical Online Articles*. Paper presented at the Proc 2008 International Conference on Data Mining. Las Vegas, Nevada, USA. July 2008;II:373-8.

Lu, Z. (2011). PubMed and beyond: a survey of web tools for searching biomedical literature. Database: the journal of biological databases and curation. *Database (Oxford), 2011*.

Ogilvie, P., & Callan, J. (2003). *Combining Document Representations for Known-Item Search*. Paper presented at the SIGIR, Toronto, Canada.

Spouge, J., Wan, H., & Wilbur, W. J. (2003). Least Squares Isotonic Regression in Two Dimensions. *Journal of Optimization Theory and Applications, 117*(3), 585-605.

Wildgaard, L. E., & Lund, H. (2016). Advancing PubMed? A comparison of 3rd-party PubMed/MEDLINE tools. *Library Hi Tech, 34*(4), 669-684. doi: https://doi.org/10.1108/LHT-06-2016-0066

Yeganova, L., Kim, W., Comeau, D. C., Wilbur, W. J., & Lu, Z. (2018). A Field Sensor: Computing the composition and intent of PubMed queries. *DATABASE*.

Zhang, Q., Cao, Y.-G., & Yu, H. (2011). Parsing Citations in Biomedical Articles Using Conditional Random Fields. *Comput Biol Med, 41*(4), 190-194.

A Framework for Developing and Evaluating Word Embeddings of Drug-named Entity

Mengnan Zhao[1], Aaron J. Masino[2], Christopher C. Yang[1]
[1]College of Computing and Informatics, Drexel University, Philadelphia, PA, US
[2]Department of Biomedical and Health Informatics, Children's Hospital of Philadelphia, PA, US
emails: mz438@drexel.edu, masinoA@email.chop.edu, ccy24@drexel.edu

Abstract

We investigate the quality of task specific word embeddings created with relatively small, targeted corpora. We present a comprehensive evaluation framework including both intrinsic and extrinsic evaluation that can be expanded to named entities beyond drug name. Intrinsic evaluation results tell that drug name embeddings created with a domain specific document corpus outperformed the previously published versions that derived from a very large general text corpus. Extrinsic evaluation uses word embedding for the task of drug name recognition with Bi-LSTM model and the results demonstrate the advantage of using domain-specific word embeddings as the only input feature for drug name recognition with F1-score achieving 0.91. This work suggests that it may be advantageous to derive domain specific embeddings for certain tasks even when the domain specific corpus is of limited size.

1 Introduction

The ability of word embeddings to capture latent, contextual information has proven useful to a variety of NLP tasks, such as named entity recognition (Santos & Guimarães, 2015), syntactic parsing (Levy & Goldberg, 2014), and question answering (Iyyer et al., 2014). Within biomedical research, word embeddings developed in most previous studies were generated from very large, generic corpora (e.g. news articles). This is appropriate for generalized language models. However, for specialized domains and tasks, it may be beneficial to generate word embeddings from a targeted corpus. We propose a biomedical domain-specific word embedding model and a novel evaluation framework, which mainly focus on representing drug names in the current stage. This framework can be expanded to other biomedical entities such as protein, gene, and chemical compound names in the future. We evaluate the developed word embeddings with a comprehensive intrinsic evaluation framework that includes relatedness, coherence, and outlier detection assessment, as well as an extrinsic evaluation that focuses on the task of drug name recognition and classification with a bidirectional long short-term memory (Bi-LSTM) RNN model.

2 Related Work

In the biomedical domain, word embeddings are primarily used for biomedical named entity recognition (BNER) with evaluations conducted on tasks such as JNLPBA (Kim et al., 2004), BioCreAtIvE (Hirschman et al., 2005), and BioNLP Shared Tasks. Tang et al. (2014) explored the impact of three different types of word representations (WR) on clustering-based representation, distributional representation and word embedding. Segura-Bedmar et al. (2015) generated word embeddings with *word2vec* and a combined Wikipedia and MedLine corpus. The results were evaluated on the SemEval-2013 Task 9.1 Drug Name Recognition dataset (Segura-Bedmar et al., 2013). Wang et al. (2015, November) used word embeddings for bio-event trigger detection. Li et al. (2015) incorporated word embedding features with bag-of-words (BOW) features for bio-event extraction and evaluated results on the BioNLP 2013 GENIA task (Nédellec et al., 2013).

Drug name recognition (DNR) in biomedical literature and clinical notes is essential for many medical information and relation extraction tasks (e.g. drug-drug interaction). Significant effort has been devoted to DNR and the common methods can be categorized as (Lu et al., 2015): (1) dictionary-based approaches (Rindflesch et al., 2000; Sanchez-Cisneros et al., 2013), (2) rule-based/ontology-based approaches (Hamon & Grabar, 2010; Coden et al., 2012), (3)

Proceedings of the BioNLP 2018 workshop, pages 156–160
Melbourne, Australia, July 19, 2018. ©2018 Association for Computational Linguistics

machine learning-based approaches (Lamurias et al., 2013; Lu et al., 2015), and (4) hybrid approaches (Korkontzelos et al., 2015).

3 Word Embeddings Training

We extracted text from PubMed and DrugBank to construct our corpus. For PubMed, we used "*drug*" as the keyword of query to broadly select drug related abstracts, which yielded 474,273 abstracts. From DrugBank [1] Release Version 5.0.5 we extracted the fields: "description" "indication" "pharmacodynamics" "mechanism-of-action" "toxicity" for 8,226 drugs.

We employed the skip-gram model in *word2vec* to generate word embeddings. Moreover, as studies have found that word embeddings have a consistent relationship with word frequencies, even after the interception of frequency-based effects by algorithms and vector length normalization (Schnabel et al., 2015), we employed correlation analysis between vectors and frequencies as the evaluation metric to tune the parameters for the word embedding model. For our final result, we trained the word embedding model in *word2vec* with parameters: *size* = 420, *window* = 5, *min_count* = 2.

4   Intrinsic Evaluation

4.1   Relatedness assessment

Relatedness evaluation is the most popular and direct intrinsic word embedding evaluation method. It is expected that high quality word embeddings will display significant correlation (e.g. Pearson's, Spearman's) between the cosine similarity of the embedding vectors for related word pairs and the human scores.

We evaluated the results on two biomedical domain inventories: UMNSRS-Rel and UMNSRS-Sim (Pakhomov et al., 2010). These datasets provide human-annotated scores of relatedness and similarity between clinical term pairs. We measured the correlation between the scores provided by the UMNSRS datasets and calculated by our model, using Spearman's correlation coefficient. We also compared our model to a publicly available word embedding set trained on about 100 billion words from Google News samples[2].

Corpora	PubMed+ DrugBank	Google News
drug-drug	**0.737**	0.430
drug-X	0.530	0.293
drug-nonDrug	0.492	0.245
whole dataset	0.555	0.345
nonDrug-nonDrug	0.565	0.368

Table 1: Relatedness assessment on UMNSRS-Rel dataset

Corpora	PubMed+ DrugBank	Google News
drug-drug	**0.764**	0.495
drug-X	0.529	0.435
drug-nonDrug	0.449	0.385
whole dataset	0.597	0.402
nonDrug-nonDrug	0.601	0.381

Table 2: Similarity assessment on UMNSRS-Sim dataset

As shown in Table 1 and 2, our model and UMNSRS show positive correlations in both relatedness and similarity assessment, with most of the correlation coefficients higher than 0.5, which means the relationship represented in vector space is consistent with human annotations. In particular, the highest consistency is achieved for the relationship of drug-drug pairs, where coefficients reach 0.737 and 0.764 for relatedness and similarity, respectively. In addition, the proposed model trained on PubMed+DrugBank shows significantly higher correlations with human scores than the model trained on a Google News corpus in all word pair types. This is important because the Google News based embeddings were trained on an extremely large dataset compared to our corpus.

4.2   Coherence assessment

Conceptually, we expect that a good word embedding should be surrounded by a coherent neighborhood of similar words. From this concept, we propose a novel intrinsic evaluation metric as a supplement to current relatedness analysis (Schnabel et al., 2015). In coherence assessment, we assess whether a given word embedding is mutually related to the word embeddings in its local neighborhood. Here we created a neighborhood for each drug name and explored the relation with the closest neighbor terms. We expect that other drug entities should be preferentially represented in the neighborhood. Setting the neighborhood size from 3 to 10, we calculated the percentage of

---

[1] www.drugbank.ca/releases/latest
[2] https://code.google.com/archive/p/word2vec/

drug names within the neighborhood of each drug, with selected results shown in Table 3.

Size of neighborhood	3	5	7	9	10
Percentage of drug/all_neighbors (%)	61.1	58.8	56.9	55.2	54.6

Table 3: Percentage of drug entities within a drug's neighborhood across all drugs.

From Table 3, we see that the percentage of drug entities declines with the expansion of neighborhood size. Noting that neighbors were arranged by the cosine similarity relative to the target word, such decline implies that drug entities tend to be the closest neighbors. Beyond that, drug entities still occupy more than half of the nearest 10 neighbors. These results suggest there is a strong coherence in the semantic space.

4.3 Outlier Detection

As a final intrinsic measure of word embedding quality, we consider a modification of a previously proposed outlier detection task. Given a group of words W, the compactness score of word $w_m \in W$ represents the compactness of the cluster $W \backslash \{w_m\}$. Performance on the outlier detection task can be evaluated by accuracy and outlier position percentage (OPP) (Camacho-Collados & Navigli, 2016). Ideally, if outliers in all the groups were identified and listed at the last position, accuracy and OPP should be 1 and 100% respectively.

In this study, the goal of outlier detection is to identify the non-drug words as outliers. We created two datasets each with 400 groups of words ($|D|$=400). Following the work of Camacho-Collado and Navigli, the first dataset, D-Manu, contains 4 to 8 drugs and 1 *manually* selected non-drug outlier ($|W| \in [5, 9]$). Additionally, we modify the previously presented work by forming a second dataset, D-Rand, in which each group contains 4 to 8 drugs and 1 *randomly* selected non-drug outlier ($|W| \in [5, 9]$). Tables 4 and 5 show the evaluation results of outlier detection on D-Rand and D-Manu. On D-Rand, outliers were identified in more than 40% of groups across different sizes, and OPP values indicate that the average outlier position was around 70% to the right end (100%) of the list arranged by compactness score. Meanwhile, for D-Manu, the accuracy values are all higher than 0.8 and the OPP values are all above 93%.

Group size-$|W|$	5	6	7	8	9
Accuracy	0.43	0.44	0.41	0.40	0.41
OPP(%)	69.2	72.0	73.6	70.3	72.4

Table 4: Accuracy and OPP of outlier detection on D-Rand

Group size-$|W|$	5	6	7	8	9
Accuracy	0.82	0.83	0.85	0.80	0.83
OPP	93.4	94.3	95.3	93.9	94.9

Table 5: Accuracy and OPP of outlier detection on D-Manu

To gain further insight on the potential correlation between the outlier task performance and the similarity distribution over the outlier term and the non-outlier terms, we calculated the average similarity between each pair of non-outlier terms and the average between non-outliers and the outlier for each group in D-Rand and D-Manu. We found that the average similarity between non-outliers was about 0.21. The average similarity between non-outliers and randomly selected outliers and manually selected outliers was about 0.16 and 0.12, respectively. This result confirmed that the greater distinction in word similarity is consistent with the better accuracies in outlier detection.

5 Extrinsic Evaluation - DNR

5.1 DNR with Bi-LSTM Model

We employ a bidirectional long short-term memory (Bi-LSTM) RNN model that is designed to process text input as a sequence of tokens (constituent parts, usually words) and predict the label for each token. The BLSTM-RNN model combines two RNNs: the forward RNN processes the sequence from left to right and the backward RNN processes it from right to left. We use a BIO scheme for the sequence labeling task. Specifically, each token is labeled as one B-X, I-X or O indicating it is at the beginning (B), inside (I), or outside (O) of the entity of type X (e.g. drug name).

In order to achieve the best results and compare the impact of the word embedding model in the labeling task, we introduced three BLSTM-RNN variants: (1) Fixed embedding (BLSTM-F): Word embedding values were provided by the pre-trained word embedding model and treated as fixed constants; (2) Varied embedding (BLSTM-V): Word embedding values were also provided by the pre-trained word embedding but treated as learnable parameters; (3) Randomly-initialized embedding (BLSTM-R): Word embedding values were initialized randomly and treated as learnable parameters.

5.2 Experiments on Drug Name Recognition

We evaluated our model on DDI-Extraction-2011 task (Segura-Bedmar et al., 2011) using two metrics: **Exact matching**-the predicted entity must have exactly the same boundary with the annotated entity and **Partial matching**-the predicted entity must have some overlap with the annotated entity. Table 6 shows the results of three BLSTM models. Regarding to the impact of pre-trained word embeddings, there is no obvious improvement when introducing the pre-trained embedding values instead of randomly initialized vector values. Moreover, the f1-score of BLSTM-V that sets embedding values as learnable parameters in RNN model is increased to 0.911 from 0.891 in BLSTM-F that treats them as fixed constants. Overall, our BLSTM models achieve very good results on DNR according to f1-scores, and treating embedding values as learnable parameters, regardless of pre-trained or randomly initialized, lead to better results than setting them fixed, indicating the great advantage of RNN models for drug name recognition task.

	Exact Matching			Partial Matching		
	P	R	F1	P	R	F1
BLSTM-F	0.89	0.90	0.89	0.91	0.92	0.91
BLSTM-V	**0.91**	0.91	**0.91**	**0.93**	0.94	**0.94**
BLSTM-R	0.90	**0.92**	0.91	0.93	**0.94**	0.93

*Bold indicates the highest score in the column.
Table 6: Evaluation results on DDI-Extraction-2011 test set.

5.3 Experiments on Drug Name Classification

In DDI-Extraction-2013 challenge (Segura-Bedmar et al., 2013), the drugs were annotated with four types instead of one type in 2011 task, including: *drug*, *brand*, *group*, and *drug_n*. Thus, it becomes a drug name recognition and classification task. We evaluated our results using four metrics provided by the organizers, with f1-scores shown in Table 7. Pre-trained word embeddings showed their advantages, for instance, f1 of strict matching were improved 16% in BLSTM-V than BLSTM-R. While updating the pre-trained embedding values did not show obvious improvement by comparing BLSTM-F and BLSTM-V.

DrugBank+MedLine	BLSTM-F	BLSTM-V	BLSTM-R
Strict matching	0.735	0.724	0.631
Type matching	0.753	0.737	0.654
Exact oundary matching	0.789	0.801	0.658
Partial boundary matching	0.816	0.823	0.688
drug	0.824	0.852	0.750
brand	0.722	0.588	0.344
group	0.722	0.702	0.697
drug_n	0.381	0.333	0

Table 7: Results on DDI-Extraction-2013 test set.

6 Conclusion

We presented biomedical domain-specific word embeddings formulated with the *word2vec* model using PubMed and DrugBank text sources and a comprehensive intrinsic and extrinsic evaluation framework for word embeddings that includes new and existing metrics. We found that our word embeddings demonstrated superior performance based on relatedness assessment, neighborhood coherence, and outlier detection. Moreover, we also found that these embeddings performed better than those generated from very large datasets such as Google News. This is significant because our training dataset is approximately two orders of magnitude smaller.

Since drug name recognition (DNR) is an important biomedical NLP task, we used DNR as the downstream task for extrinsic evaluation of the developed drug name embeddings. We utilized the pre-trained word embeddings in Bi-LSTM model for the task of drug name recognition and classification. For drug name recognition, setting embedding values as learnable parameters in RNN model has more impact on the performance than utilizing pre-trained word embeddings. For drug name classification, pre-trained word embeddings offer significant performance increases over randomly-initialized embeddings, while updating the pre-trained embedding values during the BLSTM model training has little improvement. This work provides a useful tool or framework for processing raw biomedical text and extracting drug entities, which could be helpful in processing other unstructured data and medical entities.

References

Camacho-Collados, J., & Navigli, R. (2016). Find the word that does not belong: A framework for an intrinsic evaluation of word vector representations.

In *Proceedings of the 1st Workshop on Evaluating Vector-Space Representations for NLP* (pp. 43-50).

Chiu, B., Crichton, G., Korhonen, A., & Pyysalo, S. (2016). How to train good word embeddings for biomedical NLP. In *Proceedings of the 15th Workshop on Biomedical Natural Language Processing* (pp. 166-174).

Coden, A., Gruhl, D., Lewis, N., Tanenblatt, M., & Terdiman, J. (2012, September). SPOT the drug! an unsupervised pattern matching method to extract drug names from very large clinical corpora. In *Healthcare Informatics, Imaging and Systems Biology (HISB), 2012 IEEE Second International Conference on* (pp. 33-39). IEEE.

Hamon, T., & Grabar, N. (2010). Linguistic approach for identification of medication names and related information in clinical narratives. *Journal of the American Medical Informatics Association, 17*(5), 549-554.

Hirschman, L., Yeh, A., Blaschke, C., & Valencia, A. (2005). Overview of BioCreAtIvE: critical assessment of information extraction for biology.

Hochreiter, S., & Schmidhuber, J. (1997). Long short-term memory. *Neural computation, 9*(8), 1735-1780.

Kim, J. D., Ohta, T., Tsuruoka, Y., Tateisi, Y., & Collier, N. (2004, August). Introduction to the bio-entity recognition task at JNLPBA. In *Proceedings of the international joint workshop on natural language processing in biomedicine and its applications* (pp. 70-75). Association for Computational Linguistics.

Korkontzelos, I., Piliouras, D., Dowsey, A. W., & Ananiadou, S. (2015). Boosting drug named entity recognition using an aggregate classifier. *Artificial intelligence in medicine, 65*(2), 145-153.

Lamurias, A., Grego, T., & Couto, F. M. (2013, October). Chemical compound and drug name recognition using CRFs and semantic similarity based on ChEBI. In *BioCreative Challenge Evaluation Workshop* (Vol. 2, p. 75).

Levy, O., & Goldberg, Y. (2014). Dependency-Based Word Embeddings. In *ACL (2)* (pp. 302-308).

Li, C., Song, R., Liakata, M., Vlachos, A., Seneff, S., & Zhang, X. (2015, July). Using word embedding for bio-event extraction. In *Proceedings of the 2015 Workshop on Biomedical Natural Language Processing (BioNLP 2015). Stroudsburg, PA: Association for Computational Linguistics* (pp. 121-126).

Lu, Y., Ji, D., Yao, X., Wei, X., & Liang, X. (2015). CHEMDNER system with mixed conditional random fields and multi-scale word clustering. *Journal of cheminformatics, 7*(1), S4.

Nédellec, C., Bossy, R., Kim, J. D., Kim, J. J., Ohta, T., Pyysalo, S., & Zweigenbaum, P. (2013, August). Overview of BioNLP shared task 2013.

In *Proceedings of the BioNLP Shared Task 2013 Workshop* (pp. 1-7).

Pakhomov, S., McInnes, B., Adam, T., Liu, Y., Pedersen, T., & Melton, G. B. (2010). Semantic similarity and relatedness between clinical terms: an experimental study. In *AMIA annual symposium proceedings* (Vol. 2010, p. 572). American Medical Informatics Association.

Rindflesch, T. C., Tanabe, L., Weinstein, J. N., & Hunter, L. (2000). EDGAR: extraction of drugs, genes and relations from the biomedical literature. In *Pacific Symposium on Biocomputing. Pacific Symposium on Biocomputing* (p. 517). NIH Public Access.

Sanchez-Cisneros, D., Martínez, P., & Segura-Bedmar, I. (2013, November). Combining dictionaries and ontologies for drug name recognition in biomedical texts. In *Proceedings of the 7th international workshop on Data and text mining in biomedical informatics* (pp. 27-30). ACM.

Santos, C. N. D., & Guimarães, V. (2015). Boosting Named Entity Recognition with Neural Character Embeddings. *arXiv preprint arXiv:1505.05008.*

Schnabel, T., Labutov, I., Mimno, D., & Joachims, T. (2015). Evaluation methods for unsupervised word embeddings. In *Proceedings of the 2015 Conference on Empirical Methods in Natural Language Processing* (pp. 298-307).

Segura-Bedmar, I., Martinez, P., & Sánchez Cisneros, D. (2011). The 1st DDIExtraction-2011 challenge task: Extraction of Drug-Drug Interactions from biomedical texts.

Segura-Bedmar, I., Martínez, P., & Herrero Zazo, M. (2013). Semeval-2013 task 9: Extraction of drug-drug interactions from biomedical texts (ddiextraction 2013). Association for Computational Linguistics.

Segura-Bedmar, I., Suárez-Paniagua, V., & Martınez, P. (2015, September). Exploring word embedding for drug name recognition. In *SIXTH INTERNATIONAL WORKSHOP ON HEALTH TEXT MINING AND INFORMATION ANALYSIS (LOUHI)* (p. 64).

Tang, B., Cao, H., Wang, X., Chen, Q., & Xu, H. (2014). Evaluating word representation features in biomedical named entity recognition tasks. *BioMed research international, 2014.*

Wang, J., Zhang, J., An, Y., Lin, H., Yang, Z., Zhang, Y., & Sun, Y. (2015, November). Biomedical event trigger detection by dependency-based word embedding. In *Bioinformatics and Biomedicine (BIBM), 2015 IEEE International Conference on* (pp. 429-432). IEEE.

MeSH-based dataset for measuring the relevance of text retrieval

Won Kim, Lana Yeganova, Donald C Comeau,
W John Wilbur, Zhiyong Lu
National Center for Biotechnology Information, NLM, NIH, Bethesda, MD, USA
{wonkim, yeganova, comeau, wilbur, luzh}@mail.nih.gov

Abstract

Creating simulated search environments has been of a significant interest in information retrieval, in both general and biomedical search domains. Existing collections include modest number of queries and are constructed by manually evaluating retrieval results. In this work we propose leveraging MeSH term assignments for creating synthetic test beds. We select a suitable subset of MeSH terms as queries, and utilize MeSH term assignments as labels for retrieval evaluation. Using well studied retrieval functions, we show that their performance on the proposed data is consistent with similar findings in previous work. We further use the proposed retrieval evaluation framework to better understand how to combine heterogeneous sources of textual information.

1   Introduction

PubMed is a search engine processing on average 3 million queries a day and is recognized as a primary tool for scholars in the biomedical field (M. Falagas, Pitsouni, Malietzis, & Pappas, 2008; Lu, 2011; Wildgaard & Lund, 2016).

PubMed provides access to a collection of approximately 28 million biomedical abstracts as of 2018, of which about 4.5 million have full text document available in PubMed Central. With the growing availability of full-text articles, an essential question to consider is how to leverage full text information to improve PubMed retrieval? While a number of studies have pointed out the benefits of full text for various text mining tasks (Cohen, Johnson, Verspoor, Roeder, & Hunter, 2010; Westergaard, Stærfeldt, Tønsberg, Jensen , & Brunak, 2018), combining these two resources for information retrieval is not a trivial endeavor.

Naïvely merging full text articles with abstract data, naturally increases the recall, but at a cost for precision, generally degrading the overall quality of combined search (Lin, 2009).

Research is required to understand how to best combine abstracts and full texts, examine the relative importance of different sections in full text, investigate the performance of different scoring functions, etc. A major obstacle in such efforts is the lack of large-scale gold standards for retrieval evaluation. Hence, creating such large-scale retrieval evaluation framework is the goal of this work.

Gold standards are typically assembled by using human judgments, which are time consuming, expensive and not scalable. Pioneering examples are a TREC collection (Hersh, Cohen, Ruslen, & Roberts, 2007) and a BioASQ collection (Tsatsaronis et al., 2015). Simulating test collections for evaluating retrieval quality offers a viable alternative and has been explored in the literature (Azzopardi & de Rijke, 2006; Azzopardi, de Rijke, & Balog, 2007; Kim, Yeganova, Comeau, Wilbur, & Lu, 2018). In this work we create an evaluation framework based on MeSH term assignments, and use that framework to test the performance of several classic ranking functions.

We examine the utility of MeSH terms as query surrogates and MeSH term assignments as pseudo-relevance rankings. We describe how we select a subset of MeSH terms as candidate MeSH queries and discuss the retrieval results using five different retrieval functions available in SOLR. MeSH queries are representative of real user queries. This approach allows us to create a large-scale relevance ranking framework that is based on human judgements and is publicly available. MeSH queries are available for download at:

Proceedings of the BioNLP 2018 workshop, pages 161–165
Melbourne, Australia, July 19, 2018. ©2018 Association for Computational Linguistics

2 MeSH Term Based Queries for Retrieval Evaluation

Each paper indexed by MEDLINE® is manually assigned on average thirteen MeSH terms (Huang, Neveol, & Lu, 2011) by an indexer, who has access to both the abstract and full text of articles. It is plausible to assume that MeSH terms assigned to a document are highly reflective of its topic, and the document is highly relevant to that MeSH term.

In this work we propose using a subset of MeSH terms as queries and rely on the assumption that documents with the MeSH terms assigned are relevant to the query. As queries, we aim to select MeSH terms that satisfy certain frequency requirements, and those that are correlated with real user queries. We will refer to the final set of MeSH terms that we use as queries as MeSH queries. Using MeSH terms for evaluation of various NLP tasks has been described in the literature (Bhattacharya, Ha–Thuc, & Srinivasan, 2011; Yeganova, Kim, Kim, & Wilbur, 2014). However, to our knowledge, using MeSH terms as query surrogates and MeSH assignments as relevance rankings has not been yet described.

2.1 MeSH term preprocessing

We preprocess the MeSH terms by applying several processing steps, which include lowercasing, removing all non-alphanumeric characters, and dropping stop words from MeSH term strings. We further drop tokens in the remaining MeSH term string that are pure digits.

2.2 Frequency Threshold

We apply frequency threshold to remove MeSH terms that are not likely to be useful as queries. Some MeSH terms such as *Humans*, are very general, and are not useful for evaluation of retrieval results. *Humans* is assigned to an overwhelming fraction of PubMed documents, even to those that are not directly discussing the topic. For example, an article studying *dietary experiments on rats involving the hormone "insulin"* is assigned *humans* because it studied animals to understand diabetes for humans. Another complication are ambiguous MeSH terms. With the frequency threshold, our goal is to limit the analysis to those MeSH terms that tend to carry the same meaning across the corpus.

For a single token MeSH term, we consider two frequencies: the number of PubMed documents the MeSH term is assigned to, and the frequency of the token used as a text word in PubMed abstracts. For a single token MeSH term, we required that the smaller of the two frequencies is at least half as big as the larger. For multi-token MeSH terms, the frequency with which each individual token in the MeSH term appears in the text is at most ten times as high as the frequency of the MeSH term. These requirements lead to 5,117 single-token and 1,735 multi-token MeSH terms for use as queries.

2.3 Presence in User Queries

The second essential consideration is to select MeSH terms that are likely to be used as queries. We collected PubMed queries issued in the 2017 calendar year. We normalized these user queries in the same manner as MeSH terms. We found that among the 5,117 single token MeSH terms, about half of them appeared as queries. Among the 1,735 multi-token MeSH terms 96% have been issued as a query. Based on this analysis, we decided to proceed with the multi-token MeSH queries for our experiments. We will refer to that set of MeSH terms as MeSH queries.

3 SOLR Retrieval Functions

SOLR is an open source search platform built on Apache Lucene which has been widely used in the search industry for more than a decade. It offers a number of useful features including fast speed, distributed indexing, replication, load-balanced querying, and automated failover and recovery. Lucene-based SOLR search engine is a popular industry standard for indexing, search and retrieval. SOLR provides several ranking options, and our interest is in evaluating them using MeSH queries and pseudo-relevance judgements.

We investigated most of the weighting formulas available in the native SOLR/Lucene search engine, and report the top five best performing ones: tf.idf, BM25, DFR, IBS and Dirichlet.

tf.idf is the SOLR default ranking algorithm and one of the most basic weighting schemes used in information retrieval (Robertson, 2004).

	MAP	BE
tf .idf	0.380	0.506
BM25	**0.413**	**0.532**
DFR	**0.417**	**0.536**
IBS	0.404	0.524
Dirichlet	0.305	0.454

Table 1. Retrieval results for multi-word queries, based on the top 2K retrieved documents. Presented are averages over 1,735 multi-word MeSH queries.

BM25 is the ranking algorithm described in (Robertson SE, 1995) and (Sparck Jones, Walker, & Robertson, 1998).

DFR is the implementation based upon the *divergence from randomness (DFR)* framework introduced in (Amati & Van Rijsbergen, 2002) .

IBS is based upon a framework for the family of information-based models, as described in *(Clinchant & Gaussier, 2010)*.

Dirichlet is an language model for Bayesian smoothing using Dirichlet priors from (Zhai & Lafferty, 2004).

4   Results

MeSH terms are assigned based on article abstracts and full texts, hence it is natural to include in the retrieval experiments not only PubMed articles, but also corresponding PubMed Central full text articles. To that end, we created a retrieval environment which included all PubMed articles (~27 million abstracts) and their available PMC full text counterparts (~4 million full texts) in a unified system. The search environment was created in such a way that we can distinguish PubMed and PMC records, and identify which PMC record corresponds to a PubMed abstract. The retrieval system, however, treated all PubMed and PMC documents independently. For PubMed records, we indexed the title and the abstract fields, for the PMC full text records we indexed title, abstract and full text fields. We evaluated each retrieval method available in SOLR by querying the unified database using MeSH queries. Retrieved documents (both PubMed and PMC) where scored using SOLR weighting functions and returned in the order of diminishing score.

For each MeSH query, we retrieved the top 2,000 documents. Among those, we considered only documents to which MeSH terms have already been assigned (recent documents may not have been assigned MeSH terms yet) and call them the *retrieved set*. Documents in the *retrieved set* that are assigned MeSH query as a MeSH term are treated as *positive*, while the rest are considered *negative*. Given these assignments, we can compute Mean Average Precision (MAP) and Precision-Recall Break Even (BE) (M. Falagas, Pitsouni, E., Malietzis, G., & Pappas, G., 2008) to measure the success of each retrieval function.

Table 1 presents the summary of the retrieval results from SOLR using the five different weighting formulas, averaged over the 1,735 multi-token MeSH queries. Table 1 shows that BM25 outperforms tf.idf in terms of both MAP and BE. This result is consistent with results reported in (Lin, 2009). We also observe that BM25 and DFR outperform the other three ranking methods, with DFR showing slightly better results than BM25.

A common consideration with document ranking formulas is how robust they are to document length. This next experiment examines whether different ranking formulas favor shorter PubMed abstracts to longer PMC full text documents, or the opposite. Among the top 2,000, we considered *positive* retrieved documents for which both PubMed and PMC records exist. For such articles, it is possible for both PubMed and PMC records to be included in top 2K or just one of them to be present. For each query, we counted the total number of *positive* documents as PMC articles that are ranked higher than PubMed articles (denoted as PMC > PM), as well as the number of positive documents for which PubMed articles are ranked higher (PM > PMC).

The counts are presented in Table 2. We observe that tf.idf pulls more PubMed abstracts into the highest scoring 2,000, thus favoring relatively short (PubMed) documents. Dirichlet, on the other hand favors PubMed Central full text articles. These experiments suggest that tf.idf and Dirichlet are more extreme. By contrast, BM25, DFR and IBS favor PubMed abstracts, but not as strongly.

Our next goal is to consider the value of full text articles for retrieval. We analyze the retrieval performance by computing MAP and BE measures in retrieving 1) PubMed articles only 2) PMC articles only and 3) both PubMed and PMC articles using BM25 and DFR retrieval functions. For the combined retrieval, we assign each article the maximum of its PubMed and PMC score

	Total # of positives	PMC>PM	PM>PMC
tf.idf	136K	13.5K (10%)	122.5K (90%)
BM25	190K	52K (27%)	138K (73%)
DFR	190K	59K (31%)	131K (69%)
IBS	199K	72K (37%)	126K (63%)
Di-richlet	468K	455K (97%)	13K (3%)

Table 2. Comparison of PubMed and PMC scores for multiword queries based on top 2K retrieved documents. The counts are included only for the articles for which both PubMed and PMC versions exist, and one or both are in the top 2K.

and evaluate based on that maximum. We observe from Table 3, that both BM25 and DFR performed better in retrieving PubMed articles than PMC articles. Using the maximum of the PubMed score and PMC score does not yield improved performance over the abstract-only search for both BM25 and DFR.

		BM25	DFR
PMC	MAP	0.265	0.273
	BE	0.353	0.360
PubMed	MAP	**0.305**	**0.309**
	BE	**0.390**	**0.394**
Combined	MAP	0.167	0.279
	BE	0.270	0.376

Table 3. The value of full text PMC articles in the retrieval performance. In combined retrieval, we assign each article the maximum of its PubMed and PMC score and evaluate based on that maximum.

5   Conclusion and Discussion

In this work we propose a large-scale collection for relevance testing. The collection represents a subset of MeSH terms that we use as queries and MeSH term assignments as pseudo relevance rankings. The value of this resource is significant not only in its simplicity and intuitiveness, but also in the quality of relevance judgements achieved though leveraging decades of manual curation. Moreover, by using MeSH terms we are guaranteed to include as queries significant and important PubMed topics. Many of these terms are frequently used as queries. To summarize, MeSH queries provide a reliable and high-quality collection of queries.

To further validate the feasibility of this collection, we used well studied retrieval functions on the set. In the future, we plan to use the proposed test collection to understand how to leverage full text documents for better search.

Acknowledgements

This research was supported by the Intramural Research Program of the NIH, National Library of Medicine.

References

Amati, G., & Van Rijsbergen, C. J. (2002). Probabilistic models of information retrieval based on measuring the divergence from randomness. *ACM Transactions on Information Systems, 20*(4), 357-389. doi: Doi 10.1145/582415.582416

Azzopardi, L., & de Rijke, M. (2006). *Automatic construction of known-item finding test beds*. Paper presented at the SIGIR '06.

Azzopardi, L., de Rijke, M., & Balog, K. (2007). *Building Simulated Queries for Known-Item Topics*. Paper presented at the SIGIR'07, Amsterdam, The Netherlands.

Bampoulidis, A., Lupu, M., Palotti, J., Metallidis, S., Brassey, J., & Hanbury, A. (2016). Interactive exploration of healthcare queries. *14th International Workshop on Content-Based Multimedia Indexing (CBMI)*.

Bhattacharya, S., Ha−Thuc, V., & Srinivasan, P. (2011). MeSH: a window into full text for document summarization. *Bioinformatics, 27*(13).

Clinchant, S., & Gaussier, E. (2010). Information-based models for ad hoc IR (2010). *SIGIR'10, conference on Research and development in information retrieval.*

Cohen, K. B., Johnson, H. L., Verspoor, K., Roeder, C., & Hunter, L. (2010). The structural and content aspects of abstracts versus bodies of full text journal articles are different. *BMC Bioinformatics, 11*(492).

Falagas, M., Pitsouni, E., Malietzis, G., & Pappas, G. (2008). Comparison of PubMed, Scopus, Web of Science, and Google Scholar: strengths and weaknesses. *The FASEB Journal, 22*(2), 338-342.

Falagas, M., Pitsouni, E., Malietzis, G., & Pappas, G. (2008). Comparison of PubMed, Scopus, Web of Science, and Google Scholar: strengths and weaknesses. *The FASEB Journal, 22*(2), 338-342.

Hersh, W., Cohen, A., Ruslen, L., & Roberts, P. (2007). *Genomics Track Overview.* Paper presented at the Proceedings of the Sixteenth Text REtrieval Conference (TREC 2007).

Huang, M., Neveol, A., & Lu, Z. (2011). Recommending MeSH terms for annotating biomedical articles. *J Am Med Inform Assoc, 18*(5), 660-667. doi: 10.1136/amiajnl-2010-000055

Kim, S., Yeganova, L., Comeau, D. C., Wilbur, W. J., & Lu, Z. (2018). PubMed Phrases, an open set of coherent phrases for searching biomedical literature. *Scientific Data, in press.*

Lin, J. (2009). Is searching full text more effective than searching abstracts? *BMC Bioinformatics, 10*, 46. doi: 10.1186/1471-2105-10-46

Lu, Z. (2011). PubMed and beyond: a survey of web tools for searching biomedical literature. Database: the journal of biological databases and curation. *Database (Oxford), 2011.*

Robertson, S. (2004). Understanding inverse document frequency: On theoretical arguments for IDF. *Journal of Documentation., 60*(5), 503–520.

Robertson SE, W. S., Hancock-Beaulieu M, Gatford M, Payne A. (1995). Okapi at TREC-4. *Proceedings of the Fourth Text REtrieval Conference (TREC-4)*, 73-96.

Sparck Jones, K., Walker, S., & Robertson, S. E. (1998). A probabilistic model of information retrieval: development and status (pp. 1-75): University of Cambridge.

Tsatsaronis, G., Balikas, G., Malakasiotis, P., Partalas, I., Zschunke, M., Alvers, M. R., . . . Paliouras, G. (2015). An overview of the BIOASQ large-scale biomedical semantic indexing and question answering competition. *BMC Bioinformatics, 16*, 138. doi: 10.1186/s12859-015-0564-6

Westergaard, D., Stærfeldt, H.-H., Tønsberg, C., Jensen , L. J., & Brunak, S. (2018). A comprehensive and quantitative comparison of text-mining in 15 million full-text articles versus their corresponding abstracts. *Plos Computational Biology, 14*(2).

Wildgaard, L. E., & Lund, H. (2016). Advancing PubMed? A comparison of 3rd-party PubMed/MEDLINE tools. *Library Hi Tech, 34*(4), 669-684. doi: https://doi.org/10.1108/LHT-06-2016-0066

Yeganova, L., Kim, W., Kim, S., & Wilbur, W. J. (2014). Retro: concept-based clustering of biomedical topical sets. *Bioinformatics, 30*(22).

Zhai, C., & Lafferty, J. (2004). A study of smoothing methods for language models applied to information retrieval. *ACM Transactions on Information Systems, 22*(2), 179-214.

CRF-LSTM Text Mining Method Unveiling the Pharmacological Mechanism of Off-target Side Effect of Anti-Multiple Myeloma Drugs

Kaiyin Zhou [1] **Sheng Zhang** [2] **Xiangyu Meng** [3] **Qi Luo** [2] **Yuxing Wang** [1]
Ke Ding [2] **Yukun Feng** [2] **Mo Chen** [1] **Kevin B Cohen** [4] **Jingbo Xia** [1*]

1. College of Informatics, Huazhong Agricultural University, China
2. College of Science, Huazhong Agricultural Univerisity, China
3. Center for Evidence-based and Translational Medicine, Zhongnan Hospital of Wuhan University, China
4. School of Medicine, University of Colorado, U.S.
*. Correspondence author: `xiajingbo.math@gmail.com`

Abstract

Off-target effects played a vital role in the pharmacological understanding of drug efficacy and this research aimed to use text mining strategy to curate molecular level information and unveil the mechanism of off-target effect caused by the usage of anti-multiple myeloma (MM) drugs. After training a hybrid CNN-CRF-LSTM neural network upon the training data from TAC 2017 benchmark database, we extracted all of the side effects of 16 anti-MM drugs from drug labels, and combined the results with existed database. Afterwards, gene targets of anti-MM drugs were obtained by using structure similarity, and their related phenotypes were retrieved from Human Phenotype Ontology. Furthermore, linked phenotypes to candidate genes and adverse reaction of known drugs formed a knowledge graph. Through regulation analysis upon intersected phenotypes of drugs and target genes, an off-target effect caused by SLC7A7 was found, which with high possibility unveiled the pharmacological mechanism of side effect after using combination of anti-MM drugs.

1 Introduction

Drug genetics aimed to discern the association between drugs and adverse reaction, and allowed to personalized medication (Stephen, 2011). Associating off-target effects with adverse reaction of drugs to discover the new pharmacological effect of them is a daunting task when using experimental method alone.(Eugen et al., 2012).

As a pioneer work, Lountkine et al.,(Eugen et al., 2012) explored a computational method to predict novel off-target effects of 656 marketed drugs. By using the chemoinformatics information, like ligand affinity, Similarity Ensemble Approach (SEA) (Keiser et al., 2009) was used to calculate structural similarity of drugs and targets, and the relations between drugs and targets was rebuild. In the meantime, the adverse reaction (ADR) of drug targets were curated from authoritative database including Drug-Bank, GeneGo Metabase, and Thompson Reuters Integrity. Thus, a large scale drug-target-ADR network was built, and the coincidental overlap of ADR among target gene and drugs potential gene gave illuminative explanation for the mechanism of off-targets effect.

Generally, the mechanism of off-target candidate filtering requires the prerequisite of target-drug pair indications. So far, this pair information has been widely predicted by inferring the similarity both in chemical structure and relevance info. Andreas et al., (Andreas et al., 2007) used chemical structure information to infer the drug-target pair, while Monica et al., (Monica et al., 2008) used phenotypic side effect similarities to make the inference. As a large-scale bioinformatics attempt, Mohan et al., (Mohan et al., 2008) exploited a huge training set of 10 million compounds with known in-vitro activities, predicted both primary and secondary pharmacology for 1279 molecules, and over 30 thousands possible interactions were predicted for these drugs.

Multiple myeloma (MM) is one of the most common hematological malignancies, the incidence of which ranks second just next to non-Hodgkin lymphoma. Although recent advances in MM treatment has largely improved the patients clinical outcome, it remains an incurable disease due to drug-resistance and relapse which are almost inevitable (Terpos, 2017). Common adverse drug reactions (ADRs) related to anti-MM treatment include hematologic toxic effects (eg. anemia, neutropenia and thrombocytopenia), thrombosis, impaired immune function, pe-

166

Proceedings of the BioNLP 2018 workshop, pages 166–171
Melbourne, Australia, July 19, 2018. ©2018 Association for Computational Linguistics

ripheral neuropathy, and gastrointestinal toxic effects (eg. mucositis, diarrhea), among many others. These ADRs bring harm to patients health and quality of life, and may result in premature discontinuation of treatment due to intolerance to side effects. Since the underlying mechanisms are largely unclear, currently they are mostly managed with symptomatic and/or supportive care, along with dosage reduction or treatment discontinuation (McCullough et al., 2018). A better understanding on the mechanisms will help us find ways to effectively cope with the above mentioned safety concerns in treating MM.

In this research, we proposed a novel pharmacological knowledge discovery strategy which integrated both Biomedical natural language processing (BioNLP) and medical informatics. The adverse reactions (ADRs) were trained by newly released ADR training data (Demner-Fushman et al., 2018), and were extracted on-line with large-scale of text mining upon 16 anti-MM drugs by using conditioned random field (CRF) and long short term memory (LSTM) neural networks. Subsequently, Human Phenotype Ontology (HPO) (Sebastian et al., 2017) and Ligand Similarity prediction were used to calculate the target phenotypes. Bioinformatics analysis hinted that an off-target gene, SLC7A7, played vital role in the side effect of a combination usage of anti-MM drugs.

2 Material and Method

2.1 Data Resource

Marketed drugs for MM were collected from drugs.com (Drugs). After searching anti-MM chemicals and removing drug synonyms, 16 drugs were extracted from the original pharmaceutical list, and drug targets were collected from SwissTargetPediction (David et al., 2014), as shown in supplementary table, Table S1 (Sixteen anti-MM drugs their possible targets). Meanwhile, drug labels were extracted from DailyMED database (National Library of Medicine and Services, 2005).

Human Phenotype Ontology (HPO) (Sebastian et al., 2017) provides standardized vocabulary of phenotypic abnormalities in human diseases. From HPO, matches of target genes and their corresponding phenotype terms were retrieved, as shown in table S2(Phenotype matching result for specific gene).

2.2 Sequence labeling by BioNLP Algorithm

2.2.1 Vector representation of tokens

Regarding the input form for a neural network, word embedding, controlled vocabulary - DISORDER, and part of speech (POS) are used for vector representation of tokens.

- Pre-trained Embeddings: Compared with randomly initialized word embeddings, pre-trained word embeddings generally yield better experimental results. 200 dimensional embeddings of GloVe ((Pennington et al., 2014)) was chosen, instead of word2vec word vectors, as GloVe is more preferable for named entity recognition tasks than word2vec (Ma and Hovy, 2016).

- DISO is a standardized dictionary from Metathesaurus of UMLS. The dictionary consists of the following 12 subtypes, i.e. acquired abnormality, anatomical abnormality, cell or molecular dysfunction, congenital abnormality, disease or syndrome, experimental model of disease, finding, injury or poisoning, mental or behavioral dysfunction, neoplastic process, pathologic function, and sign or symptom.

- The NLTK toolkit is taken into consideration to obtain the POS of each token. Randomly initialized feature weights was assigned to each POS type, and a lookup operation convert each sentence into a POS-embedding vector.

2.2.2 Integration of CRF and LSTM for sequence labeling

For sequence labeling task as ADR extraction, CRF is a popular mathematical method which defines the probability of the annotation of the label sequence $\mathbf{L} = (l_1, l_2, ..., l_l)$, given the observation sequence $\mathbf{O} = (o_1, o_2, ..., o_l)$: $exp(\sum_j \lambda_j t_j(l_{i-1}, l_i, \mathbf{O}, i)) + \sum_k \mu_k s_k(l_i, \mathbf{O}, i))$, where $t_j(l_{i-1}, l_i, \mathbf{O}, i))$ is a transition feature function that represents the transition distribution of label pair $\{l_{i-1}, l_i\}$ based on observation sequence $\mathbf{O}$, while $s_k(l_i, \mathbf{O}, i)$ refers to state feature function that quantify the state distribution of the label y_i given the observation sequence $\mathbf{O}$. The mechanism of CRF is to optimize the parameters λ_j and μ_k, and maximize the probability of $P(\mathbf{L}|\mathbf{O})$: $P(\mathbf{L}|\mathbf{O}, \lambda, \mu) = \frac{1}{Z(\mathbf{O})} exp(\sum_j \lambda_j t_j(l_{i-1}, l_i, \mathbf{O}, i)) + \sum_k \mu_k s_k(l_i, \mathbf{O}, i))$, where $Z(\mathbf{O})$ is for normalization (Lafferty et al., 2001).

In the meantime, LSTM is a special Recurrent neural networks(RNNs) which could capture time dynamics via cycles in the graph, and especially, is capable of capturing long-distance dependencies with the employment of a special cell and three grates, i.e. input gate, forget gate, and output gate. Supposing that t represents a time point, x_t is the input vector at time t. i_t, f_t, c_t, o_t stand for different gates state at time t. W_i, W_f, W_c, W_o are the weight matrices for hidden state h_t. U_i, U_f, U_c, U_o denote the weight matrices of different gates for input x_t. b_i, b_f, b_c, b_o denote the

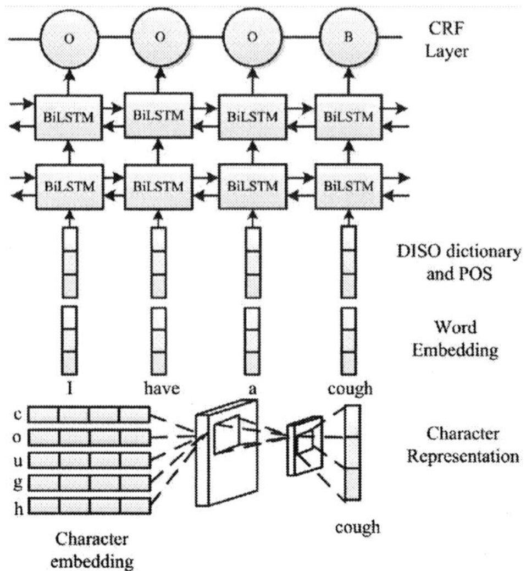

Figure 1: The idea of the CNN-LSTM-CRF sequence labeling method

bias vectors from different gates. And the formulas for LSTM unit at time t are:

$$i_t = \sigma(W_i h_{t-1} + U_i x_t + b_i)$$
$$f_t = \sigma(W_f h_{t-1} + U_f x_t + b_f)$$
$$c_t = f_t * c_{t-1} + i_t * \tanh(W_c h_{t-1} + U_c x_t + b_c) \ , \quad (1)$$
$$o_t = \sigma(W_o h_{t-1} + U_o x_t + b_o)$$
$$h_t = o_t * \tanh(c_t)$$

where σ is the element-wise sigmoid function and $*$ is the element-wise product. And h_t is the hidden state, namely the finally output of LSTM unit at time t.

To achieve a better semantic understanding in biologic domain, a combined BLSTM-CNNs-CRF neural network was put forward by Ma et al.(Ma and Hovy, 2016), where CNNs are utilized to model character-level information, bi-directional LSTM (BLSTM) is used to capture past and future information respectively, and CRF is employed to decode the best label sequence. In order to further improve the labeling accuracy for this specific task, double-BLSTM layer is taken into consideration instead of single-BLSTM layer, namely BLSTM, mentioned in Ma et al.(2016).

The detailed algorithm steps are shown in Figure 1. For each word in training text, the character-level representation vector computed by CNN, the DISO and POS feature got by lookup random initialization weights, concatenated with word embedding vector are designed as the input of the double-BLSTM network. And the output vectors of double-BLSTM are fed to the CRF layers to jointly decode the best label sequence. The flowchart of a specific labeling employment example is presented in the following.

For instance, "I have a cough." where "cough" is the target word. After the sentence being separated into words, the words are broken into letters, which can be embedded into a one-hot vector to compute the character representation vector by CNN. The character-level representation vectors of each words, their DISO and POS representation vector and word embeddings, computed by glove, are combined as the inputs of double-BLSTM, which has double-layer of two processes, i.e. the past(left) and the future(right). The past process takes information only from 'I' to 'cough' while the future process takes information only from 'cough' to 'I'. These two pieces of information was concatenated as the final outputs of double-BLSTM and, simultaneously, the inputs of CRF. With the utilization of CRF, the labels of sentence are tagged as 'O O O B'.

2.3 Phenotype matching algorithm

To decide whether two phenotype words match or not, two criteria were applied. First, both phenotypes are available in the database with the same $is\_a \cdot ID$; second, the word embedding distance of two terms are small sufficiently. The algorithm is shown in the following.

- If both phenotypes are available in the database with the same $is\_a \cdot ID$, the output will be $True$.

- If not, each target phenotype is converted into a word embedding, and if the distance of the two vectors is less than a threshold value t, the two phenotype terms are matched. Otherwise, the two terms are not matched.

Algorithm 1 Phenotype matching algorithm

Input: Term A, term B, threshold value t
Output: $True/False$
1: **if** $(A \in HPO) \bigwedge (B \in HPO) \bigwedge (A \cdot is\_a \cdot ID = B \cdot is\_a \cdot ID)$ **then**
2:     **return** $True$
3: **else if** $Cosine\ Distance(A, B) < t$ **then**
4:     **return** True
5: **else**
6:     **return** False
7: **end if**

2.4 Flowchart of the proposed strategy for off-target side effect prediction

The purpose of this research is to find the co-occurrence of phenotype from both drug and the related protein, so as to illuminate the pharmacological mechanism of the drug side effect.

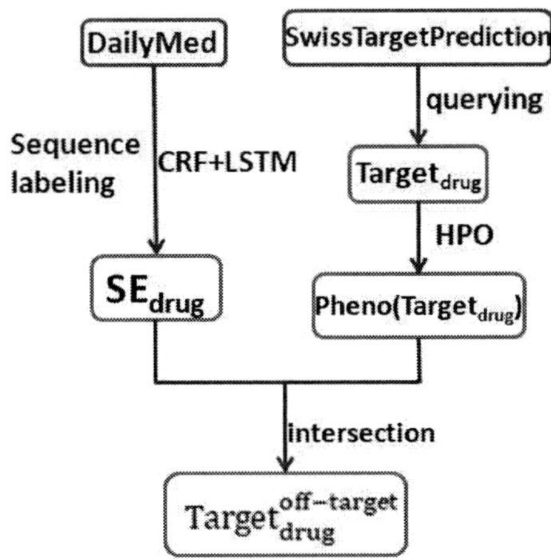

Figure 2: Flowchart of the proposed off-target mechanism discovery

By using an integration of the CRF and LSTM text mining algorithms, sequence labeling was carried on to extract side effects, SE_{drug}, of anti-MM drugs from DailyMed drug labels. Potential drug targets, $Target_{drug}$, were filtered by querying SwissTargetPredcition Database. Meanwhile, related phenotype of $Target_{drug}$, i.e., $Pheno(Target_{drug})$, was obtained by using Human Phenotyping Ontology (HPO). Subsequently, off-target gene, $Target_{drug}^{off-target}$, of corresponding drugs were filtered out by intersection analysis of SE_{drug} and $Pheno(Target_{drug})$.

3 Result

3.1 Database querying result

In total, 48 types of anti-MM drugs are collected by searching drug.com. And with the 48 drug names as searching condition, 16 different drugs and 16 corresponding labels are extracted from 27 drug labels, acquired by DailyMED. Among the 16 drugs, 2 are protein drugs, and the left 14 non-protein drugs are taken to predict their potential targets with the utilization of SwissTargetprediction, where 15 potential targets can be obtained from each drug. Searching the 15 potential targets in HPO, targets, not only our predicted targets but also targets existing in HPO, are achieved. With the application of HPO, drugs, potential target genes, and corresponding phenotype are related with each other. Meanwhile, corresponding ADRs form acquired drug labels can be collected with the strategy of sequence labeling, and improvement of the relationship between drugs and ADRs can be achieved

through drugs.com, where related drugs' ADRs are collected. Eventually, drugs, potential targets, and data of overlapping ADRs are acquired via artificial recognition. And it is revealed in the result that under the circumstance of a certain drug, its potential targets have a tight relation with its ADRs.

3.2 Phenotype matching and phenotype coincidence

For the trained samples in table S3, F-Score and Matthews Correlation Coefficient (MCC) were calculated, and a best threshold $t = 0.57$ was obtained. Here $F-score = 2\frac{Precision \times Recall}{Precision + Recall}$, and $MCC = \frac{TP \times TN - FP \times FN}{\sqrt[2]{(TP+FP)(TP+FN)(TN+FP)(TN+FN)}}$. The selection of t is shown in figure 3. The best F-score and MCC are 0.733, 0.622 separately.

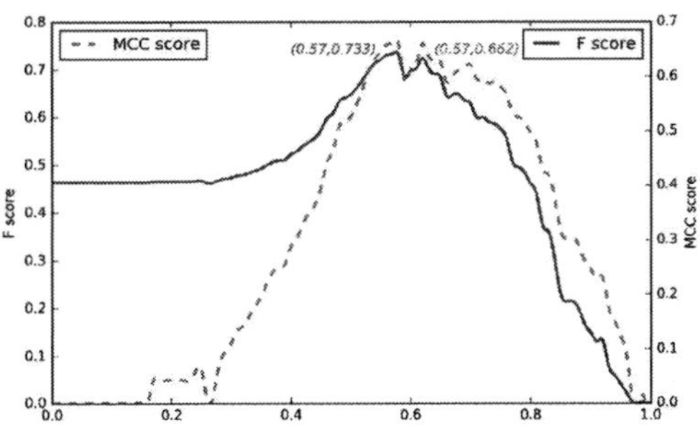

Figure 3: threshold selection for phenotype matching

By using algorithm 1, the gene whose phenotypes in HPO are highly consistent with drug ADRs were retrieved, and the coincidence were evaluated by Jaccard similarity coefficient. Among all the intersection of phenotype terms, the most prominent output pair is melphalan-SLC7A7 for Jaccard value being 0.280 and melphalan-CA2 for Jaccard value being 0.198. As shown in table 1, phenotype coincidence for melphalan and SLC7A7/CA2 is clear, that hinted that the two genes possibly play roles in the side effects of the drug.

3.3 Knowledge discovery of off-target side effect

An illuminative evidence comes from Melphalan, a common anti-MM drug. Through intersection analysis of $SE_{Melphalan}$ and $Pheno(Target_{Melphalan})$, anemia, thrombocytopenia and diarrhea were found to be the same phenotypes of the drug Melphalan and the possible target SLC7A . Observing its target genes are NR3C1, NR0B1, ANXA1, NOS2, NR1L2, and its possible target gene is SLC7A, we found that, after taking another anti-MM drug Prednisone, mRNA level of target genes goes down and that of SLC7A

Gene: SLC7A	
Known ADRs	**Off-target effect**
Sparse hair	Alopecia
Thrombocytopenia	**Thrombocytopenia**
Leukopenia	Leukopenia
Diarrhea	**Diarrhea**
Nausea	Nausea
Anemia	**Anemia**
Vomiting	**Vomiting**
Muscle weakness	Muscular parlysis
Respiratory insufficiency	Dyspnea

Table 1: Consistency of ADRs of melphalan alkeran evomela in clinical records and off-target curations

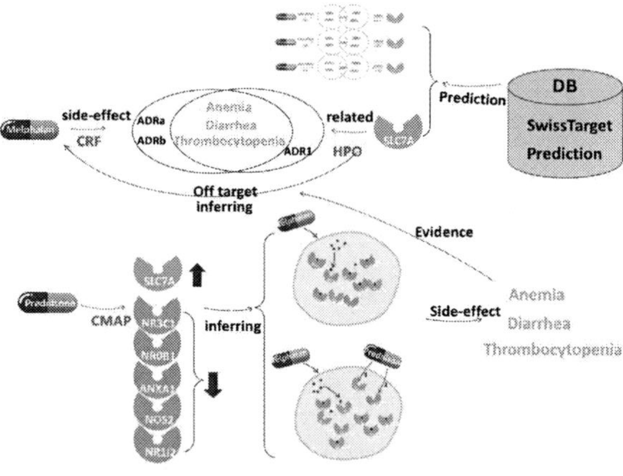

Figure 4: Mechanism of off-target side effects via functioning of SLC7A7 after MVP drug usage

goes up. That made it high chance for off-target event of SLC7A to manifest its off-target side effects: $Pheno(Target_{Melphalan})$. Thus SLC7A is with high chance the factor of the off-target effect.

4 Discussion

Mechanism of off-target effect is illustrated in this section. First, literature evidences are shown to address the side effect after anti-MM drug usage, and then the up/down regulatory mRNA-level tendency of on/off targets are shown.

4.1 Literature evidence

It was reported that a combined usage of melphalan, prednisone, and bortezomib (MPV) is regarded as common treatment for the high-risk MM patient, while neutropenia, thrombocytopenia, anemia, and gastrointestinal symptoms were common after MPV treatment (Kyle and Rajkumar., 2009).

Meanwhile, SLC7A is a heterotrimeric amino acid transporter (HAT) y+LAT-1 gene located on chromosome 14q11.2. It was reported that mutation in SLC7A caused Lysinuric Protein Intolerance. Then, delayed physical development, intestinal malabsorption, vomiting, and failure to thrive are the prominent clinical manifestations (Lawson and Loyd, 2013).

4.2 Up/Down regulation of target/off-target gene

Drug usage of Prednisone is treated as exposure in comparison analysis, and the connectivity map (CMAP) is used to unveil the up/down regulation by analyzing the before/after mRNA level of patience. We input the target genes as down regulated genes and the off-target genes as up regulated, and the output off-target gene is Prednisone, with significant P value, 0.01029.

As shown in figure 4, after taking Prednisone, as it mentioned above, the expression levels of target genes are down regulated while the off-target genes are up, the steady state is broken. In this condition, more off-target proteins lead to more combination with Melphalan than usual, which contribute to more significant side effects.

Here, we infer that the usage of Prednisone lead to an up regulation of SLC7A, and it arises competition between SLC7A and the drug targets, i.e., NR3C1, NR0B1, ANXA1, NOS2, NR1L2. The binding of SLC7A to Melphalan brings the off-target effect. Therefore, thrombocytopenia, anemia, and gastrointestinal symptoms can be easily observed after combined usage of Melphalan and Prednisone.

5 Conclusion

Sequence labeling of biomedical entities, e.g., side effects or phenotypes, was a long-term task in BioNLP and MedNLP communities. Thanks to effects made among these communities, adverse reaction NER has developed dramatically in recent years (Demner-Fushman et al., 2018). As an illuminative application, to achieve knowledge discovery via the combination of the text mining result and bioinformatics idea shed lights on the pharmacological mechanism research.

Acknowledgments

This work is funded by the Fundamental Research Funds for the Central Universities of China (Project No. 2662018PY096). We expressed our gratitude to Pierre Zweigenbaum for discussion of WAPITI and CRF, and to Köhler Sebastian et. al. for offering help in the HPO resource. We also thank anonymous reviewers for their kind suggestions.

References

Bender Andreas, Josef Scheiber, Meir Glick, John W. Davies, Kamal Azzaoui, Jacques Hamon, Laszlo Urban, Steven Whitebread, and Jeremy L. Jenkins. 2007. Analysis of pharmacology data and the prediction of adverse drug reactions and offtarget effects from chemical structure. *ChemMedChem*, 2(6):861–873.

Gfeller David, Aurlien Grosdidier, Matthias Wirth, Antoine Daina, Olivier Michielin, and Vincent Zoete. 2014. Swisstargetprediction: a web server for target prediction of bioactive small molecules. *Nucleic acids research*, 42(W1):W32–W38.

Dina Demner-Fushman, Sonya E Shooshan, Laritza Rodriguez, Alan R Aronson, Francois Lang, Willie Rogers, Kirk Roberts, and Joseph Tonning. 2018. A dataset of 200 structured product labels annotated for adverse drug reactions. *Scientific data*, 5:180001.

Drugs. Drugs.com. https://www.drugs.ca.

Lounkine Eugen, Michael J. Keiser, Steven Whitebread, Dmitri Mikhailov, Jacques Hamon, Jeremy L. Jenkins, Paul Lavan, and et al. 2012. Large-scale prediction and testing of drug activity on side-effect targets. *Nature*, 486(7403):361–367.

Michael J. Keiser, Vincent Setola, John J. Irwin, Christian Laggner, Atheir I. Abbas, Sandra J. Hufeisen, Niels H. Jensen, and et al. 2009. Predicting new molecular targets for known drugs. *Nature*, 462(7270):175–181.

Robert A. Kyle and S. Vincent Rajkumar. 2009. Treatment of multiple myeloma: a comprehensive review. *Clinical Lymphoma and Myeloma*, 9(4):278–288.

John Lafferty, Andrew McCallum, and Fernando CN Pereira. 2001. Conditional random fields: Probabilistic models for segmenting and labeling sequence data.

Guillaume Lample, Miguel Ballesteros, Sandeep Subramanian, Kazuya Kawakami, and Chris Dyer. 2016. Neural architectures for named entity recognition. *arXiv preprint arXiv:1603.01360*.

William E Lawson and James E Loyd. 2013. Interstitial and restrictive pulmonary disorders. In *Emery and Rimoin's Principles and Practice of Medical Genetics*, pages 1–22. Elsevier.

Xuezhe Ma and Eduard Hovy. 2016. End-to-end sequence labeling via bi-directional lstm-cnns-crf. *arXiv preprint arXiv:1603.01354*.

Kristen B McCullough, Miriam A Hobbs, Jithma P Abeykoon, and Prashant Kapoor. 2018. Common adverse effects of novel therapies for multiple myeloma (mm) and their management strategies. *Current hematologic malignancy reports*, pages 1–11.

Health National Library of Medicine, National Institutes of Health and Human Services. 2005. Dailymed.com. https://dailymed.nlm.nih.gov/dailymed/index.cfm.

Prerna Mewawalla and Abhishek Chilkulwar. 2017. Maintenance therapy in multiple myeloma. *Therapeutic advances in hematology*, 8(2):71–79.

Rao Mohan, Michael Liguori, Srinivasa Mantena, Scott Mittelstadt, Eric Blomme, and Terry Van Vleet. 2008. Computational prediction of off-target pharmacology for discontinued drugs. *The FASEB Journal*.

Campillos Monica, Michael Kuhn, Anne-Claude Gavin, Lars Juhl Jensen, and Peer Bork. 2008. Drug target identification using side-effect similarity. *Science*, 31(5886):263–266.

Jeffrey Pennington, Richard Socher, and Christopher Manning. 2014. Glove: Global vectors for word representation. In *Proceedings of the 2014 conference on empirical methods in natural language processing (EMNLP)*, pages 1532–1543.

Köhler Sebastian, Nicole A. Vasilevsky, Mark Engelstad, Erin Foster, Julie McMurry, Sgolne Aym, Gareth Baynam, and et al. 2017. The human phenotype ontology in 2017. *Nucleic acids research*, 45(D1):D865–D876.

Neidle Stephen. 2011. *Cancer drug design and discovery*. Academic Press.

Evangelos Terpos. 2017. Multiple myeloma: Clinical updates from the american society of hematology annual meeting 2016. *Clinical Lymphoma, Myeloma and Leukemia*, 17(6):329–339.

Zi-Hang Zeng, Jia-Feng Chen, Yi-Xuan Li, Ran Zhang, Ling-Fei Xiao, and Xiang-Yu Meng. 2017. Induction regimens for transplant-eligible patients with newly diagnosed multiple myeloma: a network meta-analysis of randomized controlled trials. *Cancer management and research*, 9:287.

A   Supplemental Material

Attached are the supplementary tables.

Table S1. Sixteen anti-MM drugs and their possible targets, (https://github.com/kyzhouhzau/crf-lstm-text/blob/master/Table%20S1.xlsx).

Table S2. Phenotype matching result for specific gene, (https://github.com/kyzhouhzau/crf-lstm-text/blob/master/Table%20S2.xlsx).

Table S3. Positive and negative samples and their distance, (https://github.com/kyzhouhzau/crf-lstm-text/blob/master/Table%20S3.xlsx)

Prediction Models for Risk of Type-2 Diabetes Using Health Claims

Masatoshi Nagata, Kohichi Takai, Keiji Yasuda[†], Panikos Heracleous, Akio Yoneyama

KDDI Research, Inc.
[†]present affiliation: Nara Institute of Science and Technology
{ms-nagata, ko-takai, pa-heracleous, yoneyama}@kddi-research.jp, ke-yasuda@dsc.naist.jp

Abstract

This study focuses on highly accurate prediction of the onset of type-2 diabetes. We investigated whether prediction accuracy can be improved by utilizing lab test data obtained from health checkups and incorporating health claim text data such as medically diagnosed diseases with ICD10 codes and pharmacy information. In a previous study, prediction accuracy was increased slightly by adding diagnosis disease name and independent variables such as prescription medicine. Therefore, in the current study we explored more suitable models for prediction by using state-of-the-art techniques such as XGBoost and long short-term memory (LSTM) based on recurrent neural networks. In the current study, text data was vectorized using word2vec, and the prediction model was compared with logistic regression. The results obtained confirmed that onset of type-2 diabetes can be predicted with a high degree of accuracy when the XGBoost model is used.

1 Introduction

The incidence of lifestyle-related diseases is increasing in many regions (WHO, 2009; Lim SS et al., 2012). Predicting the onset of lifestyle-related diseases and implementing preventive measures in advance is important for municipalities and insurers. Particularly in type-2 diabetes mellitus, not only medical cost but also indirect cost such as reduced productivity present a serious problem (American Diabetes Association, 2018), and therefore, it is very important to take preventive measures early.

From reports to date on the prediction of the onset of diabetes, it is well known that health checkup data items such as HbA1c, BMI, and ages are important indicators for estimating the onset of type-2 diabetes (Edelstein et al., 1997). Many related studies achieved accurate results by means of logistic regression and cox hazards regression models mainly based on bood test results (Droumaguet et al., 2006; Guasch-Ferré et al., 2012). These studies are aimed at predicting the onset of type-2 diabetes using a simple form. However, it is now common for machine learning and data mining methods to be used due to higher computer performance. Several studies have reported the effectiveness of using machine learning technique to improve classification accuracy (Meng et al., 2013; Tapak et al., 2013; Kavakiotis et al., 2017). Another attempt involved using clinical information such as health claims or electronic health records (EHRs). Health insurance claims data could prove to be a rich source of information for the early detection of type-2 diabetes as a previous study showed a slight improvement in prediction using such data (Krishnan et al., 2013; Razavian et al., 2015).

In this study, we aim to develop and evaluate prediction models for the risk of type-2 diabetes using health insurance claims data in addition to health checkup data.

2 Related work

Many related studies are based on conventional prediction models for early detection of type-2 diabetes (Schulze et al., 2006, Thomas et al., 2006). Some research groups use a small number of risk factors as variables as their intention is to develop a practical method. A simple risk score enables healthcare providers to evaluate patients for further intervention and treatment (Lindström et al., 2013; Kengne et al., 2014; Nanri et al., 2015). Logistic regression is one of the most effective models in these studies when compared to other machine learning models. On the other hand,

Proceedings of the BioNLP 2018 workshop, pages 172–176
Melbourne, Australia, July 19, 2018. ©2018 Association for Computational Linguistics

currently, healthcare data management systems integrate large amounts of medical information, such as diagnoses, medical procedures, lab test results, and more. Health claims and EHRs are two examples of this medical information which includes medical text data. It is suggested that there are latent factors that could improve diseases prediction models by including diagnoses and prescribed medicines (Krishnan et al., 2013; Razavian et al., 2015). In addition, some natural language processing (NLP) techniques such as word2vec have been widely used to discover novel patterns and features (Choi et al., 2017; Jo et al., 2017). It is expected that data-driven assessment of individual patient risk would provide better personalized care (Neuvirth et al., 2011).

Recently, Razavian et al. (2015) showed that using an L1-regularized logistic regression (L1LR) model with about 900 variables from health insurance claim data resulted in an area under the ROC curve (AUC) of 0.80 compared with an AUC of 0.75 when using conventional diabetes risk factors. The L1LR model is an effective method where there are many independent variables, although a recent machine learning study has suggested that a gradient boosting method (XGBoost) could achieve high performance prediction (Wei et al., 2017). Furthermore, long short-term memory (LSTM), which is based on a recurrent neural networks model, is feasible for long-range dependencies in sequential data.

In this paper, we compare multiple prediction models for diabetes incidence using health checkup and insurance claims data. In the study, three classification models (i.e. L1LR, XGBoost and LSTM) are developed, and their prediction performance is evaluated as an AUC.

3 Methods

In this section, the dataset and variables used for the evaluation of the proposed methods are described, and three prediction models are also presented.

3.1 Dataset

In the experiments, a collection of anonymized yearly health checkup and health claims at a health insurance society in Japan is used. The health checkup items consist of profile information (e.g. age, sex), lab test results (e.g. body mass index, blood pressure, HbA1c), and health questionnaire (e.g. smoking, alcohol intake, exercise level). We used 33 health checkup items as features for further experiments. The data were obtained from about 40,000 people aged 20 to 64 years. From the whole dataset, we selected those subjects who had health checkups regularly over a period of at least three years. In addition, we excluded some samples missing blood test data. After selection was complete, the final total sample size was 31,000. We used 20% of the dataset randomly sampled for test data, and the rest was used for training. Subjects were diagnosed with diabetes if they had a measured fasting blood sugar (FBS) $\geq$126 mg/dL, or HbA1c 6.5%, or a diagnosis of diabetes on a health insurance claim. Outcome was evaluated if a subject had onset of diabetes in a year in the last of dataset.

3.2 Health insurance claims

Patient records of health insurance claims include medical cost, laboratory test, medical diagnosed disease with ICD10 (International Statistical Classification of Diseases and Related Health Problems) codes and pharmacy information related to the individuals between the years 2011 and 2016. About 5% of subjects had no claim data and had never visited clinics or hospitals. We used ICD10 codes and medicine name data for additional features. To build a training data, firstly, we checked FBS level and HbA1c of health checkup data, and ICD10 codes of diabetes in health insurance claims to extract positive examples.

Our goal is to predict onset of diabetes later than next year and the after that. Thus, for training and prediction, we did not use health checkup results and health insurance claims of immediate 1 year before of diabetes diagnosis.

Since the health insurance claims are issued in monthly unit, there can be more than one ICD10 codes and medicine names in one health insurance claim. We preprocessed them by using word2vec (Mikolov et al., 2013; Rehurek R 2014; Choi et al., 2017). Here, we regarded array of ICD10 codes or medicinal ingredients of prescribed medicine as one sentence. Then we simply preprocessed by word2vec to obtain distributed expression of ICD10 codes and medicinal ingredients. In our experiments, we set both dimensions of ICD10 vector and medical ingredient vector to be 200. By the aforementioned preprocessing, a health insurance claim of one month was converted to 2 vectors (ICD10 vectors and medical ingredients vectors).

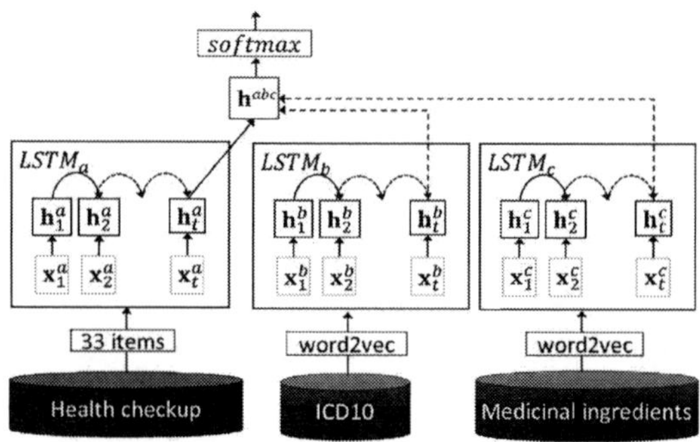

Figure 1: Diabetes prediction using LSTM.

Model	Health Claim	LSTMa	LSTMb	LSTMc
XGBoost LSTM L1LR	-	33	N/A	N/A
	ICD10	33	200	N/A
	medicine	33	N/A	200
	ICD10 + medicine	33	200	200

Table 1: Input unit of LSTM.

Characteristic	ALL subjects	Subjects with diabetes
Average age	41.63	48.23
Female ratio	0.33	0.15
Average length of data in years	3.04	3.75
Body mass index (kg/m^2)	23.07	27.18

Table 2: Characteristics of the dataset.

3.3 Prediction model

As baseline, a conventional L1LR model was used. For L1 regularization hyper-parameter, we searched over values of [0.001, 0.01, 0.1, 1, 10], and 0.1 was selected as the optimum value.

In the experiment, we compare two state of the art prediction models. One is XGBoost which is a scalable machine learning system based on tree boosting (Chen T. and Guestrin C. 2016). To train the XGBoost model, we used scikit-learn API with default parameters. For XGBoost training and L1LR models training, all features including medical checkup results, and distributed expressions of ICD10 and medical ingredients are simply concatenated.

The other prediction model is Long Short-term Memory (LSTM). Figure 1 shows the LSTM architecture used in our experiments. As shown in the figure, the LSTM method consists of two training parts. The first part is health checkup, and second is the ICD10 code, or/and medicinal ingredients of prescribed medicines. $\{\mathbf{x}_1, \cdots, \mathbf{x}_t, \cdots, \mathbf{x}_T\}$ are an array of input sequence for LSTM. For example, $\mathbf{x}_t$ could be embedded insurance claim vector at t-th month.

LSTM consist of four components comprising forget gate ($\mathbf{f}_t$), input gate ($\mathbf{i}_t$), output gate ($\mathbf{o}_t$), and memory state ($\mathbf{c}_t$). These real value vectors are calculated using the following formulas:

$$\mathbf{f}_t = \sigma\left(\mathbf{W}_f\mathbf{x}_t + \mathbf{U}_f\mathbf{h}_{t-1} + \mathbf{b}_f\right),$$

$$\mathbf{i}_t = \sigma\left(\mathbf{W}_i\mathbf{x}_t + \mathbf{U}_i\mathbf{h}_{t-1} + \mathbf{b}_i\right),$$

$$\mathbf{o}_t = \sigma\left(\mathbf{W}_o\mathbf{x}_t + \mathbf{U}_o\mathbf{h}_{t-1} + \mathbf{b}_o\right),$$

$$\tilde{\mathbf{c}}_t = \tanh\left(\mathbf{W}_{\tilde{c}}\mathbf{x}_t + \mathbf{U}_{\tilde{c}}\mathbf{h}_{t-1} + \mathbf{b}_{\tilde{c}}\right),$$

$$\mathbf{c}_t = \mathbf{f}_t \odot \mathbf{c}_{t-1} + \mathbf{i}_t \odot \tilde{\mathbf{c}}_t \qquad (1)$$

where $\mathbf{W}$ and $\mathbf{U}$ are weight matrices, and $\mathbf{b}$ are bias vectors. $\sigma\,(\cdot)$ and $\tanh\,(\cdot)$ are an element-wise sigmoid function and hyperbolic tangent function, respectively. Using these vectors, the hidden layer vector ($\mathbf{h}_t$) is calculated as follows:

$$\mathbf{h}_t = \mathbf{o}_t \odot \tanh(\mathbf{c}_t) \qquad (2)$$

Where $\odot$ is an element-wise multiplication. In our experiments, we used up to three kinds of feature sets (shows in Table 1). Each feature set is processed by individual LSTM. After processing all of feature sets by LSTMs, each of the last hidden layer vectors are concatenated as follows:

$$\mathbf{h}^{abc} = \mathbf{h}^a\mathbf{h}^b\mathbf{h}^c \qquad (3)$$

By using $\mathbf{h}^{abc}$, the output layer calculates probabilities of diabetes. The output layer calculates probability of diabetes.

4 Results

Incidence of type 2 diabetes in our dataset was 4%. The characteristics detailed statistics are shown in Table 2.

We developed three models namely XGBoost, LSTM, and L1LR. For each model, we used four patterns of health claim variables. Table 3 shows the AUC when using the three models. The results show that the performance of the XGBoost and LSTM models was superior to that of the L1LR model without health claim features. In our experiments, the highest performance was obtained

Model	Health Claim	AUC
XGBoost	-	0.81
	ICD10	0.86
	medicine	0.87
	ICD10 + medicine	0.87
LSTM	-	0.81
	ICD10	0.86
	medicine	0.82
	ICD10 + medicine	0.83
L1LR	-	0.72
	ICD10	0.74
	medicine	0.72
	ICD10 + medicine	0.74

Table 3: Performance for prediction of diabetes using health claim data

when the XGBoost with ICD10 plus medicine features was used. On the other hand, the L1LR model had the lowest AUC, though a slight improvement was obtained by incorporating health claim data.

LSTM with the ICD10 model showed a relatively high performance, however, adding prescribed medicine features did not improve its level of prediction.

5 Discussion

In this study, we compared the predictive performance of a conventional model to that of machine learning-based models using health checkup data and additional health claim features vectorized by word2vec. The results showed that the XGBoost and LSTM models achieved better performance compared to the L1LR model without using health claim information. Adding health claim features improved prediction performance in each of the three models. This is consistent with a previous study in which use of the L1LR model obtained slightly improved prediction performance (Razavian et al. 2015). These results suggest that medical information contains latent signals for risk factors associated with the onset of diabetes.

In terms of how to use health claim data, a previous study used the data as one-hot vectors. However, one-hot encoding cannot express the relationship and meaning between words. On the other hand, word2vec makes it possible to give a latent meaning to the vector. This effect was considered to be valid in the case of the XGBoost model.

In recent years, the LSTM model has been used to estimate disease name or mortality from medical information obtained from medical systems with a high degree of performance (Ayyar et al., 2016; Lipton et al., 2016; Jo et al., 2017). LSTM can embed influence over time series data across multiple layers. Therefore, although we expected this effect in our experiments, prediction performance was not improved much when ICD10 and medicine name were used in combination, compared with the case when using only ICD10. This result can probably be attributed to the difference in the quality of the information between the diagnosis disease name and prescription medicine.

Our study has several limitations. First, the vectorization from health claims data was empirically set to 200 dimensions. However, it is not clear what the optimal dimension is. Second, the duration in terms of years of the dataset is relatively short. From the standpoint of disease prevention, it may be desirable for predictive purposes to extend this period to three years or more. Finally, the dataset sample population may have been biased because our data collection depended on information from one health insurance society.

6 Conclusion and Future Work

It would be useful in terms of practicality if risk could be estimated easily with noninvasive data. However, it is also very important, from the viewpoint of personal care, to predict onset of disease with a high degree of precision with obtained from various types of medical information. In this study, we developed and evaluated several prediction models for type-2 diabetes to explore an effective means of vectorization using health claims. We used health claims, ICD10 and prescribed medicine name as variables in addition to health checkup data by vectorizing via word2vec. The results showed that the XGBoost model with health claim variables achieved a higher performance compared to the LSTM and L1LR models. Our study suggests that there are potential factors contained in large amounts of medical information which may be signals to the onset of diabetes. It is possible that the LSTM model may still be able to further improve prediction performance as well. As future work, we plan to test the effect of dimensional compression by parameter tuning.

References

American Diabetes Association. 2018. Economic costs of diabetes in the U.S. in 2017. *Diabetes Care*;41:917–928.

Ayyar S. Don' OB. & Iv W. 2016. Tagging Patient Notes with ICD-9 Codes. In *Proceedings of the 29th Conference on Neural Information Processing Systems* (NIPS 2016).

Chen T. Guestrin C. XGBoost: A Scalable Tree Boosting System. 2016. *22nd ACM SIGKDD Int. Conf.* 785 DOI: 10.1145/2939672.2939785.

Choi E. Schuetz A. Stewart WF. and Sun J. 2016. Medical concept representation learning from electronic health records and its application on heart failure prediction. *arXiv preprint arXiv*: 1602.03686.

Jo Y. Lee L. and Palaskar S. 2017. Combining LSTM and latent topic modeling for mortality prediction. *arXiv preprint arXiv*:1709.02842.

Kavakiotis I. Tsave O. Salifoglou A. Maglaveras N. Vlahavas I. Chouvarda I. 2017. Machine learning and data mining methods in diabetes research. *Comput Struct Biotechnol J.* 15:104–16.

Kengne AP. Beulens JW. Peelen LM. Moons KG. van der Schouw YT. Schulze MB. et al. 2014. Non-invasive risk scores for prediction of type 2 diabetes (EPIC-InterAct): a validation of existing models. *Lancet Diabetes Endocrinol.* 2:19-29.

Krishnan R., Razavian N., Choi Y., Nigam S. Blecker S., Schmidt A., Sontag D. 2013. Early detection of diabetes from health claims. *NIPS workshop in Machine Learning for Clinical Data Analysis and Healthcare.*

Lim SS. Vos T. Flaxman AD. Danaei G. Shibuya K. Adair-Rohani H. et al. 2012. A comparative risk assessment of burden of disease and injury attributable to 67 risk factors and risk factor clusters in 21 regions, 1990–2010: a systematic analysis for the Global Burden of Disease Study 2010. *Lancet.* 380(9859):2224–2260.

Lindstrom J. Tuomilehto J. 2003. The diabetes risk score: a practical tool to predict type 2 diabetes risk, Diabetes Care, vol. 26:725-731.

Lipton Z. Kale D. Elkan C. Wetzell R. 2016. Learning to diagnose with LSTM recurrent neural networks. *arXiv preprint arXiv: 1511.03677.*

Meng XH. Huang YX. Rao DP. Zhang Q. Liu Q. 2013. Comparison of three data mining models for predicting diabetes or prediabetes by risk factors. *Kaohsiung J Med Sci.* ;29:93–99.

Nanri A. Nakagawa T. Kuwahara K. Yamamoto S. Honda T. Okazaki H. ..., for the Japan Epidemiology Collaboration on Occupational Health Study Group. 2015. Development of risk score for predicting 3-year incidence of type 2 diabetes: Japan epidemiology collaboration on occupational health study. *PloS One.* 10:e0142779.

Neuvirth H. Ozery-Flato M. Hu J. Laserson J. Kohn MS. Ebadollahi S. Rosen-Zvi M. 2011. Toward personalized care management of patients at risk: the diabetes case study; *Proceedings of ACM international conference on knowledge discovery and data mining*; 395–403.

Razavian N. Blecker S. Schmidt AM. Smith-McLallen A. Nigam S. Sontag D. 2015. Population-level prediction of type 2 diabetes from claims data and analysis of risk factors. *Big Data.* 3:277–287.

Schulze MB. Heidemann C. Schienkiewitz A. Bergmann MM. Hoffmann K. Boeing H. 2006. Comparison of anthropometric characteristics in predicting the incidence of type 2 diabetes in the EPIC-Potsdam Study. *Diabetes Care* ;29:1921–3.

Tapak L. Mahjub H. Hamidi O. Poorolajal J. 2013. Real-data comparison of data mining methods in prediction of diabetes in Iran. *Healthcare Informat. Res.,* vol. 19, no. 3, pp. 177-185.

Thomas C. Hypponen E. Power C. 2006. Type 2 diabetes mellitus in midlife estimated from the Cambridge Risk Score and body mass index. *Arch Intern Med* 166:682–688.

Wei X. Jiang F. Wei F. Zhang J. Liao W. & Cheng S. 2017. An Ensemble Model for Diabetes Diagnosis in Large-scale and Imbalanced Dataset. *Proceedings of the Computing Frontiers Conference on ZZZ - CF17.* doi:10.1145/3075564.3075576.

WHO. 2009. Global health risks: morality and burden of disease attributable to selected major risks. World Health Organization, Geneva.

On Learning Better Word Embeddings from Chinese Clinical Records: Study on Combining In-Domain and Out-Domain Data

Yaqiang Wang[1*], Yunhui Chen[2], Hongping Shu[1], Yongguang Jiang[2]

[1] Department of Software Engineering, Chengdu University of Information Technology, Chengdu, Sichuan 610225, China

[2] School of Fundamental Medicine, Chengdu University of Traditional Chinese Medicine, Chengdu, Sichuan 610075, China

*Corresponding author: yaqwang@cuit.edu.cn

Abstract

High quality word embeddings are of great significance to advance applications of biomedical natural language processing. In recent years, a surge of interest on how to learn good embeddings and evaluate embedding quality based on English medical text has become increasing evident, however a limited number of studies based on Chinese medical text, particularly Chinese clinical records, were performed. Herein, we proposed a novel approach of improving the quality of learned embeddings using out-domain data as a supplementary in the case of limited Chinese clinical records. Moreover, the embedding quality evaluation method was conducted based on Medical Conceptual Similarity Property. The experimental results revealed that selecting good training samples was necessary, and collecting right amount of out-domain data and trading off between the quality of embeddings and the training time consumption were essential factors for better embeddings.

1 Introduction

Word embeddings, or embeddings for short, have been widely used in various natural language processing tasks, such as language modeling (Bengio et al., 2003; Sundermeyer, et al. 2012; Adams et al., 2017), syntactic parsing (Grefenstette et al., 2014; Tu et al., 2017) and part-of-speech tagging (Yang and Eisenstein, 2016). Owing to the advantage of embeddings in boosting performance, a surge of interest in applying embeddings has become increasingly evident with numerous encouraging results in the field of biomedical applications, e.g. disease prediction (Miotto et al., 2016), clinical events prediction (Choi et al., 2016a), medical concept disambigua-tion (Tulkens et al., 2016), and biomedical information retrieval (Mohan et al., 2017).

Learning embeddings from English medical texts, as a hot topic in recent years, has been extensively studied due to the efforts of open datasets, such as UMLS of NLM (Bodenreider, 2004), medical journal abstracts from PubMed (Choi et al., 2016a), and some released clinical data (Finlayson, et al., 2014; Stubbs and Uzuner, 2015). These datasets have been widely used as gold standards by the biomedical natural language processing domain for learning embeddings (De Vine et al., 2014; Choi et al., 2016b).

However, the development of learning embeddings from Chinese medical texts has fallen far behind, especially from Chinese clinical records. Due to the privacy concerns, Chinese clinical records that can be used are generally limited. Learning better embeddings based on neural network architectures, for instance the widely used skip-gram model (Mikolov et al., 2013a), usually needs a large number of training data. As a result, the learned embeddings from Chinese clinical records are not good enough.

Moreover, to the best of our knowledge, there is a limited number of studies focusing on learning embeddings from Chinese clinical records, not to mention the embedding evaluation. Many methods have been developed to learn embeddings from English medical texts, however, Chinese medical texts, especially clinical records, have their particular language features. Therefore, adaptions to the approaches of learning embeddings from English medical texts are urgently needed for learning embeddings from Chinese clinical records.

In this paper, we focused on learning embeddings from Chinese clinical records, and our major contributions were as follows:

- We proposed an in-domain and out-domain data combination method for learning better

177

Proceedings of the BioNLP 2018 workshop, pages 177–182

Melbourne, Australia, July 19, 2018. ©2018 Association for Computational Linguistics

embeddings from Chinese clinical records by the skip-gram model under the situation that we only have limited Chinese clinical records.

- Referring to the evaluation method for medical concept embeddings proposed in (Choi et al., 2016b) which is based on medical conceptual similarity property, we proposed a method for distantly evaluating the learned embeddings from Chinese clinical records using an additional standard medical terminology dataset.

- We found that selecting good training samples is necessary. Collecting right amount of out-domain data, trading off between the quality of embeddings and the training time consumption are essential factors for better embeddings.

2 Skip-Gram Model for Learning Embeddings

The skip-gram model is one of the most popular methods for learning embeddings from texts. The training objective of the skip-gram model is to find an embedding that is useful for predicting context words of one target word in a sequence. The sequence usually refers to a sentence in a specific task. In the skip-gram model, if two different target words w_k and $w_{k'}$ have (very) similar context words, then learned embeddings of w_k and $w_{k'}$ by the model would be (very) similar, because a common output weight matrix is used (Mikolov et al., 2013b). In other words, if we want to clearly distinguish two target words' embeddings, we can provide more informative context words that differentiate the target words.

The skip-gram model has been used in various domain to learn embeddings from different types of texts, and there have been also various relevant attempts to learn embeddings from medical texts by the skip-gram model. Most works directly applied the model on various medical corpora to complete this domain-specific task (Giménez et al., 2013; Liu, et al., 2016). In this paper, we continued the previous work using the skip-gram model to learn embeddings from Chinese clinical records to further explore a data combination method for improving the quality of the learned domain-specific embeddings.

3 Skip-Gram Model for Learning Embeddings from Chinese Clinical Records

3.1 Observation

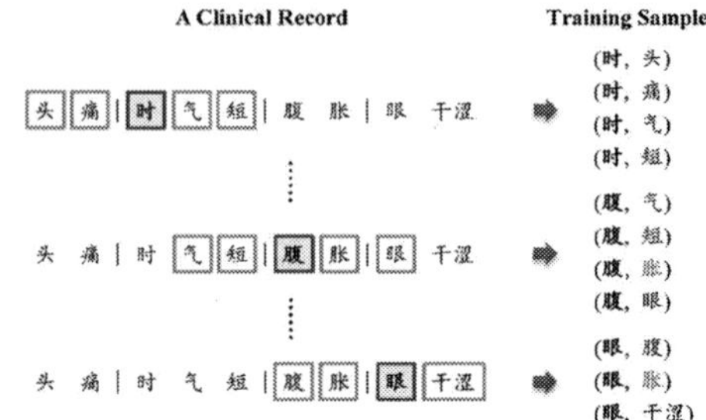

Figure 1: An example of training sample generating process of the skip-gram model.

Content of Chinese clinical records are usually brief, the occurrence of symptoms and diseases has certain correlation, and doctors have a certain habit in inquiring procedures and making records. These domain-specific characteristics challenge learning embeddings from Chinese clinical records, because it gives general domain words a high probability of having similar or even identical context words to those medical words. For example, in Figure 1, general domain word "时" (sometimes) and medical term "眼" (eye, the body part) have similar context words with medical word "腹" (abdomen, the body part), and "时" (sometimes) has more common context words with "腹" (abdomen) than "眼" (eye). Moreover, it would like to be a fixed pattern to describe certain medical problems. As a result, learned embeddings of "时" (sometimes) and "腹" (abdomen) would be more similar than embeddings of "眼" (eye) and "腹" (abdomen), although "腹" (abdomen) and "眼" (eye) belong to the same type of medical concept (i.e. the body part).

In summary, the main challenge of learning better embeddings from Chinese clinical records is to let the skip-gram model make a clearer distinction between medical words and general domain words.

3.2 Usage of Out-Domain Data

As mentioned earlier, making a clearer distinction between learned embeddings of two target words by skip-gram model requires more evidences, i.e. adding diverse context words to illustrate the difference between the two target words. Therefore, we proposed a hypothesis that adding general domain Chinese texts, i.e. the out-domain data, to Chinese clinical records, i.e. the in-domain data, would facilitate the learning of embeddings from Chinese clinical records. The intuition is that the medical words in Chinese clinical records have domain-specific usage but are not widely used in the out-domain data. However, the general domain words have a wide range of usage in the out-domain data, which is the exact opposite of using medical words. Combining out-domain data with Chinese clinical records can improve the diversity of context words of the general domain words, but without the side-effect of impairing the contexts of the medical words. Better embeddings, in turn, can be learned from the combined data.

3.3 Learning Process and Embedding Quality Evaluation Method

Chinese clinical records were segmented into words by the latest version of Stanford CoreNLP tool[1] with default settings, and adjacent words appearing in our prepared standard medical word dataset would not be segmented (Zhang et al., 2016). Punctuations were removed. Out-domain data went through a similar process but without the second process. We assume that in out-domain data there is no medical words. We directly applied skip-gram model implemented by DeepLearning4J[2] to learn embeddings. Hierarchical SoftMax is used in training process, and context window size and embedding dimensionality are set to 5 and 200 respectively (Choi et al, 2016b).

We used an intrinsic evaluation method, named Chinese Medical Concept Similarity Measure (CMCSM), to distantly measure quality of learned embeddings. CMCSM is defined below:

$$CMCSM = \frac{1}{N}\sum_{i=1}^{N}\frac{2}{C_i(C_i-1)}\sum_{j=1}^{C_i-1}\sum_{k=j+1}^{C_i}s(c_j,c_k) \quad (1)$$

where N is the number of groups of the medical words in the same level of a prepared medical word dataset $\mathbf{C}$, $C_i \in \mathbf{C}$ is one group of the medical words, and c_j and c_k are the jth and kth terms in C_i. $s(c_j,c_k)$ is any commonly used embedding similarity measure (Levy et al., 2015). In this paper, we used the cosine measure.

Dataset		Size
CCRD		25056
ODD		3010739
SMTD	Number of Terms	3617
	Number of Groups	39

Table 1: Detailed Information of the Experimental Datasets.

4 Experiments

4.1 Experimental Data

To validate performance of the proposed method, three experimental datasets were used in this paper, including a Chinese clinical records dataset (CCRD) collected from Teaching Hospital of Chengdu University of Traditional Chinese Medicine, a large scale out-domain dataset (ODD) obtained from the NLPCC 2018 Shared Task 4[3], and a standard medical terminology dataset (SMTD) gotten from WHO[4]. Medical terms in SMTD are organized into a two-layer tree structure. Index of the second layer defines the group id for medical words. Medical words in the same group are more similar. SMTD was used as the prepared medical word dataset $\mathbf{C}$ mentioned previously. The detailed information of these datasets was listed in Table 1.

4.2 Experimental Data

Firstly, we applied skip-gram model to learn embeddings from CCRD and the learned embeddings were evaluated by CMCSM. We sampled 5 sub-datasets from CCRD in order to assess effect of different size of datasets on quality of the learned embeddings. The sizes of the sampled datasets were 80%, 60%, 40%, 20% and 10% of instances in the original CCRD. The sampling process was a recursive sampling without replacement. It implied that more data means more stable learning results of embeddings. Moreover, we ran the above process 10 times to further assess the stability of the results. The results were used as the baseline, and they were shown in Table 2.

We found in Table 2 that the more Chinese clinical records were used for learning embeddings, the smaller variance of CMCSM tended to be achieved. Moreover, an interesting result was that the use of all Chinese clinical records did not nec-

---

[1] URL: https://nlp.stanford.edu/software/segmenter.shtml.
[2] URL: https://deeplearning4j.org/.

[3] URL: http://tcci.ccf.org.cn/conference/2018/cfpt.php.
[4] We filtered the terminologies which do not appear in CCRD. URL: http://www.wpro.who.int/publications/who_i-strm_file.pdf?ua=1.

essarily result in the highest quality of embeddings. It implies that if we only use in-domain data to learn embeddings, we should collect as much training data as possible and also select helpful samples from the collected data.

Secondly, we applied skip-gram model to learn embeddings from combinations of CCRD and ODD with different combination ratios. Results were listed in Table 3, indicating through combin-

should consider whether it is worthwhile to spend a lot of training time in exchange for very little quality improvement. Moreover, little quality improvement sometimes may not improve performance of downstream biomedical applications.

5 Discussion

This paper conducted only intrinsic evaluation and

	10%	20%	40%	60%	80%	100%
Time 1	0.00218	0.00254	0.00238	0.00259	**0.00268**	
Time 2	0.00210	0.00238	0.00234	**0.00269**	0.00248	
Time 3	0.00183	0.00220	0.00255	**0.00281**	0.00241	
Time 4	0.00188	**0.00254**	0.00225	0.00235	0.00232	
Time 5	0.00132	0.00218	**0.00247**	0.00226	0.00226	0.00228
Time 6	0.00229	0.00248	**0.00297**	0.00255	0.00268	
Time 7	0.00134	0.00220	0.00209	**0.00264**	0.00241	
Time 8	0.00189	0.00256	0.00261	0.00242	**0.00263**	
Time 9	0.00141	0.00213	0.00228	**0.00258**	0.00234	
Time 10	0.00199	**0.00269**	0.00255	0.00248	0.00253	
Mean	0.00182	0.00239	0.00245	**0.00254**	0.00247	-
Variance	1.11E-07	3.56E-08	5.32E-08	2.42E-08	2.08E-08	-

Table 2: CMCSM Results of the Embeddings Learned from CCRD by the Skip-Gram Model.

ing ODD into CCRD, the qualities of the learned embeddings in different conditions were improved dramatically. More ODD data is combined into CCRD, better embeddings would be learned. In the best case (combining the "Time 2-60%" dataset with the "ODD-ALL" dataset), CMCSM increased by 3.8 times.

Notably, the highest quality of the learned embeddings in each row of Table 3 was not always achieved when all data in ODD was used. This result was consistent with the result mentioned earlier, indicating that we should collect as much training data as possible and also need to pay attention to reasonably choosing training samples. In addition, the results showed that when the amount of ODD was 1000 times of the basis size of CCRD, optimal embeddings would be achieved.

Moreover, the results suggested that, in practice, the trade-off between quality of embeddings and training time consumption should be considered. Figure 2 displayed that with increasing the amount of the combined ODD, the growth rate of CMCSM of learned embeddings from basis size of CCRD decreased sharply. Furthermore, when the amount of the combined ODD was more than 50 times of the basis size, the growth rate was almost converged. While, as we know, more data were used for learning embeddings by skip-gram model, much more time would be consumed. We

requires further research involving results from extrinsic evaluations. The high quality embeddings from intrinsic evaluations is also essential for enhancing performance in downstream applications.

Experimental results in this paper casted light on the quality improvements of learning embeddings from English clinical records. Most of the existing studies about how to train good embed-

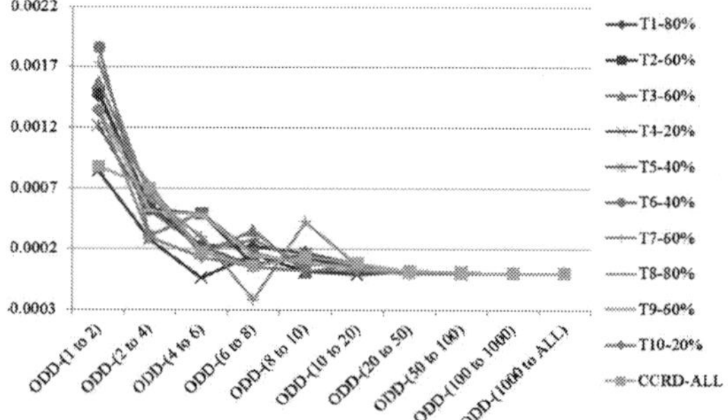

Figure 2: An Example of SMTD.

dings are based on data within the same domain (Chiu et al., 2016; Lai et al., 2016).

Further exploration needs to be continued in many aspects. For instance, how to thoroughly understand learning embeddings via complicated neural networks, which is one of current major research hotspots. Only when the complex background theory is fully interpreted, can we apply this invaluable technology in a flexible way.

	ODD-1	ODD-2	ODD-4	ODD-6	ODD-8	ODD-10	ODD-20	ODD-50	ODD-100	ODD-1000	ODD-ALL
T1-80%	0.0036	0.0050	0.0061	0.0066	0.0070	0.0074	0.0082	0.0087	0.0091	**0.0098**	0.0097
T2-60%	0.0044	0.0059	0.0070	0.0080	0.0081	0.0084	0.0089	0.0095	0.0097	**0.0102**	**0.0102**
T3-60%	0.0041	0.0057	0.0070	0.0074	0.0081	0.0081	0.0089	0.0093	0.0096	**0.0101**	0.0099
T4-20%	0.0056	0.0064	0.0070	0.0069	0.0072	0.0073	0.0072	0.0076	**0.0083**	0.0078	0.0079
T5-40%	0.0046	0.0058	0.0068	0.0073	0.0074	0.0077	0.0084	0.0089	0.0089	0.0089	**0.0092**
T6-40%	0.0050	0.0069	0.0075	0.0085	0.0088	0.0090	0.0095	0.0100	0.0101	0.0102	**0.0103**
T7-60%	0.0041	0.0058	0.0071	0.0077	0.0072	0.0081	0.0088	0.0093	0.0094	**0.0100**	**0.0100**
T8-80%	0.0036	0.0049	0.0063	0.0067	0.0073	0.0074	0.0081	0.0088	0.0092	**0.0098**	**0.0098**
T9-60%	0.0038	0.0051	0.0062	0.0071	0.0074	0.0076	0.0079	0.0088	0.0090	**0.0094**	**0.0094**
T10-20%	0.0061	0.0074	0.0080	0.0083	0.0084	0.0084	**0.0087**	**0.0087**	**0.0087**	0.0087	0.0087
CCRD-ALL	0.0035	0.0044	0.0058	0.0062	0.0063	0.0066	0.0074	0.0079	0.0083	**0.0091**	**0.0091**
Mean	0.0044	0.0058	0.0068	0.0073	0.0076	0.0078	0.0083	0.0089	0.0091	**0.0095**	**0.0095**

Table 3: CMCSM Results of the Embeddings Learned from the Combinations of CCRD and ODD by the Skip-Gram Model. "Tn-X%" means that "the dataset is the X% data of CCRD which is used for learning the highest quality of embeddings in Table 2 at Tn," and "CCRD-ALL" means that all instances in CCRD are used. "ODD-n" means that "the size of ODD currently used is 'n'×2505." "ODD-ALL" means all samples in ODD are used. 2505 is the basis size of CCRD, and it is approximately equal to the number of 10% of CCRD.

6   Conclusions

This paper presented study on how to learn better embeddings from Chinese clinical records with the supplement of out-domain data in the context of limited in-domain data. Proceeding from the Medical Conceptual Similarity Measure (Choi et al., 2016b), we applied it to distantly evaluate the quality of embeddings. The experimental results showed that a combination use of out-domain and in-domain data could potentially improve the quality of learned embeddings; collecting right amount of out-domain data, trading off between the quality of embeddings and the training time consumption, choosing the good training samples were all essential factors for learning better embeddings. Our results also proved that more data did not necessarily bring more satisfying results, which was consistent with results of Chiu et al. (2016).

Acknowledgments

Authors are pleased to acknowledge the National Natural Science Foundation of China (Grant No. 61501063), the Scientific Research Foundation of Science and Technology Department of Sichuan Province (Grant No. 2016JY0240), the Talent Introduction Project of Chengdu University of Information Technology (Grant No. 376226), and the Scientific Research Funding for Young and Middle-Aged Academic Leaders of Chengdu University of Information Technology (Grant No. J201705).

References

Oliver Adams, Adam Makarucha, Graham Neubig, Steven Bird, Trevor Cohn. 2017. Cross-Lingual Word Embeddings for Low-Resource Language Modeling. In *Proceedings of the 15th Conference of the European Chapter of the Association for Computational Linguistics*. Association for Computational Linguistics, pages 937–947.

Yoshua Bengio, Rejean Ducharme, Pascal Vincent, Christian Jauvin. 2003. A Neural Probabilistic Language Model. *Journal of Machine Learning Research*, 3:1137-1155.

Olivier Bodenreider. 2004. The Unified Medical Language System (UMLS): integrating biomedical terminology. *Nucleic Acids Research*, 32:D267-D270.

Billy Chiu, Gamal Crichton, Anna Korhonen, Sampo Pyysalo. 2016. How to Train Good Word Embeddings for Biomedical NLP. In *Proceedings of the 15th Workshop on Biomedical Natural Language Processing (BioNLP 2016)*. Association for Computational Linguistics, pages 166-174.

Edward Choi, Mohammad Taha Bahadori, Andy Schuetz, Walter F. Stewart, Jimeng Sun. 2016a. Doctor AI: Predicting Clinical Events via Recurrent Neural Networks. In *Proceedings of the 1st Machine Learning for Healthcare Conference*. PMLR 56, pages 301-318.

Youngduck Choi, Chill Yi-I Chiu, David Sontag. 2016b. Learning Low-Dimensional Representations of Medical Concepts. In *Proceedings of the AMIA Summit on Clinical Research Informatics (CRI)*. American Medical Informatics Association.

Lance De Vine, Guido Zuccon, Bevan Koopman, Laurianne Sitbon, Peter Bruza. 2014. Medical Semantic Similarity with a Neural Language Model. In *Proceedings of the 23rd ACM International Conference on Conference on Information and Knowledge Management*. pages 1819-1822.

Samuel G. Finlayson, Paea LePendu, Nigam H. Shah. 2014. Building the Graph of Medicine from Millions of Clinical Narratives. *Scientific Data*, 1: 140032.

Edward Grefenstette, Phil Blunsom, Nando de Freitas, Karl Moritz Hermann. 2014. A Deep Architecture for Semantic Parsing. In *Proceedings of the ACL 2014 Workshop on Semantic Parsing*. Association for Computational Linguistics, pages 22–27.

Siwei Lai, Kang Liu, Shizhu He, Jun Zhao. 2016. How to Generate a Good Word Embedding? *IEEE Intelligent Systems*, 31(6): 5-14.

Omer Levy, Yoav Goldberg, Ido Dagan. 2015. Improving Distributional Similarity with Lessons Learned from Word Embeddings. *Transactions of the Association for Computational Linguistics*, 3:211-225.

Yun Liu, Collin M. Stultz, John V. Guttag, Kun-Ta Chuang, Fu-Wen Liang, Huey-Jen Su. 2016. Transferring Knowledge from Text to Predict Disease Onset. In Proceedings of the 1st Machine Learning for Healthcare Conference. PMLR 56, pages 150-163.

Tomas Mikolov, Kai Chen, Greg Corrado, Jeffrey Dean. 2013a. Efficient Estimation of Word Representations in Vector Space. *arXiv*:1301.3781.

Tomas Mikolov, Ilya Sutskever, Kai Chen, Gregory S Corrado, Jeffrey Dean. 2013b. Distributed Representations of Words and Phrases and their Compositionality. *Advances in Neural Information Processing Systems 26*. pages 19-27.

Jose Antonio Miñarro Giménez, Oscar Marin-Alonso, Matthias Samwald. 2013. Exploring the application of deep learning techniques on medical text corpora. *Studies in health technology and informatics*, 205:584-588.

Riccardo Miotto, Li Li, Brian Kidd, Joel T. Dudley. 2016. Deep Patient: An Unsupervised Representation to Predict the Future of Patients from the Electronic Health Records. *Scientific Reports*, 6:26094.

Sunil Mohan, Nicolas Fiorini, Sun Kim, Zhiyong Lu. 2017. Deep Learning for Biomedical Information Retrieval: Learning Textual Relevance from Click Logs. In *Proceedings of the 16th Workshop on Biomedical Natural Language Processing (BioNLP 2017)*. Association for Computational Linguistics, pages 222-231.

Amber Stubbs, Christopher Kotfila, Ozlem Uzuner. 2015. Annotating Longitudinal Clinical Narratives for De-identification: The 2014 i2b2/UTHealth Corpus. *Journal of Biomedical Informatics*, 58:S20-S29.

Martin Sundermeyer, Ralf Schlüter, Hermann Ney. 2012. LSTM Neural Networks for Language Modeling. In *Proceedings of the 13th Annual Conference of the International Speech Communication Association (INTERSPEECH 2012)*. ISCA, pages 194-197.

Lifu Tu, Kevin Gimpel, Karen Livescu. 2017. Learning to Embed Words in Context for Syntactic Tasks. In *Proceedings of the 2nd Workshop on Representation Learning for NLP*. Association for Computational Linguistics, pages 265–275

Stephan Tulkens, Simon Suster, Walter Daelemans. 2016. Using Distributed Representations to Disambiguate Biomedical and Clinical Concepts. In *Proceedings of the 15th Workshop on Biomedical Natural Language Processing (BioNLP 2016)*. Association for Computational Linguistics, pages 77-82.

Yi Yang, Jacob Eisenstein. 2016. Part-of-Speech Tagging for Historical English. In *Proceedings of the 2016 Conference of the North American Chapter of the Association for Computational Linguistics: Human Language Technologies*. Association for Computational Linguistics, pages 1318–1328.

Shaodian Zhang, Tian Kang, Xingting Zhang, Dong Wen, Noémie Elhadad, Jianbo Lei. 2016. Speculation Detection for Chinese Clinical Notes: Impacts of Word Segmentation and Embedding Models. *Journal of Biomedical Informatics*, 60:334-341.

Investigating Domain-Specific Information for
Neural Coreference Resolution on Biomedical Texts

Hai-Long Trieu[1], Nhung T. H. Nguyen[2], Makoto Miwa[1,3] and Sophia Ananiadou[2]
[1]Artificial Intelligence Research Center (AIRC),
National Institute of Advanced Industrial Science and Technology (AIST), Japan
[2]National Centre for Text Mining, University of Manchester, United Kingdom
[3]Toyota Technological Institute, Japan
long.trieu@aist.go.jp, makoto-miwa@toyota-ti.ac.jp
{nhung.nguyen, Sophia.Ananiadou}@manchester.ac.uk

Abstract

Existing biomedical coreference resolution systems depend on features and/or rules based on syntactic parsers. In this paper, we investigate the utility of the state-of-the-art general domain neural coreference resolution system on biomedical texts. The system is an end-to-end system without depending on any syntactic parsers. We also investigate the domain specific features to enhance the system for biomedical texts. Experimental results on the BioNLP Protein Coreference dataset and the CRAFT corpus show that, with no parser information, the adapted system compared favorably with the systems that depend on parser information on these datasets, achieving 51.23% on the BioNLP dataset and 36.33% on the CRAFT corpus in F1 score. In-domain embeddings and domain-specific features helped improve the performance on the BioNLP dataset, but they did not on the CRAFT corpus.

1 Introduction

Deep neural systems have recently achieved the state-of-the-art performance on coreference resolution tasks in the general domain (Clark and Manning, 2016; Wiseman et al., 2016; Lee et al., 2017). These systems do not heavily rely on manual features since the networks automatically build advanced features from the input. Such an attribute has made deep neural systems preferable to traditional manual feature-based systems.

In the biomedical domain, coreference information has been shown to enhance the performance of entity and event extraction (Miwa et al., 2012; Choi et al., 2016a). Most of work in this domain use rule-based or hybrid approaches (Nguyen et al., 2011, 2012; Miwa et al., 2012; D'Souza and Ng, 2012; Li et al., 2014; Choi et al., 2016b; Cohen et al., 2017). These systems rely on syntactic parsers to extract hand-crafted features and rules, e.g., rules based on predicate argument structure (Nguyen et al., 2012; Miwa et al., 2012) or features based on syntax trees (D'Souza and Ng, 2012). These rules are designed specifically for each type of coreference, such as noun phrases, relative pronouns, and non-relative pronouns. Moreover, several rules are restricted to specific entities of the training corpus, e.g., protein entities for the BioNLP Protein Coreference dataset (Nguyen et al., 2011).[1]

Given the fact that deep learning methods can produce the state-of-the-art performance on general texts, we are motivated to apply such methods to biomedical texts. We therefore raise three research questions in this paper:

- How does a general domain neural system with no parser information perform on biomedical domain?
- How we can incorporate domain-specific information into the neural system?
- Which performance range the system is in comparison with existing systems?

In order to address these questions, we directly apply the end-to-end neural coreference resolution system by Lee et al. (2017) (Lee2017) to biomedical texts. We then investigate domain specific features such as domain-specific word embeddings, grammatical number agreements between mentions, i.e., mentions are singular or plural, and agreements of MetaMap (Aronson and Lang, 2010) entity tags of mentions. These features do not rely on any syntactic parsers. Moreover, these features are also general for any biomedical corpora and not restricted to the corpora we use.

---

[1]http://2011.bionlp-st.org/home/
protein-gene-coreference-task

Proceedings of the BioNLP 2018 workshop, pages 183–188
Melbourne, Australia, July 19, 2018. ©2018 Association for Computational Linguistics

We evaluated the Lee2017 system on two datasets: the BioNLP Protein Coreference dataset (Nguyen et al., 2011) and CRAFT (Cohen et al., 2017). Our experimental results have revealed that the system could achieve reasonable performance on both corpora. The system outperformed several systems on the BioNLP dataset that employed rule-based (Choi et al., 2016b) and conventional machine learning methods (Nguyen et al., 2011) using parser information, although it was not competitive with the state-of-the-art systems. Integrating in-domain embeddings and domain-specific features into the deep neural system improved the performance of both mention detection and mention linking on the BioNLP dataset, but the integration could not enhance the performance on the CRAFT corpus.

2 Methods

In this section, we briefly introduce the baseline Lee2017 system (Lee et al., 2017) and present domain-specific features to adapt the system to biomedical texts.

2.1 Baseline System

The baseline Lee2017 system treats all spans up to the maximum length as mention candidates. Each mention candidate is represented as a concatenated vector of the first word, the last word, the soft head word, and the span length embeddings. The embeddings for the first and last words are calculated from the outputs of LSTMs (Hochreiter and Schmidhuber, 1997), while those for soft head word are calculated from the weighted sum of the embeddings of words in the span using an attention mechanism (Bahdanau et al., 2014). These candidates are ranked based on their mention scores s_m calculated as follows:

$$s_m(i) = w_m \cdot \text{FFNN}_m(g_i), \qquad (1)$$

where w_m is a weight vector, FFNN denotes a feed-forward neural network, and g_i is the vector representation of a mention i.

After mentions are decided, the system resolves coreference by linking mentions back to their antecedent using antecedent scores s_a calculated as:

$$s_a(i, j) = w_a \cdot \text{FFNN}_a([g_i, g_j, g_i \circ g_j, \phi(i, j)]), \qquad (2)$$

where $\circ$ denotes an element-wise multiplication and $\phi(i, j)$ represents the feature vector between the two mentions.

2.2 Domain-specific features

We incorporate the following domain-specific features to enhance the baseline system.

In-domain word embeddings: The input word embeddings play an important role in deep learning. Instead of using embeddings trained on general domains, e.g., word embeddings provided with the word2vec tool (Mikolov et al., 2013), we use 200-dimensional embeddings trained on the whole PubMed and PubMed Central Open Access subset (PMC) with a window size of 2 (Chiu et al., 2016).

Grammatical numbers: We check mentions' grammatical numbers, i.e., whether each mention is singular or plural. A mention is singular if its part-of-speech tag is NN or if it is one of the five singular pronouns: *it*, *its*, *itself*, *this*, and *that*. A mention is plural if its part-of-speech tag is NNS or if it is one of the seven plural pronouns: *they*, *their*, *theirs*, *them*, *themselves*, *these*, and *those*.

MetaMap entity tags: We employ MetaMapLite[2] to identify all possible entities according to the UMLS semantic types.[3] In cases that MetaMapLite assigns multiple semantic types for each entity, we take into account all of the types.

The grammatical numbers and MetaMap entity tags are incorporated into the network as follows. We firstly pre-processed the input and assigned token-based values for each type of features. For example, a token may have "singular", "plural", or "unknown" as the number attribute. Meanwhile, the MetaMap entity tags are distributed to each token with their position information chosen from "Begin" and "Inside". These features are finally encoded as a binary vector of $\phi(i, j)$ in Equation 2 that shows whether two mentions i and j has the number agreement and whether they share the same MetaMap semantic type.

3 Experiments

3.1 Data

We employed two biomedical corpora: BioNLP Protein Coreference dataset (Nguyen et al., 2011) and CRAFT (Cohen et al., 2017). The BioNLP dataset consists of 1,210 PubMed abstracts selected from the GENIA-MedCo coreference corpus. CRAFT (Cohen et al., 2017) provides coref-

---

[2]https://metamap.nlm.nih.gov/
MetaMapLite.shtml

[3]https://metamap.nlm.nih.gov/Docs/
SemanticTypes_2013AA.txt

	BioNLP	CRAFT
Training set (docs)	800	54
Development set (docs)	150	6
Test set (docs)	260	7
Avg. sent. per doc	9.15	274.75
Avg. words per doc	258.00	8,060.85
Vocabulary size	15,900	27,405

Table 1: Characteristics of BioNLP and CRAFT.

erence annotations of 67 full papers extracted from PMC. While BioNLP focusses on protein/gene coreference, CRAFT covers a wider range of coreference relations such as events, pronominal anaphora, noun phrases, verbs, and nominal premodifiers corefernce. In the CRAFT corpus, coreference is divided into two types: identity chains (a set of base noun phrases and/or appositives that refer to the same thing in the world) and appositive relations (two noun phrases that are adjacent and not linked by a copula). We use only the identity chains.

The BioNLP dataset was officially divided into training, development, and test sets. Regarding CRAFT, we randomly divided it into three subsets in a ratio of 8:1:1 for training, development, and test, respectively. Detailed characteristics of the two corpora as well as these three sets are reported in Table 1. It is noticeable that CRAFT is a corpus of full papers, which makes it more challenging for text mining tools than the BioNLP dataset—a corpus of abstracts (Cohen et al., 2010).

3.2 Settings

We first directly applied the Lee2017 system to the corpora. Lee2017 used two pretrained embeddings in general domains provided by Pennington et al. (2014) and Turian et al. (2010), and all default features such as speaker, genre, and distance.

To train the Lee2017 system, we employed the same hyper-parameters as reported in Lee et al. (2017) except for a threshold ratio. Although Lee2017 used the ratio $\lambda = 0.4$ to reduce the number of mentions from the list of candidates, we tuned it on the BioNLP development set and used $\lambda = 0.7$.

We then investigate the impact of each feature on the biomedical texts by preparing the following four systems:

- Lee2017: general embeddings, speaker, genre, and distance features

BioNLP	Prec.	Rec.	F1 (%)
Lee2017	81.15	63.81	71.44
PubMed	81.01	66.12	72.81
PubMed-SG	79.23	65.73	71.85
PubMed+MM	80.41	**67.17**	73.20
PubMed+Num	**81.91**	66.31	**73.29**
PubMed+MM+Num	81.04	66.69	73.17

CRAFT	Prec.	Rec.	F1 (%)
Lee2017	70.76	48.71	57.70
PubMed	70.93	46.90	56.46
PubMed-SG	71.98	**50.24**	**59.18**
PubMed+MM	71.11	47.91	57.25
PubMed+Num	**72.79**	42.55	53.70
PubMed+MM+Num	71.60	45.00	55.27

Table 2: Results of mention detection on the development set of BioNLP and CRAFT. The highest numbers are shown in bold.

- PubMed: biomedical embeddings, same features as Lee2017
- PubMed-SG: PubMed with no speaker and genre features
- PubMed+*: PubMed with the MetaMap feature (MM) and/or the grammatical number feature (Num).

For evaluation, we calculated precision, recall, and F1 on MUC, B^3, and $CEAF_{\phi 4}$ using the CoNLL scorer (Pradhan et al., 2014). For the BioNLP dataset, we also employed the scorer provided by the shared task organisers to make fair comparisons with previous work. We reported the performance on two sub-tasks: (1) mention detection, i.e., to identify coreferent mentions, such as named entities, prepositions or noun phrases, and (2) mention linking, i.e., to link these mentions if they refer to the same thing. The result of the first task affects that of the second one.

3.3 Results

Results on the development sets of the two corpora are presented in Table 2 for mention detection and Table 3 for mention linking (see Appendix A for detailed scores in different metrics).

Regarding the BioNLP dataset, the Lee2017 system performed reasonably well even when it did not use any domain-specific features. Replacing general embeddings by the biomedical ones improved F1 score in general (Lee2017 v.s. PubMed). Removing speaker and genre features (-SG) did not help enhance the performance.

System	BioNLP	CRAFT
Lee2017	61.25	33.85
PubMed	62.51	33.92
PubMed-SG	61.47	**34.85**
PubMed+MM	**63.41**	33.91
PubMed+Num	63.16	31.28
PubMed+MM+Num	63.12	32.77

Table 3: Average F1 scores (%) of mention linking on the development set of BioNLP and CRAFT.

Adding MetaMap's tags (+MM) or the number feature (+Num) produced slightly better scores in comparison to PubMed. However, combining the two features at the same time was not as effective as expected. Among the proposed features, the agreement on MetaMap entity tags (+MM) was the strongest one on the BioNLP dataset.

The impact of the features was quite different on the CRAFT corpus. As shown in Table 2, introducing biomedical embeddings (PubMed) show slightly worse F1 score on mention detection than Lee2017 but it also show a slight improvement on mention linking. Removing speaker and genre features (-SG) boosted the performance. However, adding domain-specific features all harmed the performance. As a result, PubMed-SG showed the best score on the CRAFT development set.

Results in Tables 2 and 3 justify the fact that the CRAFT corpus is more challenging than the BioNLP dataset. The scores of the experimented systems on the CRAFT corpus were always lower than those on the BioNLP dataset. This is reasonable because (1) CRAFT consists of full papers that are significantly longer than abstracts, (2) it covers a wide range of anaphors, and (3) its identity chains can be arbitrarily long.

We applied the best performing system on each development set, i.e., PubMed+MM for BioNLP and PubMed-SG for CRAFT, to its test set, and reported the results in Tables 4 and 5 with showing the performance in previous work for comparison. Table 4 reveals that the neural system outperformed five systems that used SVM and rule-based approaches including the best system on the shared task, and the system could compete with Nguyen et al. (2012)'s. Meanwhile, on the CRAFT corpus (Table 5), we could only produce better performance than the general state-of-the-art system, especially due to the low precision.

System	Prec	Rec	F1 (%)
TEES (BioNLP ST)	67.2	14.4	23.8
ConcordU (BioNLP ST)	63.2	19.4	29.7
UZurich (BioNLP ST)	55.5	21.5	31.0
UUtah (BioNLP ST)	73.3	22.2	34.1
Choi et al. (2016b)	46.3	50.0	48.1
PubMed+MM	55.6	47.5	51.2
Nguyen et al. (2012)	50.2	52.5	51.3
Miwa et al. (2012)	62.7	50.4	55.9
D'Souza and Ng (2012)	55.6	67.2	60.9

Table 4: Results of mention linking on the test set of the BioNLP dataset. The F-scores are in ascending order.

System	Prec.	Rec.	F1
General state-of-the-art	0.93	0.08	0.14
Rule-based	0.78	0.29	0.42
Union of the two output	0.78	0.35	0.46
PubMed-SG	0.44	0.31	0.36

Table 5: B^3 scores of mention linking on the CRAFT test set in comparison with the three systems by Cohen et al. (2017). This is not a fair comparison as our system only addressed identity chains and the test set is different from theirs.

4   Conclusion

We have applied a neural coreference system to biomedical texts and incorporated domain-specific features to enhance the performance. Experimental results on two biomedical corpora, the BioNLP dataset and the CRAFT corpus, have shown that (1) the neural system performed reasonably well with no parser information, (2) the in-domain embeddings and domain-specific features did not consistently perform well on the two corpora, and (3) the system could attain better performance than several rule-based and traditional machine learning-based systems on the BioNLP dataset.

As future work, we would like to investigate feature representations to make input features useful to a target domain. We will also incorporate rules in the existing systems into the network.

Acknowledgments

This research has been carried out with funding from AIRC/AIST and results obtained from a project commissioned by the New Energy and Industrial Technology Development Organization (NEDO).

References

Alan R Aronson and François-Michel Lang. 2010. An overview of MetaMap: historical perspective and recent advances. *Journal of the American Medical Informatics Association*, 17(3):229–236.

Dzmitry Bahdanau, Kyunghyun Cho, and Yoshua Bengio. 2014. Neural machine translation by jointly learning to align and translate. *arXiv preprint arXiv:1409.0473*.

Billy Chiu, Gamal K. O. Crichton, Anna Korhonen, and Sampo Pyysalo. 2016. How to Train good Word Embeddings for Biomedical NLP. In *Proceedings of the 15th Workshop on Biomedical Natural Language Processing*, pages 166–174.

Miji Choi, Haibin Liu, William Baumgartner, Justin Zobel, and Karin Verspoor. 2016a. Coreference resolution improves extraction of biological expression language statements from texts. *Database*, 2016:baw076.

Miji Choi, Justin Zobel, and Karin Verspoor. 2016b. A categorical analysis of coreference resolution errors in biomedical texts. *Journal of biomedical informatics*, 60:309318.

Kevin Clark and Christopher D. Manning. 2016. Improving coreference resolution by learning entity-level distributed representations. *CoRR*, abs/1606.01323.

K. Bretonnel Cohen, Helen L. Johnson, Karin Verspoor, Christophe Roeder, and Lawrence E. Hunter. 2010. The structural and content aspects of abstracts versus bodies of full text journal articles are different. *BMC Bioinformatics*, 11(1):492.

K. Bretonnel Cohen, Arrick Lanfranchi, Miji Jooyoung Choi, Michael Bada, William A. Baumgartner, Natalya Panteleyeva, Karin Verspoor, Martha Palmer, and Lawrence E. Hunter. 2017. Coreference annotation and resolution in the colorado richly annotated full text (craft) corpus of biomedical journal articles. *BMC Bioinformatics*, 18(1):372.

Jennifer D'Souza and Vincent Ng. 2012. Anaphora resolution in biomedical literature: a hybrid approach. In *BCB*, pages 113–122. ACM.

Sepp Hochreiter and Jürgen Schmidhuber. 1997. Long short-term memory. *Neural Comput.*, 9(8):1735–1780.

Kenton Lee, Luheng He, Mike Lewis, and Luke Zettlemoyer. 2017. End-to-end neural coreference resolution. In *Proceedings of the 2017 Conference on Empirical Methods in Natural Language Processing*, pages 188–197, Copenhagen, Denmark. Association for Computational Linguistics.

Lishuang Li, Liuke Jin, Zhenchao Jiang, Jing Zhang, and Degen Huang. 2014. Coreference resolution in biomedical texts. In *BIBM*, pages 12–14. IEEE Computer Society.

Tomas Mikolov, Kai Chen, Greg Corrado, and Jeffrey Dean. 2013. Efficient Estimation of Word Representations in Vector Space. *CoRR*, abs/1301.3781.

Makoto Miwa, Paul Thompson, and Sophia Ananiadou. 2012. Boosting automatic event extraction from the literature using domain adaptation and coreference resolution. *Bioinformatics*, 28(13):1759–1765.

N. L. T. Nguyen, J.-D. Kim, and J. Tsujii. 2011. Overview of bionlp 2011 protein coreference shared task. In *Proceedings of the BioNLP Shared Task 2011 Workshop*, pages 74–82, Portland, Oregon, USA. Association for Computational Linguistics.

Ngan Nguyen, Jin-Dong Kim, Makoto Miwa, Takuya Matsuzaki, and Junichi Tsujii. 2012. Improving protein coreference resolution by simple semantic classification. *BMC Bioinformatics*, 13(1):304.

Jeffrey Pennington, Richard Socher, and Christopher D Manning. 2014. Glove: Global Vectors for Word Representation. In *EMNLP*, volume 14, pages 1532–1543.

Sameer Pradhan, Xiaoqiang Luo, Marta Recasens, Eduard Hovy, Vincent Ng, and Michael Strube. 2014. Scoring Coreference Partitions of Predicted Mentions: A Reference Implementation. In *Proceedings of the 52nd Annual Meeting of the Association for Computational Linguistics (Volume 2: Short Papers)*, pages 30–35. Association for Computational Linguistics.

Joseph Turian, Lev Ratinov, and Yoshua Bengio. 2010. Word Representations: A Simple and General Method for Semi-supervised Learning. In *Proceedings of the 48th Annual Meeting of the Association for Computational Linguistics*, ACL '10, pages 384–394.

Sam Wiseman, Alexander M. Rush, and Stuart M. Shieber. 2016. Learning global features for coreference resolution. *CoRR*, abs/1604.03035.

A   Detailed results

We report detailed results of mention linking on the development set of the two corpora in Table 6 and Table 7. Due to the long running time of the scorer, we were not able to report CEAF$_{\phi4}$ scores for CRAFT.

System	MUC			B^3			CEAF$_{\phi4}$			Avg. F1 (%)
	Prec.	Rec.	F1	Prec.	Rec.	F1	Prec.	Rec.	F1	
Baseline	65.31	45.03	53.30	71.50	50.27	59.03	77.06	66.54	71.41	61.25
PubMed	65.44	47.04	54.74	71.12	51.85	59.98	77.25	68.85	72.81	62.51
PubMed-SG	63.02	46.58	53.57	68.72	51.72	59.02	76.11	68.01	71.83	61.47
PubMed+MM	66.17	48.30	55.84	71.62	52.95	60.89	76.70	70.53	73.49	63.41
PubMed+Num	66.81	47.83	55.75	72.27	52.23	60.64	78.15	68.63	73.08	63.16
PubMed+MM+Num	65.73	47.37	55.06	71.68	52.66	60.72	77.53	70.04	73.59	63.12

Table 6: Results of mention linking on the BioNLP development set.

System	MUC			B^3			Avg. F1 (%)
	Prec.	Rec.	F1	Prec.	Rec.	F1	
Baseline	45.46	27.17	34.02	44.29	27.17	33.68	33.85
PubMed	47.36	27.34	34.67	44.89	26.30	33.17	33.92
PubMed-SG	46.04	28.49	35.20	43.33	28.67	34.50	34.85
PubMed+MM	46.22	27.37	34.38	43.70	27.09	33.44	33.91
PubMed+Num	47.31	23.40	31.31	48.70	23.01	31.25	31.28
PubMed+MM+Num	46.85	25.41	32.95	45.78	25.30	32.59	32.77

Table 7: Results of mention linking on the CRAFT development set.

Toward Cross-Domain Engagement Analysis in Medical Notes

Adam Faulkner [*]
Grammarly
NY, NY, USA
adam.faulkner@grammarly.com

Sara Rosenthal
IBM Research
Yorktown Heights, NY, USA
sjrosenthal@us.ibm.com

Abstract

We present a novel annotation task evaluating a patient's *engagement* with their health care regimen. The concept of engagement supplements the traditional concept of adherence with a focus on the patient's affect, lifestyle choices, and health goal status. We describe an engagement annotation task across two patient note domains: traditional clinical notes and a novel domain, care manager notes, where we find engagement to be more common. The annotation task resulted in a κ of .53, suggesting strong annotator intuitions regarding engagement-bearing language. In addition, we report the results of a series of preliminary engagement classification experiments using domain adaptation.

1 Introduction

The recent trend in medicine toward health promotion, rather than disease management, has forefronted the role of patient behavior and lifestyle choices in positive health outcomes. Social-cognitive theories of health-promotion (Maes and Karoly, 2005; Bandura, 2005) stress patient self-monitoring of life-style choices, goal adoption, and the enlistment of self-efficacy beliefs as health promotive. We call this cluster of behavioral characteristics patient *engagement*. Traditional strategies of patient follow-up have also been affected by this trend: healthcare providers increasingly employ "care managers" (CMs) to monitor patient well-being and adherence to physician-recommended changes in health behavior—i.e., engagement. In this paper, we present an annotation schema for (lack of) engagement in CM notes (CMNs) and generalize the schema to the related

domain of electronic health records (EHRs). Our high-level research questions are:

(1) Is the concept of engagement sufficiently well-defined that annotators can recognize the concept across text domains with an acceptable level of agreement?

(2) Can the annotations produced in (1) be used to classify engagement-bearing language across text domains?

In section 3, we report the results of our exploration of (1), describing an annotation task involving ~ 6500 CMN and EHR sentences that resulted in an average κ of .53. In sections 4 and 5 we address (2) and report the results of several classification experiments that ablate classes of features and use domain adaptation to adapt these features to the CM and EHR target domains.

2 Related Work

The notion of patient engagement explored here is inspired by the self-regulation paradigm of (Bandura, 2005; Leventhal et al., 2012; Mann et al., 2013), where a patient's successful completion of health-related goals is predicated on their ability to "self-regulate", i.e., to plan and execute actions that promote attaining those goals, and their ability to maintain a positive attitude toward self-care. We are also aligned with the more recent work of Higgins et al. (2017) whose definition of engagement includes a "desire and capability to actively choose to participate in care".

NLP approaches assessing doctor compliance include Hazelhurst et al. (2005) who evaluate notes for doctor compliance to tobacco cessation guidelines and Mishra et al. (2012) who assess ABCs protocol compliance in discharge summaries.

---

[*] work completed at IBM

189

Proceedings of the BioNLP 2018 workshop, pages 189–193
Melbourne, Australia, July 19, 2018. ©2018 Association for Computational Linguistics

Label	Description	Examples
Engagement with care	The patient is engaged in their well-being by describing/exhibiting healthy behavior, positive outlook, and social ties.	"Patient disappointed by lack of weight loss but is just beginning exercise regimen"; "Patient joined book club."
Engagement with CM	Adherence to a doctor or CM instruction or understanding of CM advice.	"Patient verbalized understanding"; "Patient confided that she has gaps in nitroglycerin use."
Lack of engagement with care	Lack of engagement by using language suggestive of non-adherence to guidelines, health-adverse behavior, lack of social ties, or negative impression of patient self-care.	"White female, disheveled appearance"; "Patient admits to 'sedentary' lifestyle."
Lack of engagement with CM	Non-adherence to a prescribed instruction or a negative response to interaction.	"Patient rude during call"; "Patient angrily refused further outreach."
CM Advice	CM advice or suggestion	"I suggested he watch his diet and increase exercise"
Other	Default label to be chosen when no other label fits.	"Patient has a history of atrial fibrillation on corticosteroids"; "Chest is clear with no crackles."

Table 1: Annotation labels with descriptions and anecdotal examples. We use the term CM to describe both the para-professionals interacting with patients in CM notes and the physicians in EHRs.

While there exists work dealing with sentiment in clinical notes, such as positive or negative affect (Ghassemi et al., 2015) and speculative language (Cruz Díaz et al., 2012), (lack of) engagement cannot be reduced to sentiment. Lack-of-engagement-bearing language, for example, can also contain positive sentiment, e.g., *patient is feeling better so she has stopped taking her medication*. We include sentiment in our feature set, as described in Section 4.

The most closely related work is Topaz et al. (2017) who developed a document-level discharge note classification model that identifies the adherence of a patient in the discharge note. Their annotation task differs from ours, however, as they focus only on lack of adherence, specifically, towards medication, diet, exercise, and medical appointments. We also distinguish the targets of both engagement and lack of engagement by allowing annotators to identify either the CM or the care itself as the target.

3 Annotation Task and Data

The majority of our data consists of CMNs generated by a care manager service located in Florida, USA. CMs typically contact patients via phone to inquire into the patient's status with respect to health goals and enter the resulting information into the structured sections of a reporting tool. In addition, CMs note their impressions of the patient in a note as unstructured text, which we use here. To expand the domain scope of the task, we included EHR notes from the i2b2 Heart Disease Risk Factors Challenge Data Set (Stubbs and Uzuner, 2015; Stubbs et al., 2015), which includes notes dealing with diabetic patients at risk for Coronary Artery Disease (CAD). All notes were

annotated in the same manner regardless of source.

3.1 Annotation Guidelines

Table 1 includes descriptions of the annotation labels along with anecdotal examples of each label type (original sentences are excluded due to privacy constraints[1]). Annotators were allowed to choose more than one label for each sentence, or no label at all (considered *other*). Our schema captures three different label classes: *engagement*, *lack of engagement*, and *cm advice*. We included *cm advice* because it can provide an indication that the next sentence should be classified as (lack of) engagement. We initially explored "barrier" language (e.g. *patient could not get to his appt because he didn't have a car*) as this can be indicative of lack of engagement, however, we found it to be too rare to include in the annotation tasks.

3.2 Annotation Challenges

Our first challenge was encoding a distinction between engagement and the more familiar concept of patient "adherence" (Vermeire et al., 2001; Topaz et al., 2017) in the annotation guidelines. While engagement-bearing language can include adherence-bearing language (e.g., *is monitoring blood sugar*, *made follow-up appointment*), the reverse is often not the case: Engagement-bearing language can include mentions of social ties (e.g., *discusses struggles to lose weight with sister*) and positive or negative evaluations of health-related goals (e.g., *patient was irritable when asked about efforts to reduce smoking*), neither of which involve adherence per se. By annotating such examples as engagement-bearing, we capture "self-

---

[1] All examples provided throughout the paper are anecdotal

efficacy beliefs," which theories of patient self-regulation (Bandura, 1998, 2005) have suggested are predictive of health goal attainment.

An additional distinction that emerged during the annotation process involved the target of the engagement-bearing language: Is the patient (not) engaged with the CM or with the care itself? This distinction is evident in sentences that display a lack of engagement with care but a level of engagement with the CM. For example, in the sentence *He appeared cheerful in our interactions and admitted that he has not been exercising daily*, the patient is confiding in their CM (engagement) that they are not pursuing their health goals (lack of engagement). By allowing annotators to annotate such sentences as both engaged with the CM but unengaged with care we were able to exclude sentences that contained internally inconsistent engagement-bearing language from our data.

Another challenge involved the frequent use of "canned language" in the CM data, or language that does not report the CM's interactions with the patient but is used to meet some reporting criterion recommended by the health-care provider. For example, *Patient is scheduled for follow up appointment in two weeks*, is a frequently occurring canned language. Thus, we excluded common canned language sentences from the data.

3.3 Data Statistics

After several initial pilot rounds inter-annotator agreement for our six annotators on a final pilot round of 200 sentences (100 from each source) ranged from .46 to .66 among the annotators with an overall average of .53 (using Cohen's κ), indicating moderate to substantial agreement (McHugh, 2012).

4011 CMN sentences were annotated, extracted from $\sim 10,000$ unique CMNs. In order to broaden the range of language in our data, 2561 EHR sentences were annotated, with an equal number of sentences drawn from the three patient cohorts included in the i2b2 data. For each EHR, we restricted our annotation effort to sections that were more likely to include engagement-bearing language, specifically, the *social history*, *family history*, *personal medical history*, and *history of the present illness* sections. Table 2 shows the label distribution of the annotated data relative to note source. Although we allowed the annotators to differentiate between engagement/lack of engage-

Source	Engage	No Engage	Advice	Other
EHRs	114	56	15	2376
CMNs	395	172	140	3304
Total	509	228	155	5680

Table 2: Label distribution relative to note type for all annotated sentence data.

ment with care or the CM, we ultimately conflated these two categories into one for our experiments.

4 Method

Given the small size of our data we elected to use a feature-engineering-based approach along with a discriminative classification algorithm in our experiments. Our features can be divided into five categories: *lexico-syntactic*, *lexical-count*, *sentiment*, *medical*, and *embeddings*.

Lexico-syntactic. Standard NLP features for text-classification such as n-grams and part-of-speech (POS) tags, along with dependency tuples (De Marneffe and Manning, 2008) with either the governor or dependent generalized to its POS.

Lexical-count. Frequency-based features such as sentence length, min and max word length, and number of out of vocabulary words.

Sentiment. We ran two sentiment classifiers over the data (Socher et al., 2013; Hutto and Gilbert, 2014) and included the resulting tags as features. In addition, we developed "comply word" features by inducing a lexicon based on WordNet- (Fellbaum, 1998) and Unified Medical Language System (UMLS)-based[2] synonym expansion of seed words such as "take" and "decline."

Medical. Using the MetaMap[3] tool, we generated Concept-Unique Identifiers (CUIs) for any medical concepts in the sentence. We also included both the "preferred names" and semantic types returned by UMLS for each concept

Embeddings. We extracted term-term, CUI-CUI and term-CUI co-occurrences pairs from a large medical corpus and used `wordtovecf`[4] (Levy and Goldberg, 2014) to learn embeddings from this co-occurrence dataset. We generated the mean of the embeddings for all content-words and CUIs in the sentence as a feature.

5 Experiments

All experiments were performed using an SVM classifier with a linear kernel basis function and

---

[2] http://www.nlm.nih.gov/research/umls/
[3] http://metamap.nlm.nih.gov/
[4] http://bitbucket.org/yoavgo/word2vecf

	experiment	CM			EHR		
		eng	lack	other	eng	lack	other
ALL	n-grams	18.4	25.1	91.3	7.1	3.3	95.4
	+embeddings	18.0	24.9	91.3	7.2	3.3	95.4
	+lexico-synt	22.7	20.7	90.2	7.0	3.2	94.6
	+lexical counts	21.7	21.9	89.2	9.1	3.1	93.1
	+sentiment	19.8	22.9	88.7	9.1	6.1	92.8
	+medical	22.3	24.3	89.3	8.4	8.6	93.6
	all	24.3	21.7	89.1	9.5	6.2	93.1
CM	n-grams	27.4	26.3	91.2	9.2	12.7	93.7
	+embeddings	27.5	26.8	91.3	8.5	12.7	93.7
	+lexico-synt	27.9	25.2	88.7	7.4	9.8	91.7
	+lexical counts	27.4	27.5	87.5	8.4	12.1	89.6
	+sentiment	27.6	**29.4**	87.4	7.9	11.3	89.4
	+medical	28.9	28.6	87.8	10.6	10.4	90.2
	all	**29.6**	25.4	87.1	10.8	**13.0**	89.6
EHR	n-grams	9.4	0.0	92.1	0.0	0.0	96.3
	+embeddings	9.4	0.0	92.1	0.0	0.0	96.3
	+lexico-synt	12.7	1.1	91.9	5.5	6.5	96.4
	+lexical counts	12.0	3.2	91.9	5.6	6.3	96.3
	+sentiment	10.2	3.2	91.9	6.9	6.3	96.5
	+medical	15.6	9.6	91.7	**17.0**	6.1	96.3
	all	11.1	7.2	91.7	14.4	6.3	96.2

Table 3: F-score results of ablation experiments for Engagement (*eng*) and Lack of Engagement (*lack*). Row headers refer to training sets and column headers refer to test sets. The best F-score for each test set is shown in **bold.**

experiment	CM			EHR		
	eng	lack	other	eng	lack	other
n-grams	21.0	25.7	91.6	9.2	3.3	95.8
+embeddings	21.1	25.5	91.6	10.4	3.3	95.8
+lexico-synt	23.0	22.9	89.1	9.2	3.2	93.4
+lexical count	20.2	24.2	89.1	9.9	**15.0**	91.5
+sentiment	23.0	**26.1**	88.8	7.8	2.7	93.0
+medical	**23.4**	22.9	88.8	9.9	5.1	93.2
all	22.3	23.2	89.0	**13.3**	12.8	91.4

Table 4: F-score results of DA experiments. The best F-score for each test set is shown in **bold.**

one-vs-rest multiclass classification strategy as implemented in `scikit-learn`.[5] To deal with the skew in class distribution we experimented with both over- and under-sampling but got our best performance by simply adjusting class weights to be inversely proportional to class frequencies. Given the relatively small size of our data we used 5-fold cross-validation throughout. We also conflated *cm advice* with *other* to boost performance. We show F-score results for all three classes, but our analysis will focus on (lack of) engagement since *other* is trivially high-performing due to the massive data skew.

In our first set of experiments we examined the impact of training and testing on EHRs and CMNs individually, as well as together, while ablating the feature classes described in section 4. As shown in Table 3, all feature classes seem to help the model, but sentiment helps more for predicting lack of engagement in the CMNs while medical features help more for predicting lack of engagement in the EHRs. These experiments show that a CMN-trained model can perform well on EHRs. The best result for lack of engagement occurs when training on CM notes, with an F-score of 13.0.

The results in Table 3 encouraged us to apply domain adaptation (DA) to improve the results of

---

[5]`http://scikit-learn.org`

the smaller dataset (EHRs) while also taking advantage of the larger dataset (CMNs) by considering EHRs to be "in-domain" and CMNs to be "out-of-domain". As this is still preliminary work, we started with a simple, yet effective DA strategy: the feature representation transformation procedure described in Daumé III (2007). Table 4 shows the results using DA. In the EHRs, where there is less data, on average DA provided an improvement, particularly for lack of engagement.

6 Conclusion

In this paper we presented an annotation schema that captures engagement in CMNs and EHRs. We described the challenges of developing an annotation schema for a subjective task and show that annotators achieved moderate to high agreement in our final task. We annotated 6,572 sentences for (lack of) engagement and show preliminary results of a classification experiment on our dataset using feature ablation and domain adaptation. Our results are promising, showing that both features and domain adaptation are useful. However, they remain preliminary due to the rarity of (lack of) engagement labels. In future work, we plan to explore transfer learning to increase the size of our data, which in turn will allow use to explore deep learning approaches to this task.

Acknowledgements

Deidentified clinical records used in this research were provided by the i2b2 National Center for Biomedical Computing funded by U54LM008748 and were originally prepared for the Shared Tasks for Challenges in NLP for Clinical Data organized by Dr. Ozlem Uzuner, i2b2 and SUNY. Research is based on care manager data provided by Orlando Health, a Florida-based community of hospitals.

References

Albert Bandura. 1998. Health promotion from the perspective of social cognitive theory. *Psychology and health*, 13(4):623–649.

Albert Bandura. 2005. The primacy of self-regulation in health promotion. *Applied Psychology*, 54(2):245–254.

Noa P Cruz Díaz, Manuel J Maña López, Jacinto Mata Vázquez, and Victoria Pachón Álvarez. 2012. A machine-learning approach to negation and speculation detection in clinical texts. *Journal of the Association for Information Science and Technology*, 63(7):1398–1410.

Hal Daumé III. 2007. Frustratingly easy domain adaptation. *ACL 2007*, page 256.

Marie-Catherine De Marneffe and Christopher D Manning. 2008. The stanford typed dependencies representation. In *Coling 2008: proceedings of the workshop on cross-framework and cross-domain parser evaluation*, pages 1–8. Association for Computational Linguistics.

Christiane Fellbaum. 1998. A semantic network of english verbs. *WordNet: An electronic lexical database*, 3:153–178.

Mohammad M Ghassemi, Roger G Mark, and Shamim Nemati. 2015. A visualization of evolving clinical sentiment using vector representations of clinical notes. In *Computing in Cardiology Conference (CinC), 2015*, pages 629–632. IEEE.

Brian Hazlehurst, H. Robert Frost, Dean F. Sittig, and Victor J. Stevens. 2005. Mediclass: A system for detecting and classifying encounter-based clinical events in any electronic medical record. *Journal of the American Medical Informatics Association*, 12(5):517–529.

Tracy Higgins, Elaine Larson, and Rebecca Schnall. 2017. Unraveling the meaning of patient engagement: A concept analysis. *Patient Education and Counseling*, 100(1):30 – 36.

C.J. Hutto and Eric Gilbert. 2014. Vader: A parsimonious rule-based model for sentiment analysis of social media text. In *Eighth International Conference on Weblogs and Social Media (ICWSM-14)*. *Available at (20/04/16) http://comp. social. gatech. edu/papers/icwsm14. vader. hutto. pdf*.

Howard Leventhal, Ian Brissette, and Elaine A. Leventhal. 2012. The common-sense model of self-regulation of health and illness. In *The self-regulation of health and illness behaviour*, pages 56–79. Routledge.

Omer Levy and Yoav Goldberg. 2014. Dependency-based word embeddings. In *Proceedings of the 52nd Annual Meeting of the Association for Computational Linguistics (Volume 2: Short Papers)*, volume 2, pages 302–308.

Stan Maes and Paul Karoly. 2005. Self-regulation assessment and intervention in physical health and illness: A review. *Applied Psychology*, 54(2):267–299.

Traci Mann, Denise De Ridder, and Kentaro Fujita. 2013. Self-regulation of health behavior: social psychological approaches to goal setting and goal striving. *Health Psychology*, 32(5):487.

Mary L. McHugh. 2012. Interrater reliability: the kappa statistic. In *Biochemia medica*.

Ninad K. Mishra, Roderick Y. Son, and James J. Arnzen. 2012. Towards automatic diabetes case detection and abcs protocol compliance assessment. *Clinical medicine & research*, pages cmr–2012.

Richard Socher, Alex Perelygin, Jean Wu, Jason Chuang, Christopher D Manning, Andrew Ng, and Christopher Potts. 2013. Recursive deep models for semantic compositionality over a sentiment treebank. In *Proceedings of the 2013 conference on empirical methods in natural language processing*, pages 1631–1642.

Amber Stubbs, Christopher Kotfila, Hua Xu, and Azlem Uzuner. 2015. Identifying risk factors for heart disease over time: Overview of 2014 i2b2 uthealth shared task track 2. *Journal of Biomedical Informatics*, 58:S67 – S77.

Amber Stubbs and Özlem Uzuner. 2015. Annotating risk factors for heart disease in clinical narratives for diabetic patients. *Journal of Biomedical Informatics*, 58:S78 – S91. Proceedings of the 2014 i2b2/UTHealth Shared-Tasks and Workshop on Challenges in Natural Language Processing for Clinical Data.

Maxim Topaz, Kavita Radhakrishnan, Suzanne Blackley, Victor Lei, Kenneth Lai, and Li Zhou. 2017. Studying associations between heart failure self-management and rehospitalizations using natural language processing. *Western journal of nursing research*, 39(1):147–165.

Etienne Vermeire, Hilary Hearnshaw, Paul Van Royen, and Joke Denekens. 2001. Patient adherence to treatment: three decades of research. a comprehensive review. *Journal of clinical pharmacy and therapeutics*, 26(5):331–342.

Author Index

Association for Computational Linguistics
209 N. Eighth Street
Stroudsburg, Pennsylvania 18360

ISBN 978-1-5108-6710-9